Philosophy of Science and the Occult

SUNY Series in Philosophy
Robert C. Neville, EDITOR

Philosophy of Science and the Occult

Second Edition

edited by

PATRICK GRIM

State University of New York Press

Published by
State University of New York Press, Albany

© 1990 State University of New York

For information, address State University of New York
Press, State University Plaza, Albany, N.Y., 12246

Library of Congress Cataloging in Publication Data
Main entry under title:

Philosophy of science and the occult/edited by patrick Grim. — 2nd
    ed.

        p. cm.
    ISBN 0-7914-0204-5 (alk. paper)
    1. Occultism and science. 2. Parapsychology and science.
    3. Science—Philosophy. I. Grim, Patrick.
    BF1409.5.P47   1990
    133'.01—dc20   90-9619
                                                    CIP

ISBN 0-87395-573-0 (pbk.)

10 9 8 7 6 5 4 3

Contents

Preface to the Revised Edition
General Introduction

SECTION I. ASTROLOGY

Introduction	15
Objections to Astrology: A Statement by 186 Leading Scientists	18
The Strange Case of Astrology	23
Paul Feyerabend	
On Dismissing Astrology and Other Irrationalities	28
Edward W. James	
Spheres of Influence	37
Michel Gauquelin	
Astrology: A Critical Review	51
I.W. Kelly, G.A. Dean, and D.H. Saklofske	
Suggested Readings	82

SECTION II. SCIENCE OR PSEUDOSCIENCE?:
THE PROBLEM OF DEMARCATION

Introduction	87
Winning through Pseudoscience	92
Clark Glymour and Douglas Stalker	
Science: Conjectures and Refutations	104
Karl R. Popper	
Demarcating Genuine Science from Pseudoscience	111
Daniel Rothbart	
Logic of Discovery or Psychology of Research?	123
Thomas S. Kuhn	
The Popperian versus the Kuhnian Research Programme	131
Imre Lakatos	

Kuhn, Popper, and the Normative Problem of Demarcation 140
  Robert Feleppa
The Conservatism of "Pseudoscience" 156
  Roger Cooter
Popper, Kuhn, Lakatos: A Crisis of Modern Intellect 170
  Robert F. Baum
Suggested Readings 170

SECTION III. PARAPSYCHOLOGY

Introduction 183
Parapsychology: An Empirical Science 187
  Ruth Reinsel
Coincidence and Explanation 205
  Galen K. Pletcher
Parapsychology: Science or Pseudoscience? 214
  Antony Flew
Philosophical Difficulties with Paranormal Knowledge Claims 232
  Jane Duran
Precognition and the Paradoxes of Causality 243
  Bob Brier and Maithili Schmidt-Raghavan
Second Report on a Case of Experimenter Fraud 253
  J. B. Rhine
Suggested Readings 264

SECTION IV. QUANTUM MYSTICISM

Introduction 267
Conflicting Notes from Einstein and Bohr 272
The Unity of All Things 274
  Fritjof Capra
Einstein Doesn't Like It 285
  Gary Zukav
The Yogi and the Quantum 302
  Robert P. Crease and Charles C. Mann
Quantum Mysteries for Anyone 315
  N. David Mermin
Mind, Matter, and Quantum Mechanics 326
  Marshall Spector
Suggested Readings 350

SECTION V. OTHER APPROACHES TO THE OCCULT

Introduction 353
Notes on Mysticism and Nitrous Oxide 355
  William James
Tacit Assumptions 361
  Edward Conze
Scientific and Other Values 370
  Patrick Grim
Suggested Readings 385

Index

Preface to the Revised Edition

The first edition of *Philosophy of Science and the Occult* has shown itself to be of value as a resource in informal logic and critical reasoning courses as well as a major text in courses on science and pseudoscience and a secondary source in traditional philosophy of science courses. The book has also been welcomed in its own right as a forum of debate, however—less prejudicial in tone and more serious than most—regarding, for example astrology and parapsychology. It is hoped that this double-edged use will continue to characterize the revised edition.

The crucial section on the problem of demarcation has been broadened in this edition; it now includes, for example, the work of Lakatos as well as Popper and Kuhn. I. W. Kelly's earlier "Astrology, Cosmobiology, and Humanistic Astrology" has also been replaced by a more complete and current review of research on astrology by Kelly, G. A. Dean, and D. H. Saklofske. There are numerous small changes—both mine and the original authors'—throughout.

The major change in this edition, however, is an entirely new section on quantum mysticism, using excerpts from Fritjof Capra's *The Tao of Physics* and Gary Zukav's *The Dancing Wu Li Masters* as well as critical work by Marshall Spector, Robert P. Crease and Charles C. Mann, fragments from Einstein and Bohr, and instructions from N. David Mermin for constructing a device which dramatically illustrates the genuinely puzzling phenomena of quantum mechanics.

This new section has replaced—for reasons of timeliness and space—a section on UFOs and Ancient Astronauts in the first edition. That replaced section remains of value, however, and it is hoped that the interested reader will seek it out.

1

Magic is not the "science" of the past. It is the science of the future. I believe that the human mind has reached a point in evolution where it is about to develop new powers—powers that would once have been considered magical. Indeed, it has always possessed greater powers than we now realize: of telepathy, premonition of danger, second sight, thaumaturgy (the power to heal); but these were part of its instinctive, animal inheritance. For the past thousand years or so, humankind has been busy developing another kind of power related to the instinct, and the result is western civilization. His unconscious powers have not atrophied; but they have "gone underground." Now the wheel has come the full circle; intellect has reached certain limits, and it cannot advance beyond them until it recovers some of its lost powers.

—Colin Wilson, from *The Occult: A History*

From Alexander's time to our own—indeed, probably for as long as human beings have inhabited this planet—people have discovered they could make money by pretending to arcane or occult knowledge. . . . Many of the impostures do not have a contemporary ring and only weakly engage our passions: it becomes clear how people in other times were deceived. But after reading many such cases, we begin to wonder what the comparable contemporary versions are. People's feelings are as strong as they always were, and scepticism is probably as unfashionable today as in any other age. Accordingly, there ought to be bamboozles galore in contemporary society. And there are. . . . The proponents of such borderline beliefs, when criticized, often point to geniuses in the past who were ridiculed. But the fact that some geniuses were laughed at does not imply that all who are laughed at are geniuses. They laughed at Columbus, they laughed at Fulton, they laughed at the Wright Brothers. But they also laughed at Bozo the Clown.

—Carl Sagan, from *Brocas Brain*

General Introduction

Philosophy of science is a paradigm of contemporary intellectual rigor. It offers a challenge of clarification, a promise of systematic understanding, and an invitation to innovative conceptual exploration. Such is its appeal.

The occult traditions are steeped in antiquity. They reach us with an atmosphere of mystery, a whisper of wisdom, and a hint of the beckoning unknown. Such is their appeal.

This is an attempt to bring the two together.

The collection that follows can be viewed in either of two ways: as an introduction to philosophy of science through an examination of the occult, or as a serious examination of the occult rigorous enough to raise central issues in philosophy of science. That the collection *can* be viewed in either of these ways is, I think, an indication of its two virtues. Its concern with issues in philosophy of science marks it as a more serious investigation of the occult than most. And its use of the occult as an introduction to philosophy of science makes it a more intriguing and more accessible introduction than most.

The editorial integration of these two is of course entirely intentional. But let me render asunder for the moment what has been so carefully brought together and address each of these aims as if were my only aim.

I

Much that is best in current philosophy—and in philosophy for the past fifty years—falls within the loose but useful category of philosophy of science. With the benefits of increasingly sophisticated and rigorous work, however, have come unfortunate pedagogical costs. Philosophy of science is rarely seen, by students or by others as yet uninitiated into its mysteries, for the exciting ongoing exploration it truly is. As usually presented, the standard conundrums of confirmation and enigmas of explanation too often

3

fail to jar intellectual inertia or disturb dogmatic slumbers as they should. The collection that follows is designed to change this: to make the lively excitement and intellectual fascination of philosophy of science immediate and obvious.

I think—and I think the collection demonstrates—that standard issues within the philosophy of science arise quite naturally in the context of an examination of intriguing claims concerning astrology, parapsychology, quantum mysticism, and other aspect of "the occult." An examination of these areas thus offers an appealing and accessible forum in which to discuss theoretical issues which are also of importance for astronomy, psychology, quantum mechanics, and other aspects of science.

Topics standard to philosophy of science that arise in the following pages include issues of confirmation and selection for testing, possibility and *a priori* probabilities, causality and time, explanation and the nature of scientific laws, the status of theoretical entities, the problem of demarcation, theory and observation, and science and values. That these issues are raised here in the context of a discussion of the occult in no way weakens the importance of dealing with them carefully and rigorously as quite general issues. But this context does serve, more effectively than most, I think, to make the initial introduction of these issues immediately understandable and dramatically compelling, and thus serves to motivate the careful work they require.

The collection may serve as all that is needed within philosophy of science for a major section of an introductory philosophy or humanities course. It is quite clearly suited as a core text for courses in science and pseudoscience, which have generally proven successful,[1] and as a major text in informal logic or critical reasoning courses. Within traditional philosophy of science courses it can serve as a useful auxiliary text, with some of the major issues raised here further pursued using more standard materials.

II

The collection need not be conceived, however, solely or primarily as a pedagogical tool for philosophy of science. It is also a straightforward examination of various aspects or areas of the occult.

Those areas addressed here as "the occult," for lack of a term more suitably inclusive and yet unprejudicial, are rarely met with equanimity.[2] Topics such as astrology, parapsychology, and quantum mysticism are likely to be as threatening to the calm order of polite conversation as politics and religion were once reputed to be. Some of us have blood that boils at the insufferable dogmatism of those who deny such areas of investigation their genuine integrity and their proper due. Others of us have blood that boils at the mere thought of the empty-headed gullibility of those impressed by such twaddle.

4

The theme of the collection, in most general terms, is this clash of viewpoints between impassioned defenders and icy critics. But the attempt is not to review, in tranquility and with judicial sobriety, an intellectual battle fought elsewhere. Nor is the attempt to structure debate so as to show either the defender of the occult, or the icy skeptic, triumphant. Instead, the battle itself is allowed to rage in these pages, with equally informed and sophisticated efforts on each side of the conflict and without an editorially imposed verdict. In this regard what follows is a collection of some of the most exciting arguments offered by defenders and skeptics in conflict. There are of course various shades and various forms of both defense and critique. But with that qualification duly noted, it seems safe to number Karl Popper, Thomas S. Kuhn, Clark Glymour, Antony Flew, Marshall Spector, Robert P. Crease and Charles C. Mann among the skeptics represented, with Paul Feyerabend, Michel Gauquelin, Bob Brier, Fritjof Capra, Gary Zukav, and J. B. Rhine among their opponents.

In the most general terms, then, this collection is a clash between sophisticated defenders and critics concerning topics grouped here as "the occult." But as a glimpse at the table of contents will indicate, things quickly become more complex. In anything *but* the most general terms, what is at issue is not a simple stand-off but an intricate web of challenge and response, evidence offered and disputed, argument and counter-argument. This is, of course, what is to be expected of any rigorous and systematic debate concerning these topics. Astrology, parapsychology, and quantum mysticism raise significantly different issues and call for quite different handling, whether in critique or defense. Within each topic, moreover, there are alternative lines of defense and critique and counter-critique, different types of appeal to different types of evidence, and various ways of applying various general principles at issue. To the impatient, anxious for a glib vindication or refutation, the labyrinthian twists of this complex argumentation may prove annoying. But it is precisely the intricacy and subtlety of such argument that marks a genuinely *serious* consideration of these topics, and it is a serious consideration that is intended here. It is also within the labyrinths of such careful argument that the major issues in philosophy of science, alluded to above, so clearly arise.

Let me add here a warning both to those who think of themselves as true believers and to those who think of themselves as icy skeptics with regard to the occult: one ought not expect to emerge from the philosophical discussion of the following pages, or from any serious philosophical discussion, entirely unscathed. Icy skeptics may find evidence of which they were unaware, and encounter new and unsettlingly plausible arguments against their initial stance. True believers may find that certain bodies of evidence, when actually examined, are less compelling than they are popularly reputed to be, and may discover new and strikingly sophisticated arguments on the other side. It is fine to enter the discussion as either a believer or a skeptic, as long as one is willing to take serious debate seriously, and as long as one recognizes that that offers no guarantee that one will end up with precisely the same convictions one began with.

It should also be noted that the major philosophical problems raised in this book are problems posed for both the skeptic and the believer. These are quite general and pervasive quandaries, and no particular position with regard to the occult will allow one to wave them aside.

Here the problem of demarcation, a major theme of the collection, may serve as a convenient example. Consider first the position of the skeptic. He who rejects certain topics as mere "pseudoscience" must draw a line between proper science and that which he rejects, and the problem of demarcation is the problem of precisely where, and on what justifiable basis, such a line is to be drawn. Nor can the skeptic rest content with a line of demarcation placed conveniently at the edge of undebatable contemporary scientific knowledge. Science always calls for *further* investigation, and thus one must have some grounds for distinguishing some proposed lines of investigation as properly scientific, even before they are actually pursued, from endeavors unworthy of pursuit because "pseudoscientific." For the skeptic the problem of demarcation is to draw a justifiable distinction in terms general enough that it may govern decisions regarding future work as well as judgments regarding current efforts.

But consider also the case of the true believer. No matter how sympathetic one is to the various topics that might be included as "the occult," one simply cannot, within the bounds of consistency, believe it all. The various aspects of the occult do not form a harmonious whole, and one is subjected to conflicting theories and rival explanations and contradictory claims and competing modes of investigation within occult lore, just as elsewhere. Thus even the truest of believers must distinguish good work from bad, and worthy from hopeless approaches or forms of investigation. The believer, too, is forced to draw a line, then, and to justify its placement, though his line may lie farther from recognized science and deeper within the occult. The general problem of where to draw a line, then, and why to draw it there—the general problem of demarcation—is the same for skeptic and believer. Much the same holds for general issues of causality, confirmation, probability, explanation, theory and observation, and science and values which form the major conceptual currents of the volume.

III

The two aims of the collection—as an introduction to philosophy of science through an examination of the occult, and as an exciting debate concerning the occult serious enough to broach general issues in philosophy of science—are not, I would suggest, in competition. To a large extent, the satisfactory pursuit of either aim quite naturally furthers the other aim as well. The more serious and sophisticated that debate becomes concerning astrology, parapsychology, and the like, the more clearly philosophical it becomes as well. And the more such a discussion relies on traditional work within philosophy of science, the more sophisticated it is bound to be.

Here let me offer some more complete comments regarding the structure of the volume.

Each of the major sections of the collection opens with a brief introductory overview, designed to raise the particular issues of that section, to stress points of continuity with other sections, and to emphasize crucial patterns of argument that develop in the course of that section. The last of these is perhaps the most important. Within each section pieces have been chosen and placed so as to offer a continuous thread of developing argument, critique, and response through the section. Most pieces within a section offer implicit, and sometimes explicit, replies to the pieces that precede them, or offer a further development of earlier arguments. The first selections within each section are generally the simplest and most immediately accessible and are followed by pieces of increasing sophistication as the argument develops. So although pieces may be selected from the whole for individual attention, there is also a continuity within each section that recommends that the pieces it includes be read in the order in which they appear. It is this thread of continuity that I try to make more explicit in introductory outlines.

In some ways the section that is theoretically most fundamental for the collection as a whole is not the first section but the second, which offers a thorough discussion of the issue of demarcation in general. The most common rejection of astrology, parapsychology, and the like is to brand them with the epithet "pseudoscience." But what is it that lies behind that epithet; what does it mean to call something either "scientific" or "pseudo-scientific," and what is it in general that will justify the praise of an endeavor as "scientific" or justify its vilification as "pseudoscientific"? This is Popper's "problem of demarcation," which forms a major theme within the collection as a whole.

In this second section Karl Popper, Thomas S. Kuhn, and Imre Lakatos are all represented, with comments both critical and elucidative by Daniel Rothbart and Robert Feleppa. Clark Glymour and Douglas Stalker offer a humorous lampoon of "pseudoscience" to open the section, Roger Cooter offers the view of a historian critical of the whole philosophical attempt at demarcation, and Robert F. Baum closes with some reflections on the debate and its wider cultural significance.

The first section of the collection is devoted to astrology. Its appearance before the more general theoretical discussion of demarcation is as I would recommend that it be read and as I would use it in my own teaching. The intricacies of complex argument concerning demarcation can seem twisted indeed, especially to the philosophically unitiated, without a particular problem and a specific case to keep in mind in pursuing them. Astrology is offered first as such a specific case, with its legitimacy or illegitimacy a problem the more complete examination of which calls for further work in the later section.

The section on astrology opens with a critical broadside, excoriating astrology in no uncertain terms and signed by 186 scientific notables. That attack is immediately answered, however, by Paul Feyerabend in the role of critics' critic. This section also contains an impressive report of statistical work by Michel Gauquelin, work that has quite widely been applauded as vindicating astrology. Edward W. James gives a disparaging glance at the popular literature of astrology, alluding to the work of Popper and Kuhn that appears later in the collection, and I. W. Kelly, G. A. Dean, and D. H. Saklofske offer a comprehensive review of various types of astrology in an attempt to gauge their strengths and weaknesses.

This first section does not impose, and is not meant to impose, any clear resolution concerning the legitimacy of astrology. Thus it is no weakness if some readers find the work of Gauquelin and Kelly, Dean, and Saklofske's survey of more recent developments regarding that work so impressive as to establish at least a core of legitimacy for astrology, whereas other readers find compelling the critical comments in the opening piece, in James, or in other aspects of Kelly, Dean, and Saklofske's review. The section is intended rather to evoke, in the context of lively debate concerning astrology, some broader philosophical issues that call for further pursuit. The general issue of demarcation is primary among these, and thus leads directly into the succeeding section. But Gauquelin's work also offers an opportunity for further thought regarding causality and correlation, and Feyerabend and James introduce a number of informal fallacies worthy of attention. The difficulties of deciding what does or would serve as evidence for astrology may offer quite general lessons regarding hypothesis formulation and testing.

Using the initial consideration of astrology and the further development of important conceptual tools in the general discussion of demarcation as a background, the third section—on parapsychology—renews some earlier questions and poses important new ones. Here, to begin with, is another specific case against which to judge proposals for demarcation outlined in the previous section. Is *parapsychology* to be classified as mere pseudoscience, and if so why? The readings in this section indicate major ways in which the case of parapsychology *differs* from that of astrology, and thus emphasize the many different considerations that may be at issue in any serious attempt to draw a justifiable line between science and its pretenders. But parapsychology also raises important philosophical issues not previously addressed in the collection. What does it mean to wave something aside as "mere coincidence," and on what basis do we distinguish phenomena that call for an explanation in their own right from phenomena that do not? What is the proper role in science of intuitive "basic limiting principles" such as our conviction that mental phenomena must in some way involve brains, or that all knowledge must ultimately be acquired by some sensory means? Is it possible for an effect to precede its cause?

The parapsychology section opens with a review of major parapsychological research, and a discussion of promising developments in the field, by Ruth Reinsel. This offers a helpful summary of documented work—rather than the merely anecdotal cases and merely rumored tests that abound as popular lore—against which to gauge the more general and philosophical treatments that follow. The last piece of the section is by J. B. Rhine, by far the most respected figure in the development of the field, and offers a direct response to accusations of fraud as well as addressing the discovery of experimental falsifications by Walter J. Levy, Jr. Noteworthy among the critics in this section is Antony Flew. His contribution touches on Hume's comments regarding miracles and could easily serve as an introduction to questions concerning lawlikeness and the status of scientific laws. Other issues broached in the critical pieces by Flew, Galen K. Pletcher, and Jane Duran include repeatability, units of explanation, the scientific status of *a priori* principles, theoretical entities, and causality and correlation. Duran's piece is also noteworthy for raising central questions in the analysis of knowledge and the issue of identity theory in philosophy of mind.

An issue dealt with quite thoroughly in the parapsychology section is that of "backward causation." Can an effect precede its cause? The claim that this is not possible, and thus that precognition is not possible, is first introduced by Galen Pletcher and is more fully developed as a strong critical point in the pieces by Flew and Duran. Their skeptical conclusions are quite forcefully challenged, however, by Bob Brier and Maithili Schmidt-Raghavan. Brier and Schmidt-Raghavan suggest that this standard attack rests on a confusion, and attempt to vindicate parapsychology (and their own parapsychological experiment) by arguing that backward causation is indeed possible. Here Brier and Schmidt-Raghavan rely in part on the work of John Stuart Mill.

The fourth section of the collection is importantly different from its predecessors. Here the issue is, of course, not whether quantum mechanics—the central topic of the section—should be considered "science" or "pseudoscience"; quantum mechanics' claim to full scientific status is undisputed. The question at issue here is, rather, whether a theory as firmly entrenched scientifically as quantum mechanics may not, nonetheless, have *implications* which might be considered "occult": implications that parallel Eastern mysticism, and in particular implications that consciousness plays a *creative* role in the fundamental mechanisms of the universe—in Gary Zukav's words, that "we create certain properties because we choose to measure those properties."

The claim that quantum mechanics *does* parallel Eastern mysticism, and *does* have these major implications for the role of consciousness, is clearly made by Fritjof Capra in an excerpt from *The Tao of Physics* and by Gary Zukav in an excerpt from *The Dancing Wu Li Masters*. The claim of parallelism is directly attacked by Robert P. Crease and Charles C. Mann, however, and Marshall Spector offers a careful examination of what is *physically* new about quantum mechanics in the course of emphatically criticizing the claim that consciousness here plays any special role. Controversy regarding quantum mechanics is far from new, of course: the section opens with an exchange between Albert Einstein and Neils Bohr as to whether the theory should be considered "complete". In one of the most intriguing pieces of the section, N. David Mermin leads the reader through the construction of a device which clearly illustrates the very real perplexities of the phenomena at issue. "It's not the Copenhagen interpretation of quantum mechanics that is strange," Mermin says, "but the world itself."

The final section of the collection is an attempt to consider approaches to the occult we may have neglected in the rest of the volume. In general, our concern throughout has been with aspects of the occult as claimants to scientific status or as purported implications of fundamental scientific theory. The final section addresses the possibility that this mode of procedure and the tone of our examination in general may have unfairly distorted or disparaged fully legitimate but very different human endeavors properly included within the occult. Is a mistake being made in judging some things against a scientific standard or with the model of science in mind?

William James's classic remarks on mysticism and nitrous oxide are used to open the section, and James proposes an affirmative answer to this question. There are perhaps, James suggests, different orders of experience beyond the reach of science, and these may provide the "truest of insights into the meaning of this life." Such a position is further argued by Edward Conze, who quite explicitly accuses Western "science-bound philosophers" of perversely refusing to accept other traditions as on an equal footing with science as we know it. Such a position will be familiar to readers of Carlos Castañeda and Alan Watts, among others, but Conze's work offers a more straightforward and in many ways more compelling argument. What James and Conze raise is the specter of epistemological relativism, and one of the things I attempt to do in the final piece is to offer a partial reply. I attempt to do so, however, by addressing the question of whether science is value-laden, and what this might mean in considering science as but one among rival alternatives. That my piece comes last is, of course, by no means to suggest that mine is to be taken as anything like the final word on the matter.

10

The final section is designed to broach issues regarding science and values that do not emerge so clearly elsewhere in the collection, but is also intended to offer a context in which issues of epistemological relativism, so often introduced by the students themselves, can be clearly and profitably discussed. Thus it may prove pedagogically useful to skip to the final section when the issue arises.

IV

Some of the pieces included here are reprinted from other sources. I am grateful to Paul Kurtz and the editors of *The Humanist* for permission to reprint "Objections to Astrology"; to Paul Feyerabend for "The Strange Case of Astrology" and to Michel Gauquelin for "Spheres of Influence"; to Sir Karl Popper for permission to reprint from "Science: Conjectures and Refutations"; to Thomas S. Kuhn and Cambridge University Press for permission to reprint from "Logic of Discovery or Psychology of Research?"; to Alan Musgrave and Cambridge University Press for permission to reprint an excerpt from Imre Lakatos's "Falsification and the Methodology of Scientific Research Programmes"; to Roger Cooter, Wilfrid Laurier University Press, and the editors of *Science, Pseudoscience, and Society* for permissions regarding "The Conservatism of 'Pseudoscience' "; to the editors of *The Intercollegiate Review* for permission to reprint from Robert F. Baum's "Popper, Kuhn, Lakatos: A Crisis of Modern Intellect"; to Antony Flew, the *Pacific Philosophical Quarterly*, Wilfrid Laurier University Press, and the editors of *Science, Pseudoscience, and Society* for permission to reprint "Parapsychology: Science or Pseudoscience?"; to Louisa E. Rhine and the editors of the *Journal of Parapsychology* for permission to reprint from "Second Report on a Case of Experimenter Fraud"; to Fritjof Capra and Shambhala Publications for permission to reprint an excerpt from *The Tao of Physics*; to William Morrow and Co. for permission to reprint an excerpt from Gary Zukav's *The Dancing Wu Li Masters*; to N. David Mermin and the *Journal of Philosophy* for permission to reprint "Quantum Mysteries for Anyone"; and to Professor Ninian Smart, Muriel Conze, and George Allen and Unwin Ltd. for permission to reprint from "Tacit Assumptions." The opening quotations from Colin Wilson, *The Occult: A History* (New York: Random House, 1971), p. 39, and Carl Sagan, *Broca's Brain* (New York: Random House, 1979), pp. 3 and 18, are with permission. I am also indebted to the contributors for tolerating so many requests for laborious revision in the name of coherency and effectiveness within the collection.

The illustrations at section introductions are from rough sketches by my father, Elgas Grim.

Notes

1. See Michael Martin, "The Use of Pseudo-Science in Science Education," *Science Education* 55 (1971), 53-56, and Paul Thagard and Daniel M. Hausman, "Sun

Signs vs. Science: Using Astrology to Teach Philosophy of Science," *Metaphilosophy* 11 (1980), 101-104.

2. It is intriguing that a suitably broad but unprejudicial term is so difficult to find. "Pseudoscience," of course, condemns as it classifies, and "the paranormal" seems more properly confined to parapsychology. Recourse to metaphor does little better. "The frontiers of science" seems too laudatory, as do "the edges of science" or the "borders of science," though less so. "The fringes of science," on the other hand, has precisely the opposite drawback. "Borderlands of science" is perhaps better, though its ambiguity extends not merely to normative overtones but leaves one wondering precisely what topics are at issue. Here I use "the occult" merely for lack of a better term and as the best approximation I could find to a term with an appropriately broad application but without prejudicial overtones.

ASTROLOGY

Introduction

In the next section we will consider a number of rival attempts to distinguish science from pseudoscience in general. Here, however, we will start by considering one particular controversial area in some detail: the case of astrology.

Is astrology a pseudoscience? As we will see later, many philosophers of science take astrology to be a paradigm or a prime example of pseudoscience. On this particular case they appear to be unanimous, however much they may disagree as to precisely *why* astrology is pseudoscientific or as to precisely what astrology's faults are.

Our first selection here presents a similarly unanimous condemnation of astrology. One hundred eighty-six prominent scientists, including eighteen Nobel Prize winners, have signed a statement blasting "the pretentious claims of astrological charlatans." The claims of astrology, they hold, have no scientific basis and are moreover claims against which we have strong evidence. The prominent place of astrology in the media "can only contribute to the growth of irrationalism and obscurantism."

In the piece immediately following, however, PAUL FEYERABEND attacks this statement for its religious tone, its blatant appeal to authority, and its lack of adequate argument. Those who have signed the statement, Feyerabend claims, simply "do not know what they are talking about," and he presents a number of scientific results that appear to contradict the general claims of the statement. Feyerabend is careful to note that he is not attempting to defend the practice of most contemporary astrologers. But it is clear that he thinks astrology cannot fairly be confronted, and that science in general is very poorly represented, by attacks of this kind.

In "On Dismissing Astrology and Other Irrationalities," EDWARD W. JAMES sharply criticizes attempts to make short work of astrology by quickly dismissing it as irrational. James briefly outlines Karl Popper's and

15

Thomas S. Kuhn's attempts to dismiss pseudoscience, more fully presented in the next section, and argues that neither of these attempts is successful in the case of astrology. Although a handful of other common arguments may give us good reason to disbelieve the claims of astrology, James further suggests, these arguments fail as sufficient grounds for dismissing astrology and those who uphold it as irrational. In the end James holds that we *do* know that astrology is irrational, not on the basis of any simple criterion for distinguishing between science and pseudoscience and not on the basis of any small group of telling arguments, but because we can see that astrological literature is a complex tangle of fallacy and misrepresentation. Thus any serious attempt to show the irrationality of astrology, James concludes, must take the tedious and unpleasant route of critically examining astrological literature step by step and argument by argument.

James mentions in passing the work of MICHEL GAUQUELIN, who speaks for himself in the following selection. Gauquelin does not consider himself an astrologer, and explicitly rejects zodiacal influences and the notion that external signs in the stars and mysterious astrological connections control our fates so as to "make puppets out of us." But on the basis of almost forty years of statistical research Gauquelin does claim to have shown a link between planetary movements and human character, and thus to have revealed "a golden grain of truth in astrological superstition." In the selection presented here, Gauquelin offers some of his most startling results, which seem to indicate that there is a statistically significant correlation between success in certain careers and the position of certain planets at one's birth. Gauquelin also outlines his methodology in some detail and briefly suggests an explanation for his findings. Here the reader should consider once again James's earlier critical comments in light of this more complete outline of Gauquelin's research.

Gauquelin's work is further discussed in the fifth section of I.W. KELLY, G.A. DEAN, AND D.H. SAKLOFSKE's concluding piece. These authors also offer a review of empirical research done on traditional astrology and a discussion of problems regarding the perceived validity of astrology, the comparison case of phrenology, and the 'spiritual astrology' of Dane Rudhyar and others. Kelly, Dean, and Saklofske reject traditional astrology in no uncertain terms on the basis of the relevant research and offer a critique of 'spiritual astrology' on more broadly philosophical grounds. Gauquelin's work, on the other hand, survives their generally critical survey relatively unscathed: the authors note that "the results obtained by Gauquelin remain the best factual evidence for astrology that has emerged to date." Here they do indicate that a number of important issues remain for further research, however.

Throughout this section the reader should attempt to tackle the question of the scientific status of astrology. *Is* astrology a prime example of pseudoscience, and if so why? Must we distinguish traditional astrology from 'spiritual astrology' and Gauquelin's hypothesis in order to answer such a question? With this background examination of the case of astrology in particular, the general question of what distinguishes science from pseudoscience in general will lead us into the next section.

The selections of this section also raise some other important questions. What is the relation between correlation, such as it appears in Gauquelin's data, and claims regarding causal relations between those things that are correlated? What precisely *is* the theory or hypothesis of astrology, and how might it be effectively tested? Is the attitude of "Objections to Astrology" a justifiable one? Is Gauquelin's work correctly taken as a vindication of astrology?

Here it is perhaps not out of order to add that the issue of the scientific status of astrology may also be more than merely an intellectually intriguing question. We are all familiar with astrology as it appears in the apparently harmless columns of magazines and newspapers. But it should be noted that there are also astrological medical services offering diagnosis and treatment on the basis of birth charts,[1] there are books such as T. Patrick Davis's *Sexual Assaults: Pre-Identifying Those Vulnerable*, which treats rape as a matter of astrological vulnerability to attack,[2] and there are claims by individual astrologers such as Larry Berg that we need not worry about the greenhouse effect or depletion of the ozone layer because "the dominant celestial patterns are intact."[3] If astrology *can* claim to be a science, all of these might thereby be scientifically vindicated. If astrology *cannot* claim to be a science, on the other hand, these might amount to quite dangerous forms of pseudoscientific fraud.

Notes

1. "MedScan analysis," for example, described in Eileen Nauman's *The American Book of Nutrition and Medical Astrology* (San Diego, Ca: ACS Publications, 1982) is sold in a computerized form by Cosmic Patterns of Gainesville, Florida.

2. Windermere, Fla: David Research Reports, 1978. The following is from the page facing the title page:

> The horoscope is the only tool presently known which can be used to separate those who may be vulnerable to a sexual assault from those who would not. With some horoscopes, even the general time period can be ascertained when vulnerability increases.

3. As quoted by Robert P. Crease from the 50th Anniversary Convention of the American Federation of Astrologers, Las Vegas, Nevada, July 1988.

Objections to Astrology

A Statement by 186 Leading Scientists

Scientists in a variety of fields have become concerned about the increased acceptance of astrology in many parts of the world. We, the under-signed—astronomers, astrophysicists, and scientists in other fields—wish to caution the public against the unquestioning acceptance of the predictions and advice given privately and publicly by astrologers. Those who wish to believe in astrology should realize that there is no scientific foundation for its tenets.

In ancient times people believed in the predictions and advice of astrologers because astrology was part and parcel of their magical world view. They looked upon celestial objects as abodes or omens of the Gods and, thus, intimately connected with events here on earth; they had no concept of the vast distances from the earth to the planets and stars. Now that these distances can and have been calculated, we can see how infinitesimally small are the gravitational and other effects produced by the distant planets and the far more distant stars. It is simply a mistake to imagine that the forces exerted by stars and planets at the moment of birth can in any way shape our futures. Neither is it true that the position of distant heavenly bodies make certain days or periods more favorable to particular kinds of action, or that the sign under which one was born determines one's compatibility or incompatibility with other people.

Why do people believe in astrology? In these uncertain times many long for the comfort of having guidance in making decisions. They would like to believe in a destiny predetermined by astral forces beyond their control. However, we must all face the world, and we must realize that our futures lie in ourselves, and not in the stars.

One would imagine, in this day of widespread enlightenment and edu-

© 1975, *The Humanist*. Reprinted with permission from *The Humanist* 35, no. 5 (September/October 1975): 4–6.

cation, that it would be unnecessary to debunk beliefs based on magic and superstition. Yet, acceptance of astrology pervades modern society. We are especially disturbed by the continued uncritical dissemination of astrological charts, forecasts, and horoscopes by the media and by otherwise reputable newspapers, magazines, and book publishers. This can only contribute to the growth of irrationalism and obscurantism. We believe that the time has come to challenge directly, and forcefully, the pretentious claims of astrological charlatans.

It should be apparent that those individuals who continue to have faith in astrology do so in spite of the fact that there is no verified scientific basis for their beliefs, and indeed that there is strong evidence to the contrary.

Bart J. Bok, *emeritus professor of astronomy University of Arizona*

Lawrence E. Jerome *science writer Santa Clara, California*

Paul Kurtz *professor of philosophy SUNY at Buffalo*

NOBEL PRIZEWINNERS

Hans A. Bethe, *professor emeritus of physics, Cornell*

Sir Francis Crick, *Medical Research Council, Cambridge, England*

Sir John Eccles, *distinguished professor of physiology and biophysics, SUNY at Buffalo*

Gerhard Herzberg, *distinguished research scientist, National Research Council of Canada*

Wassily Leontief, *professor of economics, Harvard University*

Konrad Lorenz, *univ. prof., Austrian Academy of Sciences*

André M. Lwoff, *honorary professor, Institut Pasteur, Paris*

Sir Peter Medawar, *Medical Research Council, Middlesex, Eng.*

Robert S. Mulliken, *dist. prof. of chemistry, U. of Chicago*

Linus C. Pauling, *professor of chemistry, Stanford University*

Edward M. Purcell, *Gerhard Gade univ. prof. Harvard Univ.*

Paul A. Samuelson, *professor of economics, MIT*

Julian Schwinger, *professor of physics, U. of Calif., Los Angeles*

Glenn T. Seaborg, *univ. professor, Univ. of Calif., Berkeley*

J. Tinbergen, *professor emeritus, Rotterdam*

N. Tinbergen, *emer. professor of animal behavior, Oxford Univ.*

Harold C. Urey, *professor emeritus, Univ. of Calif., San Diego*

George Wald, *professor of biology, Harvard University*

George O. Abell, *chmn., Dept. of Astron., U. of Cal., Los Angeles*

Lawrence H. Aller, *professor, Univ. of Calif., Los Angeles*

Edorado Amaldi, *prof. of physics, University of Rome*

Richard Berendzen, *dean, Coll. of Arts and Sci., American Univ.*

William P. Bidelman, *professor, Case Western Reserve Univ.*

Jacob Bigeleisen, *professor, University of Rochester*

D. Scott Birney, *prof. of astronomy, Wellesley College*

Karl-Heinz Böhm, *professor, University of Washington*

Lyle B. Borst, *prof. of physics and astronomy, SUNY at Buffalo*

Peter B. Boyce, *staff astronomer, Lowell Observatory*

Harvey Brooks, *prof. of technology and public policy, Harvard*

William Buscombe, *prof. of astronomy, Northwestern Univ.*

Eugene R. Capriotti, *prof. of astronomy, Ohio State Univ.*

H. E. Carter, *coord. of interdisciplinary programs, U. of Arizona*

J. W. Chamberlain, *prof. of astronomy, Rice University*

19

Astrology

Von Del Chamberlain, *Smithsonian Institution*

S. Chandrasekhar, *prof. of astronomy, Univ. of Chicago*

Mark R. Chartrand III, *chmn., Hayden Planetarium*

Hong-Yee Chiu, *NASA*

Preston Cloud, *prof. of geology, U. of Cal., Santa Barbara*

Peter S. Conti, *prof. of astrophysics, Univ. of Colorado*

Allan F. Cook II, *astrophysicist, Smithsonian Observatory*

Alan Cottrell, *master, Jesus College, Cambridge, England*

Bryce Crawford, Jr., *prof. of chemistry, Univ. of Minnesota*

David D. Cudaback, *research astron., U. of Calif., Berkeley*

A. Dalgarno, *prof. of astronomy, Harvard*

Hallowell Davis, *Central Inst. for the Deaf, Univ. City, Mo.*

Morris S. Davis, *prof. of astronomy, Univ. of No. Carolina*

Peter van de Kamp, *director emeritus, Sproul Observatory*

A. H. Delsemme, *prof. of astrophysics, Univ. of Toledo*

Robert H. Dicke, *Albert Einstein prof. of science, Princeton*

Bertram Donn, *head, Astrochem. Br., Goddard Space Cen., NASA*

Paul Doty, *prof. of biochemistry, Harvard*

Frank D. Drake, *dir., Natl. Astron. and Ionosphere Ctr., Cornell*

Lee A. DuBridge, *pres. emeritus, Calif. Inst. of Technology*

H. K. Eichhorn-von Wurmb, *chmn., Dept. of Astron., U. of S. Fla.*

R. M. Emberson, *dir., Tech. Services Inst. of E. and E. Engineers*

Howard W. Emmons, *prof. of mechanical engineering, Harvard*

Eugene E. Epstein, *staff scientist, The Aerospace Corp.*

Henry Eyring, *distinguished prof. of chemistry, Univ. of Utah*

Charles A. Federer, Jr., *president, Sky Pub. Corp.*

Robert Fleischer, *Astronomy Section, National Science Foundation*

Henry F. Fliegel, *technical staff, Jet Propulsion Laboratory*

William A. Fowler, *institute prof. of physics, Calif. Inst. of Tech.*

Fred A. Franklin, *astronomer, Smithsonian Astrophysical Obser.*

Laurence W. Fredrick, *prof. of astronomy, U. of Virginia*

Riccardo Giacconi, *Center for Astrophysics, Cambridge, Mass.*

Owen Gingerich, *prof. of astronomy, Harvard*

Thomas Gold, *professor, Cornell*

Leo Goldberg, *director, Kitt Peak National Observatory*

Maurice Goldhaber, *Brookhaven National Laboratory*

Mark A. Gordon, *Natl. Radio Astronomy Observatory*

Jesse L. Greenstein, *prof. of astrophysics, Cal. Inst. of Tech.*

Kenneth Greisen, *prof. of physics, Cornell*

Howard D. Greyber, *consultant, Potomac, Md.*

Herbert Gursky, *astrophysicist, Smithsonian Institution*

John P. Hagen, *chmn., Dept. of Astronomy, Penn. State Univ.*

Philip Handler, *president, National Academy of Sciences*

William K. Hartmann, *Planetary Science Inst., Tucson, Arizona*

Leland J. Haworth, *spec. assist. to the pres., Associated Univs.*

Carl Heiles, *prof. of astronomy, U. of Cal., Berkeley*

A. Heiser, *director, Dyer Observatory, Vanderbilt University*

H. L. Helfer, *prof. of astronomy, Univ. of Rochester*

George H. Herbig, *astronomer, Lick Observatory, U. of Cal.*

Arthur A. Hoag, *astronomer, Kitt Peak Natl. Observatory*

Paul W. Hodge, *prof. of astronomy, Univ. of Washington*

Dorrit Hoffleit, *director, Maria Mitchell Observatory*

William E. Howard III, *Natl. Radio Astronomy Observatory*

Fred Hoyle, *fellow, St. John's College, Cambridge U.*

Nancy Houk, *Dept. of Astronomy, Univ. of Michigan*

Icko Iben, Jr., *chmn., Dept. of Astronomy, U. of Illinois*

John T. Jefferies, *director, Inst. for Astronomy, U. of Hawaii*

Frank C. Jettner, *Dept. of Astronomy, SUNY at Albany*

J. R. Jokipii, *prof. of planetary sciences, Univ. of Arizona*

Joost H. Kiewiet de Jonge, *assoc. prof. of astron., U. of Pittsburgh*

Objections to Astrology: A Statement by 186 Leading Scientists

Kenneth Kellermann, *Natl. Radio Astronomy Observatory*

Ivan R. King, *prof. of astronomy, U. of Cal., Berkeley*

Rudolf Kompfner, *professor emeritus, Stanford University*

William S. Kovach, *staff scientist, General Dynamics/Convair*

M. R. Kundu, *prof. of astronomy, Univ. of Maryland*

Lewis Larmore, *dir. of tech., Office of Naval Research*

Kam-Ching Leung, *dir., Behlen Observatory, Univ. of Nebraska*

I. M. Levitt, *dir. emer., Fels Planetarium of Franklin Institute*

C. C. Lin, *professor, MIT*

Albert P. Linnell, *professor, Michigan State Univ.*

M. Stanley Livingston, *Dept. of Physics, MIT*

Frank J. Low, *research prof., University of Arizona*

Willem J. Luyten, *University of Minnesota*

Richard E. McCrosky, *Smithsonian Astrophysical Observatory*

W. D. McElroy, *Univ. of Calif., San Diego*

Carl S. Marvel, *prof. of chemistry, Univ. of Arizona*

Margaret W. Mayall, *consul., Am. Assoc. of Variable Star Obser.*

Nicholas U. Mayall, *former dir., Kitt Peak Natl. Observatory*

Donald H. Menzel, *former director, Harvard College Observatory*

Alfred H. Mikesell, *Kitt Peak Natl. Observatory*

Freeman D. Miller, *prof. of astronomy, Univ. of Michigan*

Alan T. Moffet, *prof. of radio astron., Calif. Inst. of Technology*

Delo E. Mook, *assist. prof. of physics and astronomy, Dartmouth*

Marston Morse, *prof. emer., Inst. for Adv. Study, Princeton*

G. F. W. Mulders, *former head, Astron. Section, NSF*

Guido Münch, *prof. of astronomy, Cal. Inst. of Technology*

Edward P. Ney, *regents prof. of astronomy, Univ. of Minn.*

J. Neyman, *director, statistical lab., Univ. of Cal., Berkeley*

C. R. O'Dell, *proj. scientist, Large Space Telescope, NASA*

John A. O'Keefe, *Goddard Space Flight Ctr., NASA*

J. H. Oort, *dir., University Observatory, Leiden, Netherlands*

Tobias C. Owen, *prof. of astronomy, SUNY at Stony Brook*

Eugene N. Parker, *prof. of physics and astronomy, U. of Chicago*

Arno A. Penzias, *Bell Laboratories*

A. Keith Pierce, *solar astronomer, Kitt Peak National Observatory*

Daniel M. Popper, *professor of astronomy, UCLA*

Frank Press, *professor of geophysics, MIT*

R. M. Price, *radio spectrum manager, Natl. Science Foundation*

William M. Protheroe, *prof. of astronomy, Ohio State University*

John D. G. Rather, *Dept. of Astronomy, Univ. of Calif., Irvine*

Robert S. Richardson, *former assoc. dir., Griffith Observatory*

A. Marguerite Risley, *prof. emer., Randolph-Macon College*

Franklin E. Roach, *astronomer, Honolulu, Hawaii*

Walter Orr Roberts, *Aspen Inst. for Humanistic Studies*

William W. Roberts, Jr., *associate prof., University of Virginia*

R. N. Robertson, *Australian National University*

James P. Rodman, *prof. of astronomy, Mt. Union College*

Bruno Rossi, *prof. emeritus, MIT*

E. E. Salpeter, *professor, Cornell*

Gertrude Scharff-Goldhaber, *physicist, Brookhaven Natl. Lab.*

John D. Schopp, *prof. of astronomy, San Diego State University*

Julian J. Schreur, *prof. of astronomy, Valdosta State College*

E. L. Scott, *professor, University of California, Berkeley*

Frederick Seitz, *president, The Rockefeller University*

C. D. Shane, *Lick Observatory*

Alan H. Shapley, *U.S. Dept. of Commerce, NOAA*

Frank H. Shu, *assoc. prof. of astronomy, Univ. of Cal., Berkeley*

Bancroft W. Sitterly, *prof. emer. of physics, American Univ.*

Charlotte M. Sitterly, *Washington, D.C.*

B. F. Skinner, *prof. emeritus, Harvard*

Harlan J. Smith, *dir., McDonald Observ., Univ. of Texas, Austin*

František Šorm, *professor, Inst. of Organic Chem., Prague, Czech.*

Astrology

G. Ledyard Stebbins, *prof. emeritus, Univ. of California*

C. Bruce Stephenson, *prof. of astronomy, Case Western Reserve*

Walter H. Stockmayer, *prof. of chemistry, Dartmouth*

Marshall H. Stone, *professor, University of Massachusetts*

N. Wyman Storer, *professor emeritus of astronomy, U. of Kansas*

Hans E. Suess, *prof. of geochemistry, Univ. of Cal., San Diego*

T. L. Swihart, *prof. of astronomy, Univ. of Arizona*

Pol Swings, *Institute d'Astrophysique, Esneux, Belgium*

J. Szentágothai, *Semmelweis Univ. Med. School, Budapest*

Joseph H. Taylor, Jr., *assoc. prof. of astronomy, Univ. of Mass.*

Frederick E. Terman, *vice-pres. and provost emeritus, Stanford*

Yervant Terzian, *assoc. prof. of space science, Cornell*

Patrick Thaddeus, *Inst. for Space Studies, New York, N.Y.*

Kip S. Thorne, *prof. of theor. physics, Cal. Inst. of Technology*

Alar Toomre, *prof. of applied mathematics, MIT*

Merle A. Tuve, *Carnegie Institution of Washington*

S. Vasilevskis, *emer. prof. of astronomy, Univ. of Cal., Santa Cruz.*

Maurice B. Visscher, *emer. prof. of physiology, U. of Minn.*

Joan Vorpahl, *Aerospace Corp., Los Angeles*

Campbell M. Wade, *Natl. Radio Astronomy Observatory*

N. E. Wagman, *emer. dir., Allegheny Observatory, U. of Pittsb.*

George Wallerstein, *prof. of astronomy, Univ. of Washington*

Fred L. Whipple, *Phillips astronomer, Harvard*

Hassler Whitney, *professor, Inst. for Advanced Study, Princeton*

Adolf N. Witt, *prof. of astronomy, Univ. of Toledo*

Frank Bradshaw Wood, *prof. of astronomy, University of Florida*

Charles E. Worley, *astronomer, U.S. Naval Observatory*

Chi Yuan, *assoc. prof. of physics, CCNY*

The Strange Case of Astrology

PAUL FEYERABEND

To drive the point home I shall briefly discuss the "Statement of 186 Leading Scientists" against astrology which appeared in the September/October issue 1975 of the *Humanist*. This statement consists of four parts. First, there is the statement proper which takes about one page. Next come 186 signatures by astronomers, physicists, mathematicians, philosophers and individuals with unspecified professions, eighteen Nobel Prize Winners among them. Then we have two articles explaining the case against astrology in detail.

Now what surprises the reader whose image of science has been formed by the customary eulogies which emphasize rationality, objectivity, impartiality and so on is the religious tone of the document, the illiteracy of the "arguments" and the authoritarian manner in which the arguments are being presented. The learned gentlemen have strong convictions, they use their authority to spread these convictions (why 186 signatures if one has arguments?), they know a few phrases which sound like arguments, but they certainly do not know what they are talking about. [1]

Take the first sentence of the "Statement." It reads: "Scientists in a variety of fields have become concerned about the increased acceptance of astrology in many parts of the world."

In 1484 the Roman Catholic Church published the *Malleus Maleficarum*, the outstanding textbook on witchcraft. The *Malleus* is a very interesting book. It has four parts: phenomena, aetiology, legal aspects, theological aspects of witchcraft. The description of phenomena is sufficiently detailed to enable us to identify the mental disturbances that accompanied some cases. The aetiology is pluralistic, there is not just the official explanation, there are other explanations as well, purely materialistic explanations included. Of course, in the end only one of the offered explanations is accepted, but the alternatives are discussed and so one can judge the arguments that lead to their elimination. This feature makes the *Malleus* superior to almost every physics, biology, chemistry textbook of today. Even the theology is pluralistic, heretical views are not passed over in silence, nor are they ridiculed; they are described, examined, and removed by argument. The authors know the subject, they know their opponents, they give a correct account of the positions of their opponents, they argue against these positions and they use the best knowledge available at the time in their arguments.

The book has an introduction, a bull by Pope Innocent VIII, issued in 1484. The bull reads: "It has indeed come to our ears, not without afflicting us with bitter sorrow, that in . . ."—and now comes a long list of countries and counties—"many persons of both sexes, unmindful of their own salvation have strayed from the Catholic Faith and have abandoned themselves to devils . . ." and so on. The words are almost the same as the words in the beginning of the "Statement," and so are the sentiments expressed. Both the Pope and the "186 leading scientists" deplore the increasing popularity of what they think are disreputable views. But what a difference in literacy and scholarship!

Comparing the *Malleus* with accounts of contemporary knowledge the reader can easily verify that the Pope and his learned authors knew what they were talking about. This cannot be said of our scientists. They neither know the subject they attack, astrology, nor those parts of their own science that undermine their attack.

Thus Professor Bok, in the first article that is attached to the statement writes as follows: "All I can do is state clearly and unequivocally that modern concepts of astronomy and space physics give no support—better said, negative support—to the tenets of astrology" i.e. to the assumption that celestial events such as the positions of the planets, of the moon, of the sun influence human affairs. Now, "modern concepts of astronomy and space physics" include large planetary plasmas and a solar atmosphere that extends far beyond the earth into space. The plasmas interact with the sun and with each other. The interaction leads to a dependence of solar activity on the relative positions of the planets. Watching the planets one can predict certain features of solar activity with great precision. Solar activity influences the quality of short wave radio signals hence fluctuations in this quality can be predicted from the position of the planets as well. [2]

Solar activity has a profound influence on life. This was known for a long time. What was not known was how delicate this influence really is. Variations in the electric potential of trees depend not only on the *gross* activity of the sun but on *individual flares* and therefore again on the positions of the planets.[3] Piccardi, in a series of investigations that covered more than thirty years found variations in the rate of standardized chemical reactions that could not be explained by laboratory or meteorological conditions. He and other workers in the field are inclined to believe "that the phenomena observed are primarily related to changes of the structure of water used in the experiments." [4] The chemical bond in water is about one tenth of the strength of average chemical bonds so that water is "sensitive to extremely delicate influences and is capable of adapting itself to the most varying circumstances to a degree attained by no other liquid." [5] It is quite possible that solar flares have to be included among these "varying circumstances" [6] which would again lead to a dependence on planetary positions. Considering the role which water and organic colloids [7] play in life we may conjecture that "it is by means of water and the aqueous system that the external forces are able to react on living organisms." [8]

Just how sensitive organisms are has been shown in a series of papers by F. R. Brown. Oysters open and close their shells in accordance with the tides. They continue their activity when brought inland, in a dark container. Eventually they adapt their rhythm to the new location which means that they sense the very weak tides in an inland laboratory tank. [9] Brown also studied the metabolism of tubers and found a lunar period though the potatoes were kept at constant temperature, pressure, humidity, illumination: man's ability to keep conditions constant is smaller than the ability of a potato to pick up lunar rhythms [10] and Professor Bok's assertion that "the walls of the delivery room shield us effectively from many known radiations" turns out to be just another case of a firm conviction based on ignorance.

The "Statement" makes much of the fact that "astrology was part and parcel of (the) magical world view" and the second article that is attached to it offers a "final disproof" by showing that "astrology arose from magic." Where did the learned gentlemen get *this* information? As far as one can see there is not a single anthropologist among them and I am rather doubtful whether anyone is familiar with the more recent results of this discipline. What they do know are some *older* views from what one might call the "Ptolemaic" period of anthropology when post–17th century Western man was supposed to be the sole possessor of sound knowledge, when field studies, archaeology and a more detailed examination of myth had not yet led to the discovery of the surprising knowledge possessed by ancient man as well as by modern "Primitives" and when it was assumed that history consisted in a simple progression from more primitive to less primitive views. We see: the judgement of the "186 leading scientists" rests on an antediluvian anthropology, on ignorance of more recent results in their own fields (astronomy, biology, and the connection between the two) as well as on a failure to perceive the implications of results they do know. It shows the extent to which scientists are prepared to assert their authority even in areas in which they have no knowledge whatsoever.

There are many minor mistakes. "Astrology," it is said "was dealt a serious death blow" when Copernicus replaced the Ptolemaic system. Note the wonderful language: does the learned writer believe in the existence of "death blows" that are not "serious"? And as regards the content we can only say that the very opposite was true. Kepler, one of the foremost Copernicans used the new discoveries to improve astrology, he found new evidence for it, and he defended it against opponents. [11] There is a criticism of the dictum that the stars incline, but do not compel. The criticism overlooks that modern hereditary theory (for example) works with inclinations throughout. Some specific assertions that are part of astrology are criticized by quoting evidence that contradicts them; but every moderately interesting theory is always in conflict with numerous experimental results. Here astrology is similar to highly respected scientific research programmes. There is a longish quotation from a statement by psychologists. It says: "Psychologists find no evidence that astrology is of any value whatsoever as an indicator of past, present, or future trends of one's personal life. ...." Considering that astronomers and biologists have not found evidence *that is already published, and by researchers in their own fields*, this can hardly count as an argument. "By offering the public the horoscope as a substitute for honest and sustained thinking, astrologers have been guilty of playing upon the human tendency to take easy rather than difficult paths"—but what about psychoanalysis, what about the reliance upon psychological tests which long ago have become a substitute for "honest and sustained thinking" in the evaluation of people of all ages? [12] And as regards the magical origin of astrology one need only remark that science once was very closely connected with magic and must be rejected if astrology must be rejected on these gounds.

The remarks should not be interpreted as an attempt to defend astrology *as it is practiced now* by the great majority of astrologists. Modern astrology is in many respects similar to early mediaeval astronomy: it inherited interesting and profound ideas, but it distorted them, and replaced them by caricatures more adapted to the limited understanding of its practitioners. [13] The caricatures are not used for research; there is no attempt to proceed into new domains and to enlarge our knowledge of extra-terrestrial influences; they simply serve as a reservoir of naive rules and phrases suited to impress the ignorant. Yet this is not the objection that is raised by our scientists. They do not criticize the air of stagnation that has been permitted to obscure the basic assumptions of astrology, they criticize these basic assumptions themselves and in the process turn their own subjects into caricatures. It is interesting to see how closely both parties approach each other in ignorance, conceit and the wish for easy power over minds. [14]

Notes

1. This is quite literally true. When a representative of the BBC wanted to interview some of the Nobel Prize Winners they declined with the remark that they had never studied astrology and had no idea of its details. Which did not prevent them from cursing it in public. In the case of Velikowski the situation was exactly the same. Many of the scientists who tried to prevent the publication of Velikowski's first book or who wrote against it once it had been published never read a page of it but relied on gossip or on newspaper accounts. This is a matter of record. Cf. de Grazia, *The Velikowski Affair*, New York 1966, as well as the essays in *Velikovsky Reconsidered*, New York 1976. As usual the greatest assurance goes hand in hand with the greatest ignorance.

2. J. H. Nelson, *RCA Review*, Vol. 12 (1951), pp. 26ff.; *Electrical Engineering*, Vol. 71 (1952), pp. 421ff. Many of the scientific studies that are relevant for our case are described and indexed in Lyall Watson, *Supernature*, London 1973. Most of these studies have been neglected (without criticism) by orthodox scientific opinion.

3. This was found by H. S. Burr. Reference in Watson, *op. cit.*

4. W. W. Tromp, "Possible Effects of Extra-Terrestrial Stimuli on Colloidal Systems and Living Organisms," *Proc. 5th Intern. Biometeorolog. Congress, Nordwijk 1972*, Tromp and Bouma (eds.), p. 243. The article contains a survey of the work initiated by Piccardi who started long range studies on the causes of certain non-reproducible physico-chemical processes in water. Some of the causes were related to solar eruptions, others to lunar parameters. Reference to such extra-terrestrial stimuli is rare among environmental scientists and the corresponding problems are "often forgotten or neglected" (p. 239). However, "despite a certain resistance experienced among orthodox scientists, a clear breakthrough can be observed in recent years amongst the younger research workers" (p. 245). There are special research centres such as the *Biometeorological Research Center* in Leiden and the *Stanford Research Center* in Menlo Park, California which study what once was called the influence of the heavens upon the earth and have found correlations between organic and unorganic processes and lunar, solar, planetary parameters. Tromp's article contains a survey and a large bibliography. The *Biometeorological Research Center* issues periodic lists of publications (monographs, reports, publications in scientific journals). Part of the work done at the *Stanford Research Institute* and related institutions is reported in (ed.) John Mitchell *Psychic Exploration, A Challenge for Science*, New York 1974.

5. G. Piccardi, *The Chemical Basis of Medical Climatology*, Springfield, Illinois 1962.

6. Cf. G. R. M. Verfaillie, *Intern. Journ. Biometeorol.*, Vol. 13 (1969), pp. 113ff.

7. Tromp, *loc. cit.*

8. Piccardi, *loc. cit.*

9. *Am. Journ. Physiol.*, Vol. 178 (1954), pp. 510ff.

10. *Biol. Bull.* Vol. 112 (1957), p. 285. The effect could also be due to synchronicity—cf. C. G. Jung, "Synchronicity: An Acausal Connecting Principle," in *The Collected Works of C. G. Jung*, Vol. 8, London 1960, pp. 419ff.

11. Cf. Norbert Herz, *Keplers Astrologie*, Vienna 1895, as well as the relevant passages from Kepler's collected works. Kepler objects to tropical astrology, retains sidereal astrology, but only for mass phenomena such as wars, plagues etc.

12. The objection from free will is not new; it was raised by the Church fathers. So was the twin objection.

13. On astrology see Paul Feyerabend, *Against Method*, p. 100n.

14. Cf. *Against Method* p. 208n.

On Dismissing Astrology
and Other Irrationalities

Edward W. James

How do we know that the irrational—say, astrology—is so? Or is this question peccantly complex, assuming before showing that we do know? To answer these questions we must consider the received views, which seek to dismiss astrology and its kin by appealing either to methodology or content. Consider first, then, two of our leading methodologists—Sir Karl Popper and Thomas S. Kuhn.

Popper claims that astrology is to be rejected because it fails to seek out falsifications, a task that for Popper marks the rational person. He tells us: "Astrologers were greatly impressed, and misled, by what they believed to be confirming evidence—so much so that they were quite unimpressed by any unfavorable evidence." [1] Alluring as this observation may be, however, it fails to close the door on astrology. For if Popper means that astrology, past or present, neither makes specific predictions nor seeks to revise its theories when the predictions fail, then he errs factually. For astrologers repeatedly predict on the basis of their theories and revise on the basis of their failures. We see this, for instance, in the astrological theory of subsumption. If a person whose horoscope predicts fortune instead encounters failure, the astrologer seeks to explain this by subsuming the relevant horoscope under a more inclusive one that calls for misfortune. Hence it was the horoscope of Hiroshima that proved dominant during the waning days of World War II. Yet if Popper means that astrology fails to put itself *as a whole* to the test, then he errs conceptually. For astrology claims to be a discipline or area of study, not a specific theory or law, and hence is no more to be falsified than physics or astronomy.

Let us then turn to Thomas S. Kuhn's advice, which urges us to consider what scientists do. According to Kuhn, "no process essential to scientific

I profited greatly from the encouragement and criticisms of Charlene Entwistle and Professors Steven Sanders and Robert Fitzgibbons.

development can be labeled 'irrational' without vast violence to the term." By examining the practice of science we will find what distinguishes astrology from science, namely, that astrologers "had no puzzles to solve and therefore no science to practice." [2] Kuhn argues that puzzle-solving, from recording more precise observations to determining the formal relationships of a theory, is the essential mark of a science. Kuhn argues further that astrology lacks such a mark. However, Kuhn too fails to shut out astrology. For astrology boasts of an abundance of puzzle-solving, from determining more precisely the time of birth to including other constellations besides "the twelve." [3]

Although Popper and Kuhn may be correct in seeing astrology and company as irrational, they do not show them to be so. These "disciplines" may not *rationally* falsify, or may not *rationally* solve puzzles, but they do engage in falsifying and puzzle-solving and confirming and whatever other general methodological strategies we may wish to add. To show that they do not rationally falsify or solve puzzles or whatever would demand an in-depth scrutiny of these areas, and *that* is a scrutiny neither Popper nor Kuhn cares to make.

Hence we must look to more than just a slogan if we wish to expose an area of study like astrology. So let us now consider briefly another type of critique against astrology, one that addresses its content rather than its method. We shall see immediately that this approach, dealing in carefully developed argument, does not fall into the trap of the quick dismissal. Consider then the following arguments.

1. Astrologers tell us that the stars influence our lives, especially the twelve constellations along the ecliptic—the apparent annual path of the sun through the heavens. Astrologers determined this path by observing the sun as it set and placing the sun "in" the constellation that supplied its backdrop. Why they chose sundown rather than some other time to place the sun "in" a constellation is obvious; it was easier. They chose not to place the sun at sunrise because few wished to rise that early. They chose not to place it at noon because it would be difficult to remember just where the sun was (at noon) and then to figure out the position of the stars (at noon) by calibrating their counterclockwise drift. Hence that the sun is "in" Aries, say, seems to be attributable to human sleeping habits rather than to any cosmic reality.

2. Moreover, the sun is not "in" these constellations an equal amount of time. It spends, for instance, forty-seven days in Virgo but only six in Scorpio. So to allow the poor Scorpios equal time, astrologers divided the ecliptic into twelve equal compartments of 30 degrees each, called them the "signs of the zodiac," and henceforth dealt directly not with the stars but with the signs—signs that have at best a rough-and-ready connection to the initial constellations. Apparently the signs, originated by astrologers to give equal time to the Scorpios and their neglected brethren, can mediate the powers of

the stars themselves. Or do the signs now have the Force?[4]

3. Worse still, the signs fluctuate. In the first century B.C. the sun "entered" Aries on March 20, the time of the vernal equinox, and since then astrologers as a rule have stated that the sun is "in" Aries during this time. However, as Hipparchus observed in 130 B.C. and Newton explained in 1687 A.D., the equinox itself is moving. Every 2,140 years the equinox slips back one sign, so that today the sun "enters" Aries not in the latter part of March but in the latter part of April. Hence what was a Taurus two millenia ago is now an Aries; what was a Leo is now a Cancer. Apparently there occurs a cosmic shift of character every two millenia, or the original signs (which mediated the constellations) are now themselves mediated by the "new" signs.

4. Furthermore, astrology offers no account, or even the promise of an account, of itself. To explain the infuence of the stars—or of the shifting signs—in terms of emanations, waves, or whatnot seems to be no more than obfuscation. For unlike the mathematically denoted entities of physics, such as Pauli's neutrino, the alleged emanations of astrology have never been experimentally identified. Moreover, to appeal to "the principle of correspondence," which tells us "as above so below, as below so above"—for example, as Mars above is red so Mars below means blood and war—is to commit a common fallacy of false cause and so to move from any attempt at science to mere magic.[5]

5. We can finally note that statistical studies have repeatedly shown the predictions of astrology to be false. A study of the *American Men of Science*, for instance, revealed no unusual groupings. Nor did an examination of 2,000 famous artists and painters find any preference for Libra. Nor again did a look at 7,000 persons who were connected with violent death see Mars or Saturn as ascendant.[6]

What can we say of arguments 1 through 5 above, and of similar arguments concerning content that we might present? As a group these provide excellent reasons for us to reject astrology, in the sense of rejecting astrological claims as confused, unsupported, or false. But surely we want to say more than that we have reason not to be believers. We want to say that no one today *should* adhere to astrology. We want not just to criticize it but to dismiss it, and to show that those who uphold astrology are simply irrational.

Yet no matter how devastating such a critique as the above may be, it cannot show that astrology or any kindred area is irrational. For in science and elsewhere we repeatedly find similarly severe criticisms. Consider contemporary physics, for instance, which offers not only the incoherence of the wave–particle dualism—where the fundamentally real is held to be in a specific place and interacting by local contact (like a particle) and yet to be spread out over an area and interacting by interference (like a wave)—but also a plethora of unrelated and so seemingly chaotic microparticles, red-shifts that cannot be aligned with quasars, and so on. Yet it would be silly indeed even to suggest that we dismiss physics or our various physical theories as irrational.

The point is that all positions, even our own, face severe criticisms. And all positions but our own we find to be false and even absurd. But that we find positions in such straits hardly condemns them as irrational. For to find a view irrational is to disallow argument. It is to say that argument with *them* is foolhardy or a waste of time and breath. Consequently to show that we have good reason to reject a position is not at all to show that the position is irrational. Most of the time it is an invitation for further discussion. But such an invitation is precisely what we do not want to tender astrology and its ilk. We want to say of them that it is a waste of time to discuss "the issues."

We have so far investigated two types of criticisms of astrology, of method and content, and have rejected both—the first because no formula can portray (ir)rationality, the second because major difficulties plague all positions and do not by themselves signify irrationality. At this point it seems clear that a critique of the content of astrology, no matter how damaging, will never be able to show irrationality. Indeed such a critique is a way of conferring respect; we take an area of investigation seriously enough to consider the ideas involved with careful attention. But we want to say more than that we have good reason for not heeding astrology. We want to say that astrology and company should not be taken seriously at all.

To say this we must once again turn to the methodology of astrology. Before we embrace anarchism in knowledge, where astrology and astronomy fare the same, let us once more look at how astrologers reason with regard to their study. But this time let us not look with a formula in mind, with the aim of a quick dismissal, but instead with a more careful attention to the complexities of reasoning. This means, alas, that we will have to read or listen to what astrologers say regarding astrology. But we will find that the dreariness of the study itself supplies a crucial clue to what lies behind irrationality in general and the irrationality of astrology in particular. For to listen to astrologers speak as astrologers is to reject them, not because we fail to understand them, and definitely not because what they say is bizarre—after all, contemporary physics is unmatched in the region of the strange—but instead because we understand them all too well and because what they say is so common.

Consider the following points concerning what passes for respectable reasoning—argument, falsification, puzzle-solving, self-reflection—in astrology. These points, by the way, do not merely concern streetcorner and newspaper astrologers, the bane of the "serious astrologer." They concern the allegedly respectable astrologers—the recognized leaders of the field—who hold that in a "proper, ordered society the astrologer should be in much the same position as doctor, psychologist, or priest."[7]

1'. Had there been no Common Fallacy List as supplied by logic texts, we would have had to invent one just to deal with the various fallacies endemic to astrology and kin. In just a few pages of a typical work[8] we go from the standard appeal to authority, this time to Kepler, "one of the last of the great astronomer-astrologers"—as if being an authority in the past and in one area magically carries over to being authoritative in the present and in another area—to the usual vague claim: "One of the most striking pieces of evidence in favor of astrology is the fact that those born at the same moment of time in approximately the same place have a life pattern that is very similar." What, we might well ask, is "the same moment" in an Einsteinian world? And what qualifies as "the same place" on a rotating and revolving planet: Massachusetts? Boston? Beacon Street? 325 Beacon Street? In the same few pages we meet the egregious analogy: "The construction of the atom bears a close relationship to the Solar System"—again set in a quantum-mechanical world that exploded such a notion decades ago. For the record we might note one more fallacy, a mundane reliance on false cause:

> On the day Queen Wilhelmena of the Netherlands married only one other woman in the country was allowed to marry. She was a friend of the Queen, her name too was Wilhelmena, and she was born on the same day as the Queen.

The inference here is all too easy to parody. Scotch and water makes me drunk, bourbon and water does too, and again Canadian club and water. So clearly water makes me drunk.

This is typical of the literature of astrology. Here we meet no subtleties in argument, as we do with Copernicus or Galileo, but a dull series of mismanagements of even the simplest considerations. Is it any wonder that such stalwarts as Popper and Kuhn shrink from contact with such stuff and search for an easier way—a slogan, a formula, anything?

2'. When we criticize astrology we find the replies trite and shallow, not worthy of close attention. We have already seen one attempt to reply to a criticism in the so-called principle of correspondence. Consider also this reply to the hoary objection that astrology cannot account for the twins who lead different lives. We are told that twins sometimes do

> look and act alike, marry at the same time, have the same number of children, die on the same day, etc. How can we explain this except by astrology? To explain the other possibility [where they lead different lives], we have only to take into account the differences in heredity and environment.[9]

The astrologer would have us explain the similarities by astrology, the dissimilarities by heredity and environment. But what justification could there be for this neat bifurcation of duties? Nary a word do we hear of the possibility of a unified view through genetics or even a note concerning the arbitrariness of this division—unless we count the expostulation that such a problem "is so complex and so unpromising that most astrologers prefer to admit their incompetence in the matter and get on with their work."[10] Hey, don't trouble me with *argument*, I'm busy.

3'. Furthermore, when astrologers do come upon an interesting consideration in their favor, they make no attempt to develop it—or in Kuhn's terms, to engage in "puzzle-solving." For instance, the non-astrologer Gauquelin researched 25,000 births and found a pronounced correlation between the planetary positions and general occupation. More doctors seemed to be born when Mars or Saturn was rising or had reached its culmination, for example, than would have been predicted by chance alone.[11] Since then astrologers have referred to this work as final: "Gauquelin's work proves once and for all, and incontestably, that there is *something* to astrology."[12] From the context it is amply clear that this "something" is close to if not the whole truth. Other possibilities are left unexamined and even unmentioned—for example, that children may tend to be born at the same time as their parents and also tend to follow their parents' interests (especially in Europe), or that Gauquelin may have erred in his statistics. Instead we are told emphatically that astrology has been incontestably demonstrated. That reasons bear relative weight, that statistics are multiply edged—well, most astrologers are apparently willing to admit their incompetence in the matter and get on with their work. That their work should involve knowledge of statistics and the like is of course irrelevant to raking in money.

4'. We might finally note that one of astrology's failings is its penchant for offering specific directives. Astrology sees that the social sciences do not tell us when to marry, to make a major decision, to take a chance. Thus it advertises itself as filling a crucial gap, as doing what the social sciences either cannot or refuse to do. Yet it is in this very reticence of the social sciences that we meet an important mark of rationality. For rationality does not mean that everything is explained, all tidy and neat, but instead demands a recognition of the limits of explanation and theory. These limits range from the statistical point, that our theories predict the behavior not of individuals but only of a proportion of a group, to the conceptual point, that our concepts are vague, lack sharp boundaries, and so cannot always say what is included in what. Hence the coincidental, which astrology has ostensibly declared war on, turns out not to be a failure of thought, a surrendering of explanation—as portrayed by astrology—but a consequence of the very logic of our systems. Our concepts can only be stretched so far and can only cover so much. In seeking for a life recipe, in other words, astrologers show no appreciation for the limits of reason.

Faced with 1' through 4' above as a typical collection of fallacious moves and dubious maneuvers, what can we do? Where do we begin? So much is wrong, so much is off, that it is doubtful whether any critique would help. For the critique would likely be understood only by those who do not make such a mass of errors in the first place. Do we speak of false cause? But first we must lay the groundwork by developing the notion of causality and lawful generalizations and the like. Do we speak of unifying our theories? But then we must discuss the criteria of coherence and simplicity, their relative weights, how they fit in with other criteria, and so on. Do we address the problem of developing our ideas through puzzle-solving or the notion of the limits of explanation? But think again of how much is assumed in these rather basic ideas.

To show that astrology is irrational, accordingly, is to tell a long story. It is to take what we have stated above in 1' through 4' as topic sentences and follow through these and related ideas to their full development. It is to read the woeful passages of what passes for argument in works of astrology and expose the myriad patterns of specious reasoning for what they really are. We know, perhaps, what we mean by argument, falsifying, puzzle-solving, and the like. But so much, indeed most of it, is implicit, signifying the past mastery of myriad techniques, standards, qualifications, and so on.

Astrologers thus do not miss out because they lose the "rationality race" but because they fail to qualify for and so fail to enter the race at all. Rationality involves mastering no one procedure—whether it be falsifying, puzzle-solving, or whatever—but, like qualifying for other matters, demands a mastery of a host of procedures. It does not demand that we be right, that we win, but that we be competitive—as revealed by our previous races and showings of competence. Hence to fail to do well in certain previous races does not by itself mean that we fail to qualify; for there are many races involved in the final qualifying decision. But what if one comes in last race after race, or frequently runs the wrong way? What if throughout astrological writings we meet little appreciation of coherence, blatant insensitivity to evidence, no sense of a hierarchy of reasons, slight command over the contextual force of criteria, stubborn unwillingness to pursue an argument where it leads, stark naïveté concerning the efficacy of explanation and so on? In that case, I think, we are perfectly justified in rejecting astrology as irrational.

Our initial question, How do we know that astrology is irrational? turns out after all not to be complex. We do know. Astrology simply fails to meet the multifarious demands of legitimate reasoning.

This, to be sure, is vague. But rationality is a global notion, one embracing a legion of criteria, procedures, techniques, asides, qualifications, and the like. To insist on a precise definition is to misunderstand it. Rationality could be characterized as the appropriate, correct, and self-critical openness to ideas. We reason appropriately when our reasons fit the situation, correctly when they are valid and sound, and self-critically when our ways of reasoning are themselves scrutinized. And similarly irrationality could be characterized as the pervasive lack of the appropriate, correct, and self-critical openness to ideas. But the necessary generality of any such characterization is soon seen as the real message. In the end this only tells us that rationality involves, well, legitimate reasoning—and irrationality, illegitimate reasoning.

Yet we do not lack clear ideas regarding the various constituents of rationality. It is a gross understatement to assert that we often know when an argument is valid or invalid, when it should be or has not been qualified, what sort of considerations are or are not suitable.

We can be said to know that astrology is irrational not because of our advertised critiques but in spite of them. We know for so many reasons—or for so many failures of reasons. But we hardly want to be troubled to chronicle all of these failures. For as we have seen this would involve developing so many elementary ideas and rudimentary techniques and methods—all of which has been done before, by others—that it would in no way be time well spent. Hence we economize and advertise a neat and quick dismissal, one that appears to do the job as long as we do not examine it too carefully.

Nor is this to say that rationality is just a matter of common sense. Rationality must be learned—if we are to do it well. Rationality may have no essence, a formula to learn and practice as we did the multiplication tables, but it still must be taught and learned. Indeed, even more so. For mastering the full complexity of rationality is a task that is inherently incomplete-able—as well as one that astrologers and their kin have hardly begun. For their arguments are as if Plato and Aristotle and Lucretius and Aquinas and Galileo and Locke and Hume and Planck and Einstein and Russell and Wittgenstein and, indeed, Popper and Kuhn, and so on, had never been. Moreover, that most astrologers would grasp this list with trembling and irate hand, charging that Plato and Aristotle and Aquinas were not adverse to their star study, only betrays how far they have to travel. If the simple fallacies like appeal to authority escape them, what can we expect when the analysis gets tough?

If this sounds harsh, well, so be it. For there are not that many areas of study that can be rejected as irrational. Since rationality is such a complex notion, it admits of varying degrees and forbids sharp lines (as found, say, in slogans). To commit a fallacy or two, to be caught in a contradiction, to fail to pursue a consequence, and the like do not by themselves condemn one to irrationality. Howlers happen to us all. It is only when we encounter one piece of bogus reasoning after another, as we do with astrology, that we have a clear case of irrationality.

Of course none of these considerations will succeed in abolishing astrology and its companions. Although astrologers do appear silly when compared to thinkers who reason well, not all care to compare, let alone strive to attain such standards. Moreover, since rationality lacks an essence, the one standard that must be met, it is easily counterfeited. When there is no one lack that we can point to as decisive, when we cannot dismiss astrology's gibberish by a word, few care to tell or listen to the required tale. For then we must read the astrological literature. Unless we are slumming or looking for another instance of fallacious reasoning, wading through such muck offers no pleasure. And woe, even after we do take the time to point out so many foolish flaws, other astrologers arise, like mythic armies of yore, claiming to have swept out their houses. Although a cursory look on our part reveals the same dreary deal, the thought of grading such remedial reasoning again—and again and again—is overwhelming.

Hence it should be no surprise that we tend to follow Popper and Kuhn and look for a quickie dismissal, one that suffices for the already converted but does not approach what we really need to say. A wave of the hand, a clever aside, may protect a tired psyche but hardly suffices to reject the irrational. For that requires an extended, nay, an indefinitely long series of analyses, war after war, and all along we have been treating it as a skirmish.

Notes

1. Karl R. Popper, "Science: Cinjectures and Refutations," in *Conjectures and Refutations* (New York: Basic Books, 1962). Part of "Science: Conjectures and Refutations" is included in this volume.

2. The first quotation is from Thomas S. Kuhn, "Reflections on My Critics," in *Criticism and the Growth of Knowledge*, ed. Imre Lakatos and Alan Musgrave (New York: Cambridge University Press, 1970), p. 235; and the second is from Kuhn, "Logic of Discovery or Psychology of Research?" *Criticism and the Growth of Knowledge*, p. 9 and included in this volume.

3. For instance, for a discussion on new constellations see Owen Rachleff, *Sky Diamonds* (New York: Hawthorne Books, 1973); or for the theory of subsumption, see Henry Weingarten, *A Modern Introduction to Astrology* (New York: ASI, 1974), p. 65.

4. For more on 2 and 3 see Isaac Asimov, *Matters Great and Small* (New York: Doubleday, 1975), Chs. 3 and 4.

5. Cf. Lawrence E. Jerome, "Astrology: Magic or Science?" *The Humanist* 35, no. 5 (September/October 1975): 10–17.

6. Cf. Michel Gauquelin, *Astrology and Science*, trans. James Hughes (London: Peter Davies, 1969), pp. 132–146.

7. Ronald C. Davison, *Astrology* (New York: Bell, 1963), p. 169.

8. Ibid., pp. 9–13.

9. Weingarten, p. 66.

10. John Anthony West and Jan Gerhard Toonder, *The Case for Astrology* (Baltimore: Penguin Press, 1973), p. 148.

11. Gauquelin, pp. 161–171.

12. West and Toonder, p. 172.

Spheres of Influence

Michel Gauquelin

"You shouldn't dismiss as incredible the possibility that a long enough search might reveal a golden grain of truth in astrological superstition," was the advice of the great Renaissance astronomer Kepler. In all modesty we might claim to have proved Kepler right when we describe our work and the surprising results we have obtained. For I believe that I have shown a link between the movements of the planets and human character and career.

In 1953, we had just finished a major inquiry into astrology. We had subjected its laws to statistical tests and all the results remained negative. There seemed to be no truth in the idea that signs of the zodiac determined anything or that horoscopes could be used to predict the future.

When, however, we came to examine the position of the planets at the hour of birth of 576 members of the French Academy of Medicine, something very odd showed itself. Among these distinguished doctors, certain planetary positions appeared at the moment of birth far more than they should. This result could not be dismissed as "mere chance." Any statistician would have judged it to be very significant—which is what we did.

What we observed did not fit any traditional law of astrology exactly. Famous doctors seemed to have a peculiar preference for being born when either Mars or Saturn had just risen above the horizon or, second, when either planet had reached its high point in the sky (its culmination) for that day. Ordinary mortals seemed to differ. We took care to examine the hours of birth of perfectly ordinary people taken at random from electoral rolls. They showed no preference for timing their birth according to either Mars or Saturn.

All this needed to be examined in greater detail and, if possible, explained using a sound methodology. It is worth stressing again that these findings clearly had nothing to do with signs of the zodiac.

Reprinted from *Psychology Today* (Brit.), No. 7, October 1975, pp. 22–27.

A Day in the Life of a Planet

Because the earth rotates round its axis every 24 hours, the heavenly bodies—the sun, moon, planets, constellations and stars—seem to rise in the east. Each body then climbs to its culmination, its high point in the sky for that day, before coming down, like the sun, to set in the west. The planets follow this pattern. Every day, each planet rises above the horizon close to the time it did the day before, climbs in the sky and then sets.

To apply probability theory to the positions of the planets we divided the circle of the daily orbit of each planet into twelve, eighteen and thirty-six sectors. The more sectors, the more detail one can obtain, but the methodology remains the same in all three cases. In this article the results we present are based entirely on division into twelve sectors. We have at our disposal, therefore, a kind of cosmic roulette with numbers running from one to twelve. When a child is born, each planet in the solar system is in one of these twelve sectors. If we examine 100 or 1,000 births, we can count the number of times that Mars lies, for example, in sector one, sector two and so on, to sector twelve.

20,000 Famous Births

Our data on doctors who were members of the French Academy of Medicine encouraged us to gather data on the times of birth of more successful people in various fields. We looked at scientists, actors, writers, sports champions and other professionals. Success may seem an arbitrary criterion but it is an objective one. It is relatively easy to agree whether or not a person is successful. This also allowed us to examine people in each profession who clearly showed the psychological characteristics that marked them for success in it.

Celebrities are also useful because they are well documented in dictionaries of national biography, *Who's Who* and a host of other reference works. In these we found many dates of birth, although the hour of birth is not—alas—usually noted.

We therefore got in touch with the Registrar's office at the place of birth of each of our subjects. In many countries, including France, Italy, Belgium, Holland and Germany, the hour of birth is automatically recorded. (In England and Wales the time is noted only in the case of twins. It is different in Scotland!) Our total number of subjects was over 20,000 European notables including politicians, artists, scientists and actors. In 1970, our laboratory undertook to publish in six volumes the hours of birth and corresponding planetary positions of each of these 20,000. This will enable all interested scientists to verify our data and conclusions.

Timing Your Destiny

The more calculations we performed, the more the initial trend was confirmed. Surprising as it may seem, a more and more precise statistical relationship appeared to emerge between the time of birth and great men and their professional success. It was no longer just a matter of the doctors of the French Academy of Medicine. Each professional group seemed to have a planetary "clock" all its own. This basic observation seemed to hold especially good for the moon. Mars, Jupiter, Saturn and Venus showed the trend to a lesser degree.

It was almost as if, for certain occupations, the presence at birth of a planet which had either just risen over the horizon, or reached its high point in the sky, seemed to "provoke" or "cause" success. Among 3,647 scientists, 704 were born after the rise (sector one) or after the culmination (sector four) of Saturn. By chance alone one would expect 598. The odds against the number we found are 300,000 to one. Mars dominates a sample of 2,088 sports champions. 452 were born either after it rose or culminated. Chance predicts 358 should be. The odds against: five million to one.

Soldiers and Jupiter

Among 3,458 soldiers, Jupiter is to be found 703 times, either rising or culminating when they were born. Chance predicts this should be 572. The odds here: one million to one. There is a lesser, but similar, trend observed for births to occur when the planets are in sectors seven and ten—just after the setting and at the lower culmination, the low point of its path.

The effect does not always seem to be extended on the *plus* side. Sometimes, the presence of a planet just after it had risen over the horizon, or just after it had culminated seemed to "prevent" success in certain professions. Only 203 of 1,473 great painters were born with Mars in sectors one or four. The odds against this: 200 to one. Only 287 writers and journalists were born with Saturn in the same position. The odds against this: 300 to one. Table I shows the essential data; figure 1 gives details for Jupiter.

Other Research

All manner of scientists responded sceptically to these astonishing results. Statisticians, psychologists, astronomers, demographers, biologists and even midwives took pains to explain it away. They argued that we might have been biased in choosing our births, mistaken in our use of statistics, incompetent in handling astronomical details. But the evidence didn't budge.

The "resistance" of these inexplicable relationships intrigued the grandly titled Belgian Committee for the Scientific Investigation of Phenomena Reputed to be Paranormal. This "Para-Committee," based in Brussels, decided to repeat our observations. The Committee is made up of thirty

TABLE I. Professional groups, with an abnormal frequency (higher or lower) in the number of births before the rising and culmination of the planets (sectors 1 and 4):

Planet	Profession	No. of births	Frequency observed at rising and culmination	Theoretical frequency at rising and culmination	Divergence	Possibility of chance 1 in:
Mars	Science and Medicine	3647	724	626	+ 98	500,000
	Sport	2088	452	358	+ 94	5,000,000
	Army	3438	680	590	+ 90	1,000,000
	Painting	1473	203	253	− 50	200
	Music	866	120	149	− 29	30
Jupiter	Army	3438	703	572	+ 131	1,000,000
	Politics	1003	205	167	+ 38	100
	Theatre and Cinema	1409	283	235	+ 48	1,000
	Journalism	903	185	150.5	+ 34.5	100
	Science and Medicine	3647	602	540	+ 62	30
Saturn	Science and Medicine	3647	704	598	+ 106	300,000
	Painting	1473	188	238	− 50	200
	Literature and Journalism	2255	287	338	− 51	500
Moon	Literature	1352	292	225	+ 67	100,000
	Politics	1003	189	167	+ 22	20

For the calculation of theoretical frequency, the exact astronomical definition of the positions of the zones after rising and culmination, and the actual values obtained, please refer to M. and F. Gauquelin, vol. 1, Series C, Statistical Results of the Series A and B (Laboratoire d'etude des Relations entre Rythmes Cosmiques et Psychophysiologiques, Paris, 1972.)

scientists of different disciplines. Since it was set up 20 years ago, it had often denounced false correlations especially where the influence of the stars was being studied. The Committee is feared by those who believe they have discovered new facts since, often, the Committee's attempts to verify a new fact have revealed nothing more novel than a mistake. The Committee was very hostile to our conclusions but decided in 1968 to repeat one of our most significant studies on a new set of new subjects. They chose sports champions. They assembled a different group of 535 champions and a computer was duly fed all the necessary details about the hours of birth of these sportsmen and with all the complex movements of Mars.

The statistical evaluation of the position of Mars at the time of birth of these 535 sportsmen showed, once more, that the planet was either rising or culminating when they were born far more often than one would expect by chance. The Committee noted the phenomenon with surprise and judged that it posed a real scientific problem. It also expressed doubts as to what

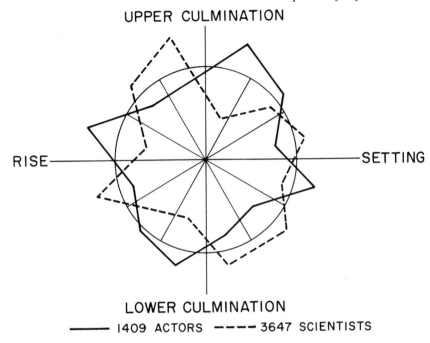

UPPER CULMINATION

RISE —— SETTING

LOWER CULMINATION

—— 1409 ACTORS ———— 3647 SCIENTISTS

Figure 1. The position of Jupiter at the birth of actors is very different from its position at the birth of scientists. The times favouring the birth of successful actors (continuous lines) correspond to the times that "prevent" the birth of scientists (broken lines). The circle indicates the theoretical frequency for ordinary births. The effect is statistically remarkable.

might have caused it. Despite a welter of critical studies and the use of exceptionally refined controls, the fact persists. (Table II)

TABLE II. The planet Mars in its daily movement at the moment of birth of sports champions.

A Sectors	1	2	3	4	5	6	7	8	9	10	11	12	Total
Observed	172	126	128	160	117	104	126	125	132	112	118	133	1,553
Expected	138	135	133	128	126	124	122	122	127	130	134	134	1,553

B Sectors	1	2	3	4	5	6	7	8	9	10	11	12	Total
Observed	68	47	35	52	35	31	36	51	53	53	40	34	535
Expected	48	47	45	44	43	43	42	42	43	45	46	47	535

Mars is much more frequent than usual in sector 1 (after rising) and in sector 4 (after culmination):
A. Our investigation of 1553 sports champions (332 against 266, significant to a degree of .001)
B. The investigation carried out by the "Para-Committee" on 535 other sports champions (120 against 92, significant to a degree of .01)

Personality and Success

How does one explain these facts? Psychological factors underlie everybody's success. Success at the highest level in a profession reveals both personal taste and a powerful vocation. There is a psychological connection between personality of individuals and their success. Many psychologists have noted this. Character is an important part of success. And each profession has a typical psychological profile.

We sent out a questionnaire to examine these characteristics of professional groups. University-educated subjects were asked to specify which characteristics or personality traits seemed to favour and also to handicap those who sought success in science, art, letters, politics and the military. The analysis of replies allowed us to frame a psychological profile for each profession. Scientists are, for instance, scrupulous, self-absorbed, timid, quiet, discreet, prudent, cautious, stubborn and methodical. Sports champions are persistent, courageous, energetic and bear hardship well. "Captains of Industry" are realistic, hard, dominating, authoritarian and well organized.

This image of traits is confirmed by psychologists who have looked at the link between personality and professions. Anne Roe studed a small number of selected scientists, all of whom were distinguished. She found that the physicists and biologists tended to have had an isolated childhood. They dated little, had a restricted social life and few contacts with others.

Skilled and Supple Sportsmen

In his work on the psychology of sport, Michel Bouet has written of champions: "Sport does not merely call on their physical attributes of speed, endurance, skill and suppleness, but also calls on their will, courage and persistence." The French psychologist, Claude Levy Leboyer, writing of important business men in his study of success, argued: "Successful businessmen reveal traits of character which one does not find among famous writers or scientists. These include the ability to make rapid judgements, a high sense of realities and a need for constantly driving themselves (a new goal is set the moment an old one is reached)."

To return to the planets. Two factors set scientists apart. They are usually introverted and their times of birth are unusually linked to various positions of the planet Saturn. Hypothesis: there is a relationship between Saturn and the classic personality of the scientist. Sports champions, for their part, are energetic, brave and their times of birth seem unusually linked to Mars. Is there, in this case, a relationship between the planet Mars and the personality traits that go to make a sports champion? One could repeat the hypothesis for other groups—namely that what these planetary oddities reveal is not luck or fate but the workings of various personality factors which have been influenced by the planets.

Collecting Individual Traits

To establish this hypothesis, one has to examine many individual cases to show the link between the planets and personality.

We proceeded like this. We used the same set of people as before. This time, we collected their biographies. In 1968 we devised a way of using biographies objectively. We based our method on the idea that there are permanent traits of character such as that defined by J. P. Chaplin in his *Dictionary of Psychology*: "a relatively persistent and consistent behaviour pattern manifested in a wide range of circumstances." A short example may be useful. This appreciation of the Nobel prizewinning writer, François Mauriac, appeared the day after his death:

> One of the dramatic features of his long career was his impetuous curiosity which made him turn to journalism where he soon made his name. Irritable—and irritating for others—he wrote ferociously, impulsively and indefatigably. He made his *Bloc-Notes* [a literary diary that appeared weekly in *The Figaro*] into an institution. He expressed in them his enthusiasms and his hates directly. His political activity was the product of his whims and passions. Such a career made François Mauriac a 'character' of letters. His books were awaited, his sayings quoted. His attacks were themselves attacked and he responded to these and the whole battle received much official notice. The long emaciated silhouette of Mauriac became one of the images of Parisian life. He liked the theatre, the music hall and the atmosphere of premieres. His body had become bent with age. His voice became thick and harsh as a result of a painful illness, but this seemed to point rather than lessen the words which poured out, and the enthusiasm or anger that he expressed lacked no clarity.

This small, vivid portrait on its own is useless for a statistical analysis. We therefore underlined all the traits of character and behaviour attributed to Mauriac. This impoverishes the portrait but it does make it possible to pool it with others in a large statistical analysis. For, by reducing the portrait to each mention of a trait, we obtain the following version of the writer: "Curious—impetuous—irritable—ferocious—impulsive—indefatigable—enthusiastic—cruel—expresses himself directly—prey to whims—passionate—'a character'—coins sayings—lively—loves theatrical life—easily angered—enthusiastic." For each trait we opened a file, carrying on one side Mauriac's name, and on the other the position of the planets when he was born.

What we did for Mauriac, we repeated for other celebrities, using published accounts of their personality. This gives us files on thousands of personality traits taken systematically from biographies. To ensure the objectivity of the study we have to:

(i) achieve a certain unity in all this biographical material.

(ii) never eliminate a biography because it fails to fit our hypothesis.

(iii) consider every trait mentioned in every biography.

A computer programme at the University of Paris's Faculty of Science allowed us to analyse the data collected as objectively and minutely as possible. Table III gives a short extract of one of the listings obtained. The study looked at one professional group after another. The data is published by our laboratory in our *Psychology Monograph Series* and each monograph gives, in English, a description of the method used, the statistical results, the catalogue of all the personality traits and, in full, all the biographical references from which these were culled.

TABLE III. Relationships between the position of the planets and character traits.

Name	Character	Frequencies	Moon	Venus	Mars	Jupiter	Saturn
Busnel, R.	Active	1	5	28	30	20	6
Codos, P.	"	1	30	20	24	10	35
Faucheux, L.	"	2	11	2	32	30	27
Fournier, A.	"	1	5	27	12	29	14
Gauthier, B.	"	2	24	21	2	11	16
Gerardin, L.	"	1	24	23	22	17	35
Gerbault, A.	"	1	33	36	11	26	12
Graule, V.	"	2	1	11	2	1	4
Hilsz, M.	"	1	33	21	3	26	25
Anglade, H.	Adaptable	1	8	22	18	19	3
Batteux, A.	"	1	18	19	26	22	19
Bastien, J.	Skillful	1	33	9	11	16	7
Batteux, A.	"	1	18	19	26	22	19
Deglane, H.	"	1	23	9	8	19	22
Drigny, G.	Affable	1	23	25	29	22	28
Rimet, J.	"	1	3	11	35	12	34
Dewaele, M.	"	1	33	22	9	1	29
Thoret, J.	Nervous	1	6	11	21	7	24
Michard, L.	Agreeable	1	18	19	9	3	6
Bigot, J.	Aggressive	1	21	36	10	25	12
Famechon, E.	"	1	5	32	2	9	6
Codos, P.	Bitter	1	30	20	24	10	35

Examples for sports champions. From left to right: the name of the champion, the character trait attributed to him, the number of times that the trait has been attributed to the champion, the position (from 36 sectors of the daily movement) at birth of the moon, Venus, Mars, Jupiter, and Saturn.

The Martian Personality

We assembled many biographies of sports champions from works like *The Dictionary of Sports* and sporting weeklies edited by sports specialists. They give not just information about performances and records but also indications as to personality. The 6,184 character traits attributed to these individuals were each filed on a card alphabetically. With each trait is also filed Mars's position at the birth of the individual who has the trait.

We then looked at the literature on the psychology of sport. We used especially the work of M. Vanek and B. Cratty, *Psychology and the Superior Athlete* (Macmillan 1970), and M. Bouet, *Les Motivations des Sportifs* (editions Universitaries 1969). These allowed us to establish a character image of a typical champion. Here are some of his attributes: energetic, dynamic, strong-willed, brave, hard, tireless, lively, iron-willed, persistent.

We examined our list of champions. Each time we noted a trait which had been listed, we saw where the planet Mars was at the time of birth of that particular champion. If Mars is over the horizon, and after culmination is linked with becoming a sports champion, then one can argue that the influence of Mars ought to be even stronger on those sportsmen whose personality approaches the typical sportsman's character.

This prediction was confirmed. Take the traits *brave* and *strong-willed*. Mars is either rising over the horizon or, more often, just culminating at the birth of champions who have these traits. The excess of Mars, so to speak, is so great that the odds against it happening by chance are ten million to one. If one chooses sportsmen whose characteristics are untypical—calm, lazy,

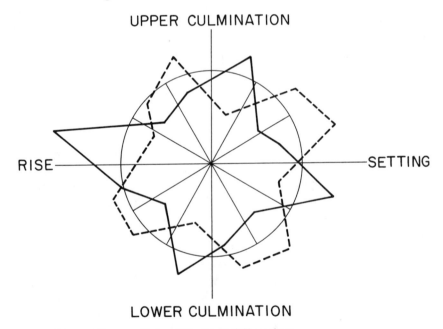

UPPER CULMINATION

RISE

SETTING

LOWER CULMINATION

Figure 2. Mars and Strong-willed and Weak-willed Champions
Circle: the mean number of champions born with Mars in each part of its daily orbit.
Continuous lines: strong-willed champions are born more often just after Mars has risen or culminated.
Broken lines: weak-willed champions are born less often with Mars in these positions.

nonchalant—not many of them were born when Mars was either rising or culminating. The position of Mars at the birth of strong-willed champions is utterly different from the position of Mars at the birth of weak-willed champions (see figure 2). One last statistic: Mars appears in sectors one, four, seven and ten in 64 percent of the births of strong-willed champions as against 33 percent for the average man. For sports champions lacking the iron will Mars is present at 29 percent of births. The position of Mars is the expression of a personality factor. Only incidentally is it related to professional choice.

The Saturn and Jupiter Temperaments

We conducted a similar study on well-known actors. This time we examined the personality traits connected with the planet Jupiter, which is linked with actors' births more than expected by chance (see Table I). Psychologists have listed the following among the most noted characteristics of actors—elegant, immodest, exuberant, funny, at ease, wastrels, boasters, theatrical, eccentric and egocentric. Not all actors share this motley assortment of defects. Louis Jouvet, the great actor, suggested a distinction between exhibitionistic and restrained actors. The latter were rarer and possessed quite opposite traits. They were modest, shy, scrupulous and quite untheatrical. Thanks to the 17,960 traits that we amassed we classed actors into these two categories. The results were clear. Exhibitionist actors were born twice as often when Jupiter had just risen over the horizon or had reached its culmination.

We also divided scientists into two personality groups according to their biographies, which had yielded 9,547 traits. The first were introverted scientists who were self-absorbed and modest, and the second, extroverted scientists who were more worldly and more vain. Saturn, the planet linked to scientists, should appear more clearly among the group of introverted scientists, the model of the true scientist. On the other hand, it should have no relationship with less typical scientists. The correlations confirm this (see figure 3).

The pattern of our results is that the passage of a planet just over the horizon, or just after its culmination, is the expression of a combinant factor in the personality. We can speak of a Mars factor, Jupiter factor, or Saturn factor with some precision, thanks to the method of analysing character traits. (Table IV). We are researching the moon factor and Venus factor.

The Planets and Modern Psychology

It is a fascinating project to try and compare planetary positions with personality factors. Hans Eysenck's work has established that the personality

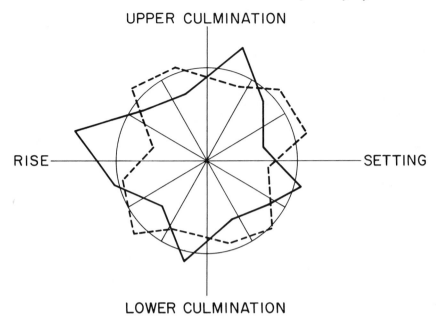

UPPER CULMINATION

RISE

SETTING

LOWER CULMINATION

Figure 3. Saturn and Introverted and Extroverted Scientists
Circle: shows the mean number of scientists born with Saturn in each part of its daily orbit.
Continuous lines: introverted scientists are born more often just after Saturn has risen or culminated.
Broken lines: extroverted scientists are born less often with Saturn in those positions.

is structured along two bi-polar orthogonal axes: the axis of introversion/extroversion and the axis of neuroticism or stability/instability.

Our discussions with Eysenck have made it possible to devise an experimental method for linking his work and ours. We are beginning a project in which we will compare the replies of 2,000 persons on the Eysenck Personality Inventory with the position of the planets at the time of their birth. Eysenck's descriptions of introverts and extroverts make it possible to formulate precise hypotheses. Introverts should tend to be born after the rising or culmination of Saturn; extroverts should tend to be born with Jupiter in those positions. If the results are positive, they show that one can extend the findings from famous to ordinary people. It also implies that one could predict the behaviour and character of a newborn baby when one knows where the planets are at its birth—providing the birth takes place under certain conditions. We are eagerly awaiting the results.

Like Father, Like Son

These planetary influences are hard to explain and what follows may seem very controversial but one must attempt some explanation. Do rays of some

TABLE IV. Twenty character traits describing the planetary ingredients of personality:

Keywords

Mars	Jupiter	Saturn
Active	At ease	Formal
Ardent	Ambitious	Concentrated
Aggressive	Thrusting	Conscientious
Reckless	Authoritarian	Cold
Pugnacious	Talkative	Methodical
Courageous	Exhibitionist	Meticulous
Dynamic	Comedy (a sense of)	Modest
Energetic	Communicative	"Looker on"
Fiery	Coquettish	Organized
Indefatigable	Spendthrift	Speaks little
Battling	Gay	Precise
Offensive	Gestures a lot	Reflective
Afraid of nothing	Good Humoured	Withdrawn
Ingenuous	Independent	Reserved
Hardy	Joyful	Sombre (temperament)
Bold	Proud	Wise
Brave	Extravagant	Timid
Vitality (full of)	Joker	Hard-working
Vital	Sympathetic	Taciturn
Strong-willed	Vain	Sad

This provisional list of traits is an often convincing image of the personality of a person born when the planet had just passed the horizon or the meridian of the place of birth. These portraits, of course, do not exclude other influences (family, culture, and so on) which combine to make up the individual's total personality.

sort affect the infant's personality? The child is born, fully formed, with all his or her inherited potentialities. It cannot be that the planet affects the whole genetic structure at the point of birth. It seems inconceivable that planets should "add" anything that is not already innate in the child. The effect of the planets would seem more possible if its role were linked to heredity.

Work by biologists and psychologists has suggested that we inherit much of the temperamental fundamentals of our personality. The research of Kretschmer and Sheldon indicated that different kinds of physical body builds went with different personality types. Sheldon classified body builds into three main categories—endomorphism, mesomorphism and ecto-morphism: Kretschmer into leptosomic, pyknik and athletic. They claimed that with each body build went a particular kind of personality. (It should be noted that their work remains controversial and not all psychologists accept it.)

In his work, Professor Hans Eysenck deals with the same theme. There is even talk of a biochemistry of behaviour with its origin in the genetic endowment each person has, since the research at the Michigan Institute and Edinburgh University showed a connection between levels of uric acid in the blood and social behaviour. If the planets give an indication of character and personality, this must involve an inherited element. And so children must be born when the heavens are in the same shape, so to speak, as at their parents' births.

We verified this hypothesis by looking at 30,000 times of birth of parents and children. Children tend to be born when a particular planet is rising above the horizon or culminating—if that same planet was doing the same at the birth of one of their parents. Observations on this show what seems a weak effect. But given the large number of births examined, the chances of that effect appearing at random are still 500,000 to one. The chance of a child being born thus is doubled if both parents had the same planet rising or culminating when they were born. Only the moon, Venus, Mars, Jupiter and Saturn show this effect, however.

The child seems gifted with a "planetary sensitivity" which triggers the moment when he is born. The planet does not alter the genetic or biochemical make-up of the infant. It acts as a trigger for the birth when it is in a favoured position and this position in the sky serves to indicate the psychobiological temperament of the newborn. But this brings a host of new problems, for how is one to justify the effect of the planets on heredity? There is both a biological problem and an astronomical problem.

The Biological and Astronomical Unknowns

First, the biological unknown. Medical science knows that when a foetus is ready to be born, nothing is needed to trigger its birth. This propensity the foetus has to be born when it is ready might be linked with subtle cosmic factors. It is arguable that the child could react to a "sign" to which those of his heredity were more sensitive than others. This idea has the merit of fitting in with contemporary medical thinking. C. G. Liggins of the National Women's Hospital in Auckland, New Zealand, has shown that it is the foetus that cues in the moment of birth through a hormonal mechanism. Liggins' work has been discussed at a number of international symposia.

What happens, therefore, when doctors interfere with the normal process of birth, either through a Caesarian section or through induction? Here we have an obvious control group, for in these cases any planetary effect on birth disappears. Medical influence rules out cosmic influence. The position of the planets at the time of birth then gives no indication as to personality.

There still remains the astronomical unknown. How could a planet take it upon itself to affect the actual times of birth? It might be something like a catalyst, acting perhaps through a screen in the solar field. A recent experiment has allowed us to confirm that the planetary effect on heredity appears twice as marked on a day of considerable disturbance in the solar field.

In the last 20 years, technical progress has made it possible to record more than merely the light coming from planets. Radio sources, gravitational changes and magnetic variations have been recorded. Some of these could cause subtle hormonal changes, though it has yet to be shown how.

It is no longer impossible to imagine that scientists will find that man's biology and psyche have invisible connections with the clocks of the cosmos. And these connections may not be mysterious, astrological ones which would make puppets out of us—our fate controlled by external signs in the stars—but empirical ones, based on such forces as magnetism.

Greek philosophers noted long ago that man's character makes his destiny. The planets affect this, though they do not allow us to predict the future. And since the word *astrological* is associated with predictions and horoscopes, I think we should stop using the word and coin a new one once we have established the nature of the influence of the planets upon man's personality.

Astrology: A Critical Review

I.W. KELLY, G.A. DEAN, AND D.H. SAKLOFSKE

The claims of astrology are not knowledge.
—Reuben Abel, *Man is the Measure*

Introduction

The fundamental claim of astrology is a simple one: whatever is born at a particular moment, be it a person, a company, an event, a nation, a question, or anything with a distinct moment of beginning, will manifest the quality of that moment. The quality of that moment will exist everywhere but can be seen most conveniently in the heavens, since they are clearly visible. So there will be a correlation between the heavens and terrestrial affairs. As above, so below.

This basic claim translates into an astrology that exists today on many levels from the trivial to the profound and which can embrace science, art, philosophy, religion, and entertainment. Most newspapers and almost all women's magazines carry sun-sign columns. According to various opinion polls a quarter of the population of Western countries believes in astrology, while about half the population reads sun-sign columns at least occasionally (Fullam 1984). Given the vagueness of the terms 'believe' and 'astrology' such polls can be misleading, but the more precise polls suggest that about 1 person in 20 is a strong believer and acts on what their horoscope indicates. However, other indicators suggest that active support for astrology may not be as strong as popular belief may lead us to expect.

The *New Scientist* (1985) asked a representative sample of 950 United Kingdom adults to allocate priorities for government spending in 12 research areas such as agriculture, defense, medicine, pollution, new forms of energy . . . and astrology. Astrology came last with a vote of 1%, with

Thanks are due to Suitbert Ertel (Germany), Michel Gauquelin (France), Arthur Mather (Scotland), and Maureen Perkins (Australia), for helpful comments on an earlier draft of this chapter.

robotics next at 4%. When asked which areas should receive reduced spending, astrology came first with a vote of 42%, just ahead of space exploration.

As for practitioners, in Western countries roughly 1 person in 10,000 is practicing or studying serious astrology, which is about the same as for psychology (Dean and Mather 1977:7). In Western languages serious astrology is currently the subject of more than 100 periodicals and about 1000 books in print (1 in 2000 of all books, or about the same as for astronomy), of which about half are in English. Since 1960 the annual output of new titles has doubled every ten years, at which rate the year 2000 could conceivably see ten new astrology books every week, excluding almanacs and sun-sign books (Dean 1987:167).

Despite public interest, however, astrology has generally been ignored by scientists. To a large extent this is due to (1) advances in relevant fields such as astronomy, biology, and the social sciences, which seem to be incompatible with astrological claims (the difference between identical and fraternal twins should not exist if astrology were true, for example), and (2) a tendency to mistake the popular astrology of fairground tents and newspaper columns for the serious astrology of consulting rooms and learned journals. Over the last few decades, however, a small but increasing number of scientists have investigated the claims of astrology. In the last ten years several major critical surveys of astrology have appeared. Studies made since the first edition of this book (1982), in particular, have greatly clarified the emerging picture. Because the material is now far too voluminous to examine in detail here, however, we can give only a brief overview. We have divided our survey into the following sections:

—Astrology and science.
—The research to date.
—The salutary lesson of phrenology.
—Problems of validity.
—The Gauquelin work.
—Spiritual astrology.
—The future of astrology.

Astrology and Science

A number of excellent chapters in this book attempt to explain what good science is, so we will not repeat that topic here. Because astrology covers so many areas, science may not always be relevant, as will be apparent from the following simplified breakdown:

Where the aim of astrology is to	The area is	Where symbols stand for	Example *
Convey information	Science	The concrete	10g of NaC1
Stir the emotions	Art	What the artist feels	Red red rose
Reveal meaning	Religion	The unknowable	Everywhere but nowhere

* A spoonful of salt, Burns, the Tao

The information to be conveyed by astrology can be anything from an ancient aphorism like "the Dragon's Tail on the Descendant indicates a dwarf," to modern 'psychological insight', like "Moon-Saturn suggests early problems in childhood with your mother." Emotions can be stirred by the beauty of astrological imagery, or by its expression in cartoons and poetry (for example:

> Your plump white hand selects
> From the open drawer alongside you
> A green plain chocolate peppermint cream
> For your moist waiting mouth.
>
> —*Mrs. Taurus*
> Joe Cooper, 1981)

or by assurances that "women who have Mars with the Moon are all right" (an indication personally guaranteed by the early Italian astrologer Jerome Cardan). The esoteric side of existence can be revealed through symbols that, like Linear A, point to things that exist but are unknowable.

Clearly science is relevant only to those parts of astrology that can be tested. What can be tested, however, is constantly expanding as techniques advance; the initial untestability of Freudian ideas and phrenology, for example, steadily disappeared as techniques improved. We can now test the idea that Leos are more generous than other signs, but as yet cannot test the idea that Leos were Cancerians in their previous life.

Having isolated an astrological idea that is testable, a scientist tests it by conducting an experiment. Suppose for example that the idea at issue is that astrology is explained by the principle "as above, so below." The questions that a scientist might ask about this claim, together with the answers as far as we presently know them, might be:

1. Is this explanation correct?	Don't know
2. Is it compatible with existing knowledge?	No
3. Are there other explanations?	Yes
4. What would prove each explanation wrong?	Scientific research

Here the research would consist of tests of hypotheses based on each explanation. The results would then be critically examined by other scientists, who would conduct further experiments, and so on, until consensus was reached. Thus the scientific approach to astrology (as to anything else) has three features: it is collaborative, it is impartial, and it is self-correcting. We might also add that since research regarding astrology in particular requires expertise in so many disciplines (at least astronomy, psychology, sociology, and statistics, as well as astrology), it has endless pitfalls for the unwary. Research into astrology, like life, was not meant to be easy.

Since 1900, many of the basic ideas of astrology have been tested by scientific methods. In the next section, we look at some of the results.

The Research to Date

The first comprehensive critical review of research into astrology was compiled by Dean and Mather et. al. (1977), who surveyed material from over 700 astrology books and 300 scientific works. Since then further critical reviews have appeared, notably by Kelly (1979, 1982), Eysenck and Nias (1982), Gauquelin (1983), Startup (1984), Dean (1986a, 1987), Niehenke (1987) and Culver and Ianna (1988). These authors have diverse backgrounds: Eysenck and Nias are psychologists, for example, while Culver and Ianna are astronomers and Niehenke is an astrologer and psychotherapist. Nonetheless their conclusions are largely in agreement and can be summarized as follows under discussions of effect size and non-astrological factors.

1. Effect sizes are too small to be useful. How strong is the relationship between astrological indications and reality? Traditionally, research in astrology has been preoccupied with statistical significance, i.e., whether the results can be explained by chance, which tells us nothing about how big the effect is. Significance levels can also be misleading. For example, as Nias and Dean (1986:374) point out, "one's astonishment at a test of 40,000 coin tosses that produced evidence for psi at the .001 level might well evaporate on our learning that for every 100 tosses it required averaging 50.8 heads instead of the 50 expected by chance." Startup (1984) was the first to submit the results of astrological research to power analysis, which looks at the power of an experiment to detect a real effect and estimates the size of that effect. The results that follow are an enlargement of Startup's pioneering work.

The results obtained by combining available studies are shown in Figure 1. Here the effect size is expressed as a correlation and can be interpreted via the examples shown in Figure 2. Three things emerge from Figure 1, namely: (1) that in most cases the number of studies is reasonably large, so we can have some confidence in the results, (2) that the supposed superiority of the whole chart over individual factors is not supported, and (3) that in each case the effect size is far too small to be useful.

Because this point about effect size is crucial, let us examine it more closely. The minimum effect size for psychological tests generally accepted as being useful is around .4, equivalent to getting 5 hits in every 7 tries instead of the 3.5 hits expected on average by chance. This may not sound like much of an improvement, but if such a system were available for winning at roulette it would break the bank in an hour. In fact many established tests do considerably better. The correlation between IQ and achievement results is around .6, for example, as is the correlation between extroversion scores on a self-report personality questionnaire and ratings by independent judges. This is the sort of act that astrology has to follow, but with correlations consistently under .1 it is clearly not even in the race.

Figure 1.

EFFECT SIZES FROM CONTROLLED TESTS OF ASTROLOGY USING ORDINARY PEOPLE

INDIVIDUAL FACTORS		Effect size as a correlation
Signs	(18 studies)	<.05
Diurnal position	(15 studies)	<.05
Aspects	(11 studies)	<.10

WHOLE BIRTH CHART

Accuracy of chart judgments (35 studies, 570 astrologers, 1500 charts)	.05
Accuracy of predictions (4 studies, 3400 predictions)	No better than chance or shrewd guesses
Ability to pick own chart interpretation (10 studies, 290 subjects)	.03
Agreement between astrologers judging the same birth chart (14 studies, 400 astrologers)	.11 (Minimum acceptable .80)

— —

The above table does not include Gauquelin studies, for which see Figure 4. Each value is the mean effect size except for individual factors, where <.10 means that, for the most sensitive study, the observed effect size was less than .10 when tested against personality (usually measured by personality tests). If the true effect size had been .10, then in 4 out of 5 tests it would have been detected at the .05 significance level. For individual factors many studies exist, but only those with proper controls for social, demographic and astronomical influences are included here.

Sources: Eysenck and Nias (1982), Startup (1984), and articles through September 1988 from the main journals reporting scientific studies of astrology, namely Astro-Psychological Problems , Correlation , NCGR Journal , and Skeptical Inquirer .

Figure 2.   EXAMPLE CORRELATIONS

1.00	Perfect, eg inches and cm.
.95	Length of right and left arms.
.9	Adult IQ tested a year apart.
.8	Height at ages 2 and 4.
.7	Adult height and weight.
.6	Height at ages 2 and 18.
.5	IQ of husbands and wives.
.4	Minimum for tests recognised as useful.
.3	Height of husbands and wives.
.2	Ink blots or graphology vs independent tests.
.1	IQ vs head size.
.0	Useless. Same as tossing coins.

Corresponding average hit rates:

1.00	5 in 5	A perfect score, no misses.
.7	5 in 6	vs 3 average out of 6 by tossing coins.
.4	5 in 7	vs 3.5 average out of 7 by tossing coins.
.25	5 in 8	vs 4 average out of 8 by tossing coins.
.0	5 in 10	Same average as tossing coins.
-1.00	0 in 5	No hits, a perfect inverse score.

A value of one, at the top of the list, means the correlation is perfect, as between inches and centimetres. Zero, at the bottom of the list, means there is no correlation at all, as between the tosses of two coins. Negative correlations (not shown here) mean that two things vary inversely, like more day means less night. In between one and zero lie the diversity of real-life correlations. At .95 is the near-perfect correlation between right and left arm lengths, which supports our everyday observation that any difference is small. At .7 is the less-perfect correlation between height and weight—heavyweights tend to be tall while light-weights tend to be short, but there are individual exceptions. At .5 we have the extent to which bright men prefer bright women and vice versa. Perhaps surprisingly, it is more marked than the preference of tall men for tall women at .3 and vice versa. At .1 we have the almost negligible correlation between IQ and head size, which seems to sum up the problems of phrenology in a nutshell. Because few effect sizes in astrology exceed .1, it also sums up the problems of astrology.

The most devastating result in Figure 1 is the last, which concerns only the agreement (i.e. reliability) between astrologers, not their accuracy. Thus if astrologers agreed that all cats were black, they would show perfect *agreement* (correlation = 1.00), even if half the cats were in fact white. Similarly, if they measured the length of each cat, we would expect good agreement if they used a steel ruler, and poor agreement when they used a rubber ruler. The minimum acceptable agreement for psychological measures when applied to individuals is around .8, below which the measure is too rubberlike to be useful. Yet for astrology the average agreement displayed in Figure 1 is only .11, showing that there is almost no agreement between astrologers on what a birth chart means. This is surprising in view of the uniformity of most astrology books. But it is consistent across all 14 studies, so there is no reason to suppose it is atypical. Furthermore, most of the studies were conducted not by hostile critics but by astrologers anxious to demonstrate the value of their craft, so they cannot be dismissed as biased. The result is devastating: if astrologers cannot even agree on what a birth chart *means* then their entire practice is reduced to absurdity.

The results in areas other than those covered by Figure 1 are just as negative. Nelson's system for predicting radio propagation quality has long been quoted in support of astrological ideas, but a critical evaluation of 7000 forecasts found the correlation between forecast and outcome to be a mere .01, almost exactly chance level and which indicates essentially no relationship (Dean 1983). Jonas's claim that an astrological ovulation cycle exists has been negated by *in vitro* fertilization studies, which found no ovulation cycle other than the normal one (Dean 1984c). A survey of 37 studies of lunar-lunacy effects found that alleged effects can be explained by artifacts, the correlation with moon phase being below the limit of detection of <.1 (Rotton and Kelly, 1985). A survey of 21 studies of lunar birthrate effects was equally negative (Martens, Kelly, Saklofske, 1988).

Sun-sign columns are condemned by critics and serious astrologers alike—and with good reason. All the half-dozen scientific studies examining sun-sign columns collected by us have found them to be neither valid nor in agreement, and therefore acceptable only as entertainment. Watson (1988) notes that some are ridiculous. For example, the horoscope in the Australian women's magazine *Cosmopolitan* for January 1988 said Librans could expect to visit France next summer—meaning France's 54 million population would have over 400 million visitors. Libran women could also expect to visit their gynecologist on 22 February. This was checked by a reporter from Perth who counted 104 visitors to Perth gynecologists on that day, of whom 8 were Librans—some 20,000 short of the number predicted (Channel 9 news, Perth, 22 February 1988). In 1984, The Committee for Scientific Investigation of Claims of the Paranormal (CSICOP) urged American newspapers and magazines to label their horoscopes with a disclaimer saying they were for entertainment only and had no basis in fact. But by 1988 only 2% of newspapers had done so. In New Zealand no newspapers adopted the disclaimer when urged by New Zealand scientists, but one did add the caveat "for entertainment," and two major dailies changed the title to "Stars for Fun," suggesting that disclaimers will not be adopted unless brief and to the point (Dutton 1987).

These results do not deny the possibility that some as yet untested areas of astrology (e.g. mundane astrology, for which dramatic claims have been made by Lynes (1987), and horary astrology), or certain people (either individually or as a group), may produce adequate effect sizes, or that some entirely new and valid astrological technique may be discovered. Nor do they deny the possibility that genuine effects may exist, too weak to be of use (like the bending of light by gravity), but nevertheless of scientific importance. However the sheer volume of available results, in what probably amounts to well over 200 man-years of research by scientists (and which excludes a much larger amount of non-scientific work by astrologers), and the consistent failure of those results to show adequate effect sizes, raise questions that astrologers must answer.

2. Alleged effects can be explained by non-astrological factors. Here the word 'effects' needs qualification. In general, astrologers judge astrology on how *helpful* it is, whereas scientists judge it on how *true* it is. Thus astrologers see that astrological consultation seems to help people and conclude that astrological effects are real, whereas scientists see that tests are negative (Figure 1) and conclude the opposite. The issue boils down to this: to be credible in a healing role, astrology must produce benefits beyond those produced by non-astrological factors. But does it?

Non-astrological (i.e. psychological) factors that can contribute to belief in astrology are surprisingly numerous (Figure 3). Yet astrologers show no awareness of them. Consequently, when faced with negative results "they rush to invoke way-out explanations such as synchronicity, the horrors of science and paradigm differences, rather than simple human failings. This is rather like invoking the end of the world to explain . . . a power cut" (Dean 1984b). Non-astrological factors are also generally far more *potent* than the astrological factors shown in Figure 1. The acceptance of tarot interpretations, for example, increases with their generality and social desirability, the correlation being about .3 in each case (Blackmore 1983). On this basis it seems reasonable to conclude that astrology does not produce effects *usefully* beyond those produced by non-astrological factors (Figure 3).

Figure 3. TWENTY WAYS TO CONVINCE CLIENTS THAT ASTROLOGY WORKS

(from Dean, 1987 P. 263)

Twenty Ways to Convince Clients that Astrology Works

Principle	Factor	How it works.
Cues	Cold reading.	Let body language be your guide.
Disregard for reality	Illusory validity.	Sound argument yes, sound data no.
	Procrustean effect.	Force your client to fit the chart.
	Regression effects.	Winter doesn't last forever.
	Selective memory.	Remember only the hits.
Faith	Predisposition.	Preach to the converted.
	Placebo effect.	It does us good if we think it does.
Generality	Barnum effect.	Statement has something for everybody.
	Situation dependence.	Everybody has something for statement.
Gratification	Client misfortune.	The power of positive thinking.
	Rapport.	Closeness is its own reward.
Invention	Non-falsifiability	Safety in numbers.
Packaging	Dr. Fox effect.	Blind them with science and humor.
	Psychosocial effects.	The importance of first impressions.
	Social desirability.	I'm firm, you're obstinate, he's —
Self-fulfilling prophecies	Hindsight bias.	Once seen, the fit seems inevitable.
	Projection effects.	Find meaning where none exists.
	Self-attribution.	Role-play your birth chart.
Self-justification	Charging a fee.	The best things in life are not free.
	Cognitive dissonance.	Reduce conflict-see what you believe.

This table shows that there are many nonastrological reasons why clients should be satisfied by an astrological consultation, none of which require that astrology be true. But if clients are going to be satisfied with the product offered, then astrologers can hardly fail to believe in astrology. In this way a vicious circle of reinforcement is established whereby astrologers and their clients become more and more persuaded that astrology works. An astrologer typically spends years learning to read charts and thus has ample chance to respond to such reinforcements.

An interesting check on this point was conducted by Dean and Mather (1985). They persuaded various astrological groups to sponsor a $5000 prize for "evidence that the accuracy of chart interpretations cannot be explained by non-astrological factors." Over 60 intentions to enter were received from a total of 14 countries, with a breakdown of topics roughly as follows: one third concerned events, one third concerned personality, and one third concerned other areas such as discrimination, synastry, and horary. Of these, 34 individuals sent in entries (totaling over 1500 pages) that were then evaluated by a panel of independent judges. Only one entry qualified, but this was a fake study entered as a control to evaluate a particular criticism, namely that the prize was unwinnable because appropriate tests could not be designed and the judges were not impartial.

Of course this does not deny the possibility that certain astrologers may succeed where others have failed. But the onus is on astrologers to demonstrate it. In the meantime we can gain further insight from an examination of belief in phrenology.

The Salutary Lesson of Phrenology

Phrenology is a system of intellectual and moral philosophy that is based on reading character from brain development as shown by head shape. Phrenology is now effectively dead, but in the 1830s its popularity exceeded that of astrology today, with over 30 phrenological societies in the United Kingdom alone. At that time one quarter of the United Kingdom population was illiterate, and a phrenology book cost half a week's wages. Yet roughly 1 person in 3000 was practicing or studying phrenology, a number greater than "of persons equally advanced in geology, entomology, botany, astronomy, or similar sciences" (Watson 1836:223).

Phrenology provides a salutary lesson because of its similarity to astrology. Like astrology, it is concerned with individual potential and the philosophy of existence. That is, it encourages one to (1) think about oneself in phrenological terms, (2) assess oneself via phrenological principles, and (3) act on the findings to achieve a physical, mental, and spiritual whole and thus harmony with the world. Except for prediction, then, astrology and phrenology cover the same ground.

But the important point is this: like astrology, phrenology persisted because day after day its practitioners and clients could see that it 'worked'. They were therefore unmoved by what the critics said, and for the best of reasons. For example in his book *Popular Phrenology,* the phrenologist J. Millott Severn (1913) cites the following remarkable testimonies:

Thomas Alva Edison:	"I never knew I had an inventive talent until phrenology told me. I was a stranger to myself until then."
Andrew Carnegie:	"Not to know yourself phrenologically is sure to keep you standing on the Bridge of Sighs all your life."
William E. Gladstone:	"I declare that the phrenological system of mental philosophy is as much better than all other systems as the electric light is better than the tallow dip."
Alfred Russel Wallace:	"The phrenologist has shown that he is able to read character like an open book . . . with an accuracy that the most intimate friends cannot approach."

Similarly, Edgar Allen Poe wrote that "Phrenology is no longer to be laughed at . . . It has assumed the majesty of a science and as a science ranks among the most important" (quoted in Hungerford, 1930:209).

The fundamental postulates of phrenology are now known to be wrong, however. Character is *not* indicated by brain size and shape because the brain does not work like that. Nor does character break down into the entities required by phrenological theory. So a bulge here or a depression there *cannot* mean what it is supposed to mean, even though the underlying philosophy of "know thyself" has undeniable appeal.

As Flugel (1965:37) says in *A Hundred Years of Psychology*, "The failure of phrenology, with the implied immense amount of misdirected effort and ill-informed enthusiasm, was the price that had to be paid for this neglect of scientific caution." Unfortunately this neglect of scientific caution is also raging out of control among astrologers and their teaching institutions today, a point to which we shall return in the final section. As noted by Dean and Mather (1985:71):

> Astrologers are like phrenologists. Their systems cover the same ground, they apply them to the same kinds of people, they turn the same blind eye to the same lack of experimental evidence, and they are convinced for precisely the same reasons that everything works. But the phrenologists were wrong. So why shouldn't critics conclude for precisely the same reasons that astrologers are wrong?

This is an honest question that astrologers have yet to answer. At this point it is worth taking another look at why invalid systems can nonetheless seem to be valid.

Problems of Validity

In a study that found no evidence of a link between handwriting and occupational success, Ben-Shakhar et. al. (1986:652) asked the same question that we can ask of astrology, namely "why are its users and clients so satisfied with [handwriting analysis] and so often prepared to swear by it?" Ben-Shakhar notes that, since there is a lack of *empirical* support, the answer has to lie in the area of *non-empirical* support, of which there are at least two kinds. The first is *face validity*, which means that the system *looks* as if it should work. For example, a new transistor radio *looks* functional even if its batteries are flat. Astrology certainly has face validity—a long history, complex techniques, impressive jargon, computerized charts, national associations, international conferences in Europe and the United States that can attract over a thousand people, and a vast literature base totaling some

200 shelf-metres of Western-language books and periodicals (Nias and Dean 1986:357). To the unwary the effect is dazzling.

Kelly et. al. (1989) note that many of the arguments put forward to support belief in astrology rely on face validity, as shown by the following examples (with the counter-argument in parentheses):

Astrology has great antiquity and durability. (So has murder.)

Astrology is found in many cultures. (So is belief in a flat earth.)

Many great scholars have believed in it. (Many others have not.)

Astrology is based on observation. (Its complexity defies observation.)

Extraterrestrial influences exist. (None are relevant to astrology.)

Astrology is not science but art/philosophy. (Not a reason for belief.)

The second form of non-empirical support is *personal validation*, or how one feels about a system following personal exposure to it. Here the unwary can easily mistake artifacts (in this case nonastrological factors) for the real thing. For example one may find the astrologer to be so friendly, so sensible, and such a good listener, that one ascribes to astrology benefits that in fact arise from other sources.

Face validity and personal validation can be so persuasive that other more important kinds of validation are not sought out. If the latter cannot be found even when sought, as seems to be the case with astrology (see Figures 1 & 3), then a serious problem arises. Hergenhahn (1982:18) takes the position that

> astrology is a highly developed formal system that has little or no relationship to actual events . . . it sounds good but adds practically nothing to our understanding of human behavior.

He then points out that a theory is only of value if it has a better explanatory power than alternative theories. Now many astrologers see this as being precisely true of astrology. For example Elwell (1986) claims that astrology is an alternative model of reality, whose true value lies in

> its power to give us a completely new slant on everything, not just tell us what we can find out by more conventional means. . . . if astrology is only part-way true, the universe is very different from what it seems to be.

The problem with such claims is *vagueness*. Astrologers never describe precisely what their model predicts, what the criteria are by which it could be tested, and what sort of evidence they would accept as proving it wrong. When this argument was put to Elwell (Dean 1986b), he replied that it was not necessary to justify one model in terms of another, just as "proponents of the wave concept of light are not required to prove it in terms of particle theory, and vice versa" (Elwell 1987a).

Similar challenges to other astrologers have been equally unproductive (Dean 1984a). But perhaps that is the whole point: being too specific, as with the emperor's new clothes, could be disastrous. Given the reluctance of astrologers to produce testable theories and the slow but steady improvement in alternative theories as knowledge increases, the prospects for astrology as a useful theory do not seem promising (see also Carlson, 1988).

The Gauquelin Work

Michel Gauquelin has been tirelessly researching astrology since 1949, for much of that time assisted by his former wife, Francoise. Gauquelin was not the first to investigate astrology, but he was the first to do so rigorously with a large database that currently totals over 100,000 cases, all with birth certificates giving the time, date, and place of birth. This work is described in many publications, the most up-to-date being Gauquelin (1983). Together with the earliest work, originally published in French and recently available in English (Gauquelin 1988a,b), it is essential reading for anyone interested in the scientific investigation of astrology. Because the results have been so widely published we give here only a brief summary, which necessarily fails to capture the full rigor and extent of the investigations. We will try to outline recent developments, however, with greater detail.

Gauguelin's tests of zodiac signs and planetary aspects were negative, the mean effect size in each case being around .01 or less, as were tests of astrologers themselves, whom Gauquelin says are "sometimes so disillusioned that they accuse me of rigging the cases" (Gauquelin 1983:139). These negative findings have generally been confirmed (Figure 1), but Gauquelin's test of planetary diurnal position (equivalent to house position) produced positive results that appeared to replicate. At the birth of eminent professional people such as actors and scientists, certain planets tended to concentrate in certain sectors of the sky more than they should (Figure 4). As shown in Figure 4, the effect sizes were very small, but they were statistically significant, that is, unlikely to be due to chance.

In 1960 these results led Gauquelin to formulate two general hypotheses that, with minor exceptions discussed later, were subsequently confirmed (see Gauquelin 1988a:101, 150-153). The first hypothesis stated that the planetary effect was linked to *occupation*, different occupations showing different effects. The second hypothesis stated that the effect was linked to *eminence*: the lower the eminence the less the effect. Later work on the first hypothesis suggested that the real link was with *traits*, not occupation (see Gauquelin 1983:47); the stronger the trait the greater the effect. Interestingly, in many cases the relevant traits were those predicted by

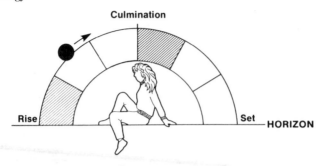

Figure 4.   THE GAUQUELIN EFFECT

At the birth of eminent professional people, Gauquelin found that certain planets seemed to prefer or avoid the shaded sectors (called key sectors) more than they should, while the rest showed no effect. To a much lesser extent they also preferred or avoided the opposite sectors (not shown) below the horizon. No such effect was found at the birth of ordinary people. Astronomic and demographic influences were meticulously controlled to avoid artefacts. For comparison with Figure 1, the effect sizes which are .03 or more are as follows:

Group	Effective planets					Ineffective planets				
	MO	VE	MA	JU	SA	SO	ME	UR	NE	PL
24961 Ordinary people	.	.	.	.	.	.	.	.	.	.
2008 Sports champions	.	.	.10	.	.	.	.	.	.03*	.
2552 Physicians	.	.	.08	.	.08	.	.	.04	.	.
1095 Scientists	.	.	.	-.07	.05*	.	.	.	-.05	.
3046 Military men	.	.03	.06	.09	.	.	.	.	.	.
1003 Politicians	.05	.	.	.10	.	.	.	.	.	.
1473 Painters	.	.	-.08	-.04	-.08	.	.	.	.	.03*
866 Musicians	.	.04*	-.08	.	.	.	-.04*	.	.	-.07
1409 Actors	.	-.03*	.	.07	.	.	.	.	.	-.03*
1352 Writers	.12	.03*	-.03*	.	-.04*	-.03*	.	.	.	.
1654 Mean eminent**	.03	.02	.05	.05	.04	.01	.02	.02	.02	.02

-Indicates key sectors are avoided. *Not significant at the .10 level. **Ignoring sign. Thus the mean of .08 and -.08 is 0 . Each value is the mean of all 9 eminent groups.

In the above table the effective planets are the Moon, Venus, Mars, Jupiter and Saturn. The ineffective planets are the Sun, Mercury, Uranus, Neptune and Pluto. The effect size (calculated by us) is the phi correlation coefficient, calculated via chi-squared with one degree of freedom, using observed vs expected frequencies in the two key sectors combined and in the remaining sectors combined. The frequencies were obtained from Gauquelin (1972, 1978, 1984).

The main features are as follows: (1) Most of the values are below the minimum detectable effect size of .05 (N=3000) to .08 (N=1000). As a result there is little overall difference between effective and ineffective planets, the means ignoring sign for the eminent groups being .04 and .02 respectively.   (2) Even the largest effect sizes are very small. Thus the most consistently effective planets (Mars, Jupiter and Saturn) have effect sizes between .04 and .10, albeit much larger than their mean effect size with ordinary people of .003.   (3) Nevertheless the results are very significant, with ten results significant at the .01 level vs .01×100 = 1 expected by chance. For such results the surplus or deficit is typically 10-20% of the expected frequency. For example the effect increases the chance of an eminent soldier having Jupiter in key sectors from 1 in 6 to 1 in 5.   (4) Similarly, at the .10 level, effective planets have 15 significant results whereas ineffective planets have 3 significant results, the number expected by chance being .10×50 = 5 in each case. (5) At first sight Venus seems too weak to be classed as an effective planet. But in heredity studies, where key sector positions were compared between parents and children, Venus was as prominent as the other effective planets (but see text), whereas the ineffective planets remained ineffective. In subsequent trait studies a stronger effect for Venus was confirmed.

astrology (and by no other theory), but Gauquelin (1979:153) considers this to be a coincidence. He points out that the overall fit is poor, and that contrary to astrological predictions

> The sun does not appear among the leaders (military, political etc.); nor Venus for artists (musicians, painters, actors); nor Mercury for writers and business-men . . . And none of the more distant planets—Uranus, Neptune, and Pluto—justifies the symbolism that modern astrologers have lightly attributed to them.

These hypotheses suggest that planetary effects might be linked to heredi-tary factors. A subsequent study of 24,961 parents and children showed that children tended to be born with the same particular (i.e. effective) planet in key sectors as their parents. The effect sizes were even smaller than those shown in Figure 4, being typically .02 for effective planets and .00 for ineffective planets, but they suggested that planetary effects applied (albeit weakly) to ordinary people and not just to eminent people. Ten years later the results were successfully replicated (Gauquelin 1983).

The results of these and many other studies suggested that the planetary effect on eminent and ordinary people has many baffling features. It seems to work for the Moon, Venus, Mars, Jupiter, and Saturn, but not for the Sun or other planets, so it cannot be due to light or gravity. It works best if the person is eminent. It disappears if the birth is induced or surgically assisted, so it contradicts the doctrine held by many astrologers that deliv-ery conditions have no effect on birth chart validity. It is also enhanced if geomagnetic activity is high or if both parents have the same planet empha-sized. In short, it seems to behave like a genuine physical effect. So is there an explanation?

One possible explanation would be that the fetus tends to trigger its own birth only if the planetary signals are appropriate. But as Gauquelin (1983) points out, even this poses baffling problems. None of the obvious signals—electromagnetic radiation, magnetism, or gravity—fit what we know about planets and their solar-terrestrial effects. If the planet helps to trigger labor, why is the effect observed at *birth*? Why birth, when character was presum-ably formed months before? And what possible natural advantage could a planetary effect bestow? Elwell (1987b:3) adds a further point: "if such forces existed . . . on what would the stars imprint their nature in the case of inanimate things, like ships and nations, which also have horoscopes?"

The usual explanation in such a situation is the self-fulfilling prophecy: people adjust their self-images to fit their birth charts, thus inflating the apparent hit rate. Self-fulfilling prophecy is known to apply to sun signs (Eysenck and Nias 1982:57), where the effect size can reach .15 or more (Startup 1984:256), but it seems unlikely to apply to planetary positions of which most people are unaware—especially as the prophecies to be fulfilled tend to contradict astrology, which predicts weakness rather than strength for the position observed.

Despite these baffling features, and despite the small effect sizes, these results obtained by Gauquelin remain the best factual evidence for astrology that has emerged to date. As noted by Mather (1979:96), in a now-famous quote,

> Both those who are for and against astrology (in the broadest sense) as a serious field for study recognize the importance of Gauquelin's work. It is probably not putting it too strongly to say that everything hangs on it.

It is against this background that we now move to some more recent developments.

Recent Development in the Gauquelin Work

The Gauquelin occupation hypothesis findings have survived independent replications for sports champions (Comité Para 1976, Ertel 1988) and eminent physicians (Müller 1986). A replication using United States sports champions ended in dispute (see Curry et al (1982) for an independent account). A computer recalculation of the entire Gauquelin database gave results for eminent professionals that were essentially unchanged from the original hand calculations, although effect sizes were generally 15% lower with a corresponding reduction in significance levels (Gauquelin 1984). For ordinary people and planetary heredity, the effect sizes in the first replication were reduced by a similar amount, but those in the second replication were reduced almost to non-significance (mean effect size .01), while a third huge replication involving 50,942 parents and children fared even worse with a mean effect size of zero (Gauquelin 1988b). This leaves the reality of the Gauquelin heredity effects in some doubt.

Other researchers using the same database have extended Gauquelin's work. For example the British astrologer John Addey (1979, 1981a,b) found that, as the planetary position at birth changed, the personality traits (determined from biographies) varied smoothly as in a spectrum (Figure 5). In Figure 5 the spectrum repeats itself every quadrant, but Addey found groups where (1) the spectrum repeated in other divisions of the circle such as thirds, sixths, and fifteenths, which he called harmonics, and (2) the traits forming the spectrum varied with harmonic. For example the third harmonic spectrum for Mars was modest, then influential and rigorous, then creative, then kind and hardworking. In this case the first meaning was *centered* on the rising point, whereas for the fourth harmonic (Figure 5) the first meaning *began* at the rising (Addey 1981a). From Figure 4 this suggests there are two types of sports champion, one modest (third harmonic) and the other aggressive (fourth harmonic). Sadly, this intriguing work was cut short by Addey's untimely death in 1982.

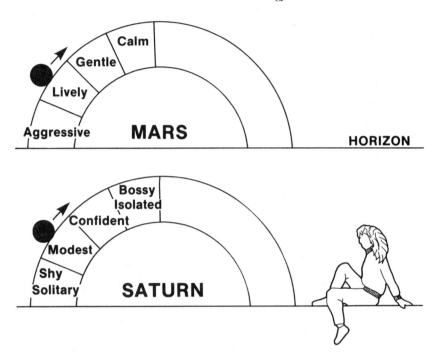

Figure 5.   THE HARMONIC EFFECT

Addey asked the question: as a planet changes its position from the birth of one eminent professional to another, how does the personality change? From Figure 4 one might predict that, if the personality in shaded areas is black, then elsewhere it should be white, with shades of grey in between. But Addey found no evidence for this. Instead the personality (determined from biographies) seemed to exhibit a SPECTRUM of changes. Thus as Mars rises above the horizon, the professionals tended to be successively aggressive, then lively, then gentle (which is the opposite of aggressive), then calm (which is the opposite of lively), after which they became aggressive again, and so on, the cycle repeating itself in each quadrant back to the rising point. This variation was observed not only for sports champions but also for scientists, actors and writers. Much the same thing seemed to happen with Saturn and the other Gauquelin planets. As Addey points out, planets with variable meanings are quite contrary to what astrology predicts. From Addey (1979, 1981a).

Other researchers such as Startup (1984) have attempted to detect planetary effects via personality tests applied to ordinary people, but so far without clear results, due possibly to a mismatch between test and planet. Tests describing personal traits have been devised that fit each planet more exactly, but so far they have been too confounded by artifacts to give clear results (Schneider-Gauquelin 1988). This is disappointing because, as noted earlier by Eysenck and Nias (1981:206), eminence effects pose a severe problem. If the link is with traits and not with occupation the effect should apply to all who have a particular trait, not just to the eminent. For example Mars seems to be linked to extroversion, which is a major dimension of personality shared by all, yet mediocre sportsmen can be just as extroverted as the elite. This point is further addressed below.

The most important recent developments are based on the work of Suitbert Ertel, working at the University of Gottingen with an enlarged Gauquelin database. His first development was to quantify eminence in terms of citation frequency (Ertel 1987a, 1987b, 1988). Thus a person of high eminence will be cited in more biographical dictionaries than a person of low eminence. Provided enough dictionaries are available, and each citation is given equal weight, eminence can be objectively determined.

The Gauquelin eminence hypothesis predicts that the planetary effect (excess or deficit in key sectors as in Figure 4) should increase with increasing eminence. Ertel's results confirmed this for Mars at the birth of sports champions (excess), and for Mars and Saturn at the birth of musicians, painters and writers (deficit). The increase in planetary effect with increasing eminence was remarkably regular, which Ertel interprets as ruling out the possibility of data manipulation, since this would have destroyed the regularity. But the Venus effect for musicians, painters, and writers *decreased* with increasing eminence, as did the Moon effect for writers and the Saturn effect for scientists. This does not necessarily weaken the eminence hypothesis, however, because Gauquelin had never claimed that it should apply to Venus (which performs only weakly in Figure 4), and had previously noted that the Moon effect for writers "scarcely seem[s] sensitive to [eminence]" (Gauquelin 1988a:149). Indeed, it could be argued that agreement on these details adds to the trustworthiness of the data. The new finding may mean that the eminence effect is not always positive, however, as has hitherto been assumed.

In a second development, Ertel considered further problems with the trait hypothesis. Why do certain groups with distinctive traits, such as priests and psychotics, show no planetary effect? How realistic are trait effects now that heredity effects seem doubtful? And in those cases where the effect decreases as eminence increases, why should reducing a *typical* trait increase success? Gauquelin had tested the trait hypothesis by counting traits in the biographies of eminent sports champions, actors, writers, physicians, and scientists. He found that, for example, sports champions born with Mars in key sectors tended to have more traits typical of Mars (such as energy and aggression) than champions born with Mars elsewhere. Artifacts due to eminence and its effect on biography length were controlled by Gauquelin by expressing the counts as a proportion of all traits.

In his search for possible artifacts, Ertel (1988) looked at the biographies. He found the following:

(1) As eminence increases so does biography length and the total number of traits. This means that a relation between planetary effect and eminence (the eminence hypothesis) must necessarily affect total traits—hence Gauquelin's controls.

(2) As biography length increases, it is likely that biographers will devote an increasing proportion of space to traits that are typical of the group, such as energy and aggression for sports champions.

(3) Typical traits tend to be more positive and socially desirable than non-typical traits, for example "perceptive" (typical of successful writers) versus "arrogant" (non-typical). So the natural tendency for biographers to be more positive as eminence increases (the halo effect) will reinforce the tendency to increase the proportion of typical traits.

These 3 points suggest that eminence may artificially inflate Gauquelin's controls, in which case the trait effect may be due to eminence rather than traits.

Ertel (1988) tested this view using 43 pairs of the most eminent writers (10% of the parent sample) and 58 pairs of the most eminent painters (8% of the parent sample). Each pair had the target planet (Moon, Venus, Mars, Jupiter, or Saturn) in positions of maximum effect (i.e. in key sectors) or minimum effect respectively, while the positions of the other four planets were matched as closely as possible. This maximized the planetary differences while keeping everything else constant. So unlike Gauquelin, who collected traits and looked at planetary position, Ertel collected planets and looked at traits. When 14 professors of literature and 7 professors of art were asked to decide which subject best matched the typical traits, the results showed no consistent trends in support of the trait hypothesis, but this may have been due to sampling variations associated with the small sample sizes. If better tests with larger samples produce positive results, then the Gauquelin trait hypothesis will be strengthened. But if the results are still negative, it will indicate that *trait* effects in the Gauquelin data (and by extension the findings of Addey) are probably due to artifacts, at which point the support for *astrology* will largely collapse. This will not of course explain *occupation* and *eminence* effects, so the results of Figure 4 will be unaffected. Such tests are currently in progress, so further comment at this stage would be premature.

In a third development, Ertel (1989) examined the influence of planetary physical variables such as planetary distance, apparent magnitude, solar elongation, geomagnetic activity, and so on, the size of the Mars effect. No relationships were uncovered. Even Gauquelin's geomagnetic findings did not hold up. Ertel points out that this seriously weakens Gauquelin's planetary trigger hypothesis, and indeed any hypothesis based on physical causation. What remains is a choice between something like synchronicity (an acausal connection between events) and artifacts as an explanation for the occupation and eminence effects (as in the case of trait effects where artifacts have, it seems, already been uncovered).

Over the past few decades a spiritual form of astrology has become increasingly popular, and is examined next.

Spiritual Astrology

As we show below, the main difference between everyday astrology and spiritual astrology is one of emphasis. The former is concerned with *practical* matters such as relationships and vocation ("you have career difficulties"), the latter with spiritual matters such as potentials and transcendent meaning ("your task is to find your ego—the father within rather than the father outside"). In contrast to astrologers who see astrology as an art or science, the spiritual astrologer sees it as a philosophy or occult system.

There are numerous variations, some with particular names such as 'transpersonal astrology' and 'esoteric astrology'. To simplify discussions we exclude those requiring a particular spiritual belief, such as Christian astrology, and those requiring specialized symbolism such as the Kabbalah, and concentrate on the broad spiritual approach.

To the philosopher, spiritual astrology seems to raise the same problems as does religion. Does it say anything that could be shown to be true or false? Or is it rather an appeal to 'super-personal' values that neither requires nor is capable of rational test?

Some astrologers do indeed restrict astrology to super-personal values. Like Paracelsus, they may be committed to a "metaphysick" or 'grand scheme of things' which happens to be conducive to astrology, yet may reject its material claims, just as Hindus can be committed to a concept of God and yet reject certain claims of Christianity. But most astrologers claim that astrology has the unique advantage of encompassing both the material and spiritual planes. For example, the astrologer Dennis Elwell (1987b:4) says "astrology is the best and maybe the last hope of religion, because it offers a meeting-ground for the scientific and religious views of reality, reconciling many of their differences." That would mean, as previously mentioned, that spiritual astrology is a matter of emphasis rather than something new.

As far as everyday astrology is concerned, there is no doubt that many astrological statements are intended to be true ("people with a lot of Air grow old prematurely"), and we have already seen that the evidence justifies disbelief in such statements. If we link a scale of 0 to 10 (from highly improbable to highly probable) to the correlations of 0 to 1 shown in Figure 2, such statements would emerge with a rating close to zero, or highly improbable. So what about the statements of spiritual astrology? Blackstone (1963, see also Nielsen, 1985) argues that religious statements don't qualify as knowledge on the grounds that they are not capable of being tested (and possibly disconfirmed) and so cannot be supported by reliable evidence. A similar claim might be made against spiritual astrology, in support of our introductory quote from Abel (1976:23) that "the claims of astrology are not knowledge." Blackstone also points out that the religious believer seeks spiritual fulfillment rather than truth, so inconsistencies and falsifying data are simply ignored, and this too applies to spiritual astrology. For example, the astrologer Gregory Szanto (1985) claims that the astrological symbols in our birth chart help to align our outer physical expression (where we have free will) with our inner spiritual nature (which is set by God) to achieve harmony with the universe. He then claims that only intuition can reveal the inner nature shown by the birthchart, and therefore that the meaning of the chart must be allowed to rise spontaneously from the unconscious (where meaning resides) by using the birthchart as a crystal ball. Note the problem: there is nothing here that could be true or false. Note also how the unreliability of intuition (Abel 1976:208-211), and the problem of resolving opposing intuitions, are ignored.

A revealing case of ignoring negative evidence is provided by Niehenke (1984), who for his Ph.D. in psychology obtained over 3000 responses to a massive 500-item questionnaire that included almost 300 items relevant to astrology. The results were negative. For example, subjects with as many as four Saturn aspects (which are supposed to indicate heavy responsibility and depression) felt no more depressed than those with no Saturn aspects. But Niehenke is also an astrologer, and every day he finds that the horoscope "provides deep insights into the real being of a client . . . such feelings of evidence cannot be simply washed away." It doesn't seem to matter to Niehenke that his "evidence" may have a simple non-astrological explanation (Figure 3). So he carries on regardless: "A world in which astrology exists is surely a more enjoyable world than one without it. The need that astrology be a reality is much stronger than all the rational demonstrations against it" (Niehenke 1984:15).

Another revealing case of ignoring falsifying data is provided by the American astrologer Zipporah Dobyns, who like Niehenke has a Ph.D. in psychology. Despite her scientific training she ignores the results of controlled tests in favor of uncontrolled anecdotal observations, which leads her to assert that "There is no way to support a world of chance and to label as coincidence the incredible correspondences which I see daily in my work with astrology" (Dobyns, 1986:45). Because she works with birthcharts to which up to 369 asteroids have been added, the probability of a hit due to chance is greatly increased—a case of safety in numbers, as indicated in Figure 3. So the lack of controls makes her observations meaningless. Yet this does not deter her from continuing, "I am convinced that the underlying reality of the world is closer to 'mind' than to 'matter', that minds can connect across time and space without the involvement of matter, that minds can and do influence matter directly." Apart from the problem of how she could know this, note that if it were true then it seems illogical for astrologers to bother with the 'matter' of astrological charts and horoscopes when their minds could do it all directly so much better. If their minds cannot deliver, on the other hand, then the claim is irrelevant.

Further problems arise when we look at the claims made by the late Dane Rudhyar, perhaps the best-known proponent of the broad spiritual approach to astrology, 'spiritual' being his preferred term (Rudhyar 1975:49). What follows is largely a condensation of a detailed analysis by Kelly and Krutzen (1983) and Kelly, Culver, and Loptson (1989):

1. Astrology deals with individuals, science does not. According to Rudhyar (1936/1970:460), "modern science is obliged to ignore the individualness of every living entity" whereas astrology "deals essentially with the individual." But science does not ignore individuals. In fact, a major focus of psychology is the study of individual differences. Individual differences were central to the investigations of Galton in the late 1800s and are manifest today in such recent publications as the journal, *Personality and Individual Differences* or the book on individual differences coedited by one of the authors of this paper (Saklofske & Eysenck, 1988). Hence Rudhyar has set up a straw man.

2. Astrology's truths are not empirical. Rudhyar (1936/1970:459) denies that astrology gives empirical truth, that is, truth which can be confirmed by observation. Rather the birth chart reveals a person's potentials but does not indicate what will actualize these potentials. To paraphrase Eysenck's (1985:195) critique of Freudian theory, this presents us with a difficulty; what if the potentials are merely speculations that are actually untrue, and how can we tell? There is no doubt that astrologers would disagree on what a given chart indicated (Figure 1), so how can we tell who is right? Thus even if we accept astrology as a source of insight, we still need criteria for deciding its truth or falsity—and none are provided.

The explanation given by Rudhyar (1979a) is that the issue is not whether the astrological indication is correct but whether it is seen to be valid. Thus it is sufficient if, after having studied his birthchart, a person "is able to feel a direction and purpose in his life" (Rudhyar 1936/1970:7). But if correctness is irrelevant, the system becomes meaningless, so why bother to learn astrology? Why not just give everyone the same horoscope or the same check list of potentials? After all, everyone uses the same Bible, Koran, or Bhagavad-Gita. Ironically, Rudhyar seems to be saying that astrology need not be true, which is the same conclusion reached empirically by Dean (1987).

3. Astrology sees man and the universe as a whole. In the way Rudhyar uses this notion it requires the assumption that it is meaningless to examine any part in isolation. Just as a note has meaning only as part of a melody, so life can be understood only as part of the universe, and astrology "if it is properly used" is the best way of finding your place (Rudhyar 1977). If astrology doesn't seem to you to help you find your place, of course, then astrology "is not being used properly." Compare this with faith healing—if you have faith you will be healed, but if you are not healed then you have insufficient faith! By invoking holism, astrology is thus made nonfalsifiable and is elevated beyond criticism. Note how the impossibility of examining the whole is conveniently ignored. Furthermore, even the supposedly "whole chart" of 10 planets is still a microscopic "part" of the universe of over 10^{22} stars, so by Rudhyar's own holistic argument the birth chart would seem to be necessarily meaningless.

Rudhyar's use of holism involves other problems as well. Rudhyar compares, for example, the web of relationships revealed by modern particle theories with those revealed by astrology (Rudhyar 1979b). Many other astrologers have done the same, such as Elwell (1987b:3) who states: "astrology is entirely compatible with the leading edge of modern physics, however strenuous the protestations that it is fundamentally unscientific." But such a comparison is invalid. Modern theories are subjected to much testing before they are adopted, and are discarded if they fail. A nonfalsifiable astrology, by contrast, could not have been arrived at by the testing of anything.

Ultimately Rudhyar (1980) defines astrology as being beyond inquiry: "To reduce astrology to a practice susceptible of . . . [objective analysis] is to me to repudiate its very special character as a discipline of understanding—a path to broad psycho-spiritual wisdom." This would seem to absolve astrology from any responsibility to establish the truth of its claims. On this type of reasoning, moreover, it seems that no belief about anything must be rejected as false—anything could be defended against criticism by making it nonfalsifiable and nonverifiable. The problem with that type of reasoning, of course, is that if you believe it is all foolish nonsense then by its own rules you are right.

The Future of Astrology

We have seen how research to date has consistently failed to demonstrate adequate effect sizes. This is of no consequence if astrology is used as entertainment or as a mere conversation piece. But if astrology is presented as being not merely helpful but also true (and most astrology books do so present it), then the astrologer is reduced to a charlatan and dealer in lies, a true DFAstrolS (Disseminator of the Fallacies of Astrological Superstition). The remedy is simple:

Astrologers wishing to be taken seriously must become more responsible. They must become aware of relevant research findings, they must desist from making claims at variance with the known facts, and they must label their product honestly so that the public is not misled. . . . Until this happens, the professional astrologer will remain a contradiction in terms. (Dean 1987, see also Carlson, 1988).

Based on past performance, however, there seems to be little hope of this happening, at least not until a new generation of astrologers replaces the current practitioners. So how do astrologers view the future of astrology?

In 1986 a survey of 19 well-known American astrologers gave a somewhat gloomy view of the state of astrology (*Astro*Talk* May/ June 1986). According to them, American astrology is: at a dead halt, in a stormy situation of uncertain outcome, generally of decreased quality, in a very sorry state, plagued by bickering, too commercial, not accepted by society, maturing, often a waste of time, insufficiently person-centered, too person-centered, making progress, too ingrown, in trouble, in chaos, ignorant of relevant disciplines, and best in the world for its sensitive understanding of the human condition. The main need, they felt, was for: a theoretical basis, more facts and better theories, qualified people to do research, wider horizons such as application to ecological issues, reintegration of the sacred and scientific, rigorous scientific testing, more person-centeredness, investigation of underlying mechanisms, proper accreditation, new ideas, more professionalism, better accreditation, more sophistication, thorough testing, and scientific research.

Here the majority view is that astrology is in trouble and in need of proper testing. Quite a different view emerges from *The Future of Astrology*, a recent book consisting of 14 essays from "some of the foremost international astrologers in the world today" (Mann 1987). Here there is nothing but unbounded enthusiasm. Curiously, half the essays say nothing about the future of astrology (Figure 6). And not one of them gives the slightest hint that hundreds of research studies exist—some of them not inconceivably pertinent to the future of astrology. The result reads like a weather forecast in which forecasters resolutely refuse to look out the window. Clearly, if this is the best that foremost international astrologers can do, then their deliberations are not helpful in predicting the future of astrology.

Nevertheless some essays are of interest. Harvey comes nearest to fulfilling the promise of the title. He predicts that fairly soon full-time astrology courses will appear in universities while the profits to be made from political, economic, medical, and weather forecasting will stimulate a great upsurge in mundane astrology. But the biggest changes will come via computers. Huge banks of computer-analysed charts complete with portraits and case histories will be available on compact disk (some already are) to provide instant large samples of a particular aspect or event or data on close time twins. Harvey predicts that this will lead to a much improved understanding of astrology and ultimately to an adequate model for integrating the various physical, psychological, and spiritual aspects of a person. This will in turn "demand that the student has a real grasp of psychology, philosophy and theology as well as a good grounding in all the sciences and arts" (in Mann, 1987:74). This is in stark contrast to American astrologer George Noonan's (1976) description of the typical American astrologer, whose "ignorance of astrology is exceeded only by his lack of knowledge of just about every other intellectual discipline." Harvey bravely notes that astrological effects "can be measured," and agrees with the point made in the essay by Elliot that the future will depend on the demonstration of these effects. But nothing so pragmatic as an actual observed effect size is allowed to sully these sublime visions.

Figure 6.   THE FUTURE OF ASTROLOGY (from Mann 1987)

AUTHOR	VIEW
Rudhyar	Astrology will bring order to the chaos of human life.
Rael	Rudhyar was a true seed man.
Hand	We must see the universe as a living conscious entity.
Addey	Astrology will become central to scientific thought.
Gauquelin	During 40 years I got positive and negative results.
Harvey	Astrology will be woven into the fabric of our lives.
Mann	Astrology will be part of medicine, education, religion.
Elwell	Every individual is a cosmic deed directed to some end.
Lewis	Growth to late 1988, then consolidation, then flowering.
Hamaker	Astrology will regain its accepted place in science.
Oken	Astrology is the purest form of occult truth.
Huber B&L	We must create new methods for spiritual delineation.
Lynes	Unless mundane astrology is recognised, tragedy looms.
Elliot	No future unless astrology can be shown to work.

The same defect applies to the essay by Hand. He argues that science is mechanist-materialist, that is, that it sees the universe as a machine and consciousness as a mere by-product of chemistry and physics. (This is *wrong*! To be successful, science requires only some order and observability; if there is neither, then science is unworkable—but so is astrology.) By contrast, astrology sees consciousness and the universe as a unity, and therefore cannot be investigated by a science which holds the contrary view. (Again, this contention is *wrong*! To test, say, whether a person fits his own birth chart better than a control person requires no mechanist-materialist assumptions whatever. Hand is setting up straw men.) He claims that the astrological world view is correct, citing as evidence the Gauquelin work, which "is one of the strongest threats to mechanist-materialism in existence" because it "has no known mechanistic explanation and it will strain the possibilities of mechanism to find one" (in Mann, 1987:37). In view of Ertel's work described earlier, which shows that straightforward explanations may be easy to find once you start looking, and in light of the negligible size of the Gauquelin effect (zero for such mainstream concepts as zodiac signs), Hand's claim is anything but convincing. Nevertheless he argues that astrology has no future unless astrologers "overthrow the mechanist-materialist world view" and "reestablish the idea of the universe as a living conscious entity" (in Mann, 1987:38). This may well be so, but Hand conveniently fails to specify the tests by which this revolution can be achieved and the evidence he would accept as proving his ideas wrong. Thus everything stays in the blue beyond, secure from rational discussion — just one more example of the *vagueness* to which astrologers habitually reduce crucial issues, as noted in the preceding section of this chapter on problems of validity. The same defect applies to the essay by Elwell, who argues that tests of astrology involving groups of, say, physicians or doctors are likely to fail because "the cosmos seldom operates in the categories which seem important to us" (in Mann, 1987:106). He claims that scientists behave like Procrustes, stretching or chopping their subject to fit the test bed, which explains the negative results. But the same criticism applies to Elwell, who stretches and chops the cosmos so that signs become "a set of imperatives to be obeyed," while planetary configurations show "those activities, problems" and the like necessary for self-development. How does he know *these* are the right categories? How could they be shown to be wrong? On such crucial issues Elwell remains silent.

In several other essays there is agreement that astrology will resume its central role in science. For example Addey (whose essay was written in 1971) points out that science, by addressing itself to matter, has resulted in a huge neglect of spirit. Astrology alone can embrace the two. Therefore:

> From being an outcast from the fraternity of sciences, it [astrology] seems destined to assume an almost central role in scientific thought. . . . its impact will be felt in the next twenty years (in Mann, 1987:43,53).

Today, almost two decades later, no such impact is apparent. Nevertheless the same sentiments are confidently repeated by others such as Elwell (1987:3) who states that "mainstream science will eventually be obliged to embrace the astrological if it is to unify its picture of the universe," and in the program notes for the 1987 London astrological research conference "Astrology promises to revolutionise science and contemporary thought". Although a poll of the conference audience of 80 or so showed that only 25% agreed (*Correlation* 1987), later sentiments expressed by the Urania Trust (one of the conference's organizing bodies) were even more extreme:

Astrology is in fact one of the great primary sciences. Once developed, astrology holds a key to many of humanities [sic] most pressing problems: in holistic medicine, in psychology, in economics, in meteorology, sociology, in international affairs and in many other areas [including] the potential to help prevent famines, cancer and countless other scourges of mankind." (Urania Trust 1988).

Similar claims of revolutionary status have been made for parapsychology, which as noted by Nias and Dean (1986:363) has testable hypotheses, a much larger body of research than astrology (perhaps 100 times as many studies), and many more competent researchers. Yet even here, as Hövelmann and Krippner (1986) point out, such claims are premature. So they are even more premature for astrology. To paraphrase their points: (1) despite half a century for research, astrology has had no appreciable impact in any field of science, nor has it initiated any changes; (2) it has not succeeded in persuading scientists to reconsider their theories and methods; (3) it cannot represent a new way of doing scientific research that will emerge after a future paradigm shift, because paradigms are overthrown by better paradigms, which astrology does not have—and without convincing positive findings never can have; (4) the positive findings of astrology are far too scanty, weak, and unreliable to warrant confidence that it will lead anywhere.

To these we can add that (5) the same revolutionary status was claimed for phrenology! One prominent supporter of phrenology, as noted above, was Alfred Russel Wallace, FRS, one of the most eminent scientists of his time, the author of 24 books and numerous scientific papers, co-founder with Charles Darwin in 1858 of the Theory of Evolution, and the recipient of the Royal Society's Royal Medal, Darwin Medal, and Order of Merit. In 1898 he wrote that

in the coming century phrenology will assuredly attain general acceptance . . . and one of the highest places in the hierarchy of the sciences. (Wallace 1898:19)

Ironically, Addey's prediction about astrology is almost identical, and on present evidence seems unlikely to escape the same fate. To paraphrase Hövelmann and Krippner (1986:2), if the revolutionary nature of astrology "is used to tempt others into the field, who may be unsatisfied by the way ordinary science works, then this propaganda is clearly made under false pretenses."

References

Abel, R. (1976). *Man is the Measure: A Cordial Invitation to the Central Problems of Philosophy*. New York: Macmillan.

Addey, J.M. (1979). Personal communication to Dean.

Addey, J.M. (1981a). The basic harmonics of Mars and their interpretive value in the horoscope. *Astrological Journal*, 23(4) 208-216 and 1982, 24(1), 12-20. His earlier work, based on 20 years of painstaking research, is described in *Harmonics in Astrology*, Fowler, 1977.

Addey, J.M. (1981b). The true principles of astrology and their bearing on astrological research. *Correlation* 1981, 1(1), 26-35. Related work is reported by Roberts, P., Harmonic analysis of the diurnal distribution of Gauquelin's professional groups. *Correlation* 1987, 7(1), 18-25.

Astro*Talk (1986). The state of astrology: where are we headed? *Astro*Talk (Bulletin of the Matrix User's Group)*, 3(3), pages 1, 6-7, 11. A poll by Michael Erlewine of some top U.S. astrologers.

Ben-Shakhar, G. et. al. (1986). Can graphology predict occupational success? Two empirical studies and some methodological ruminations. *Journal of Applied Psychology*, 71, 645-653.

Blackmore, S.J. (1983). Divination with tarot cards: an empirical study. *Journal of the Society for Psychical Research*, 52, 97-101.

Blackstone, W.T. (1963). *The Problem of Religious Knowledge: The Impact of Philosophical Analysis on the Question of Religious Knowledge*. Englewood Cliffs, N.J.: Prentice-Hall.

Carlson, S. (1988) Astrology. *Experientia* (Switzerland), 44, 316-326. A critical review with 40 references.

Comité Para (1976). Considerations critiques sur une recherche faite par M.M. Gauquelin dans le domaine des influences planétaires. *Nouvelles Brèves*, 43, 327-343. Important details that support Gauquelin but which are omitted from the original account appeared in *Zetetic Scholar*, 1982, 10, 67-71.

Cooper, J. (1981). *Astroverse*. Leeds: Duffield.

Correlation (1987). Report on the 6th London Astrological Research Conference, 20-22 November. *Correlation*, 7(2), 2-8.

Culver, Roger G., and Ianna, Philip A. (1988). *Astrology: True or False?—A Scientific Evaluation*. Buffalo, N.Y.: Prometheus Books. 1010 references.

Curry, P. (1982). Research on the Mars effect. *Zetetic Scholar* 9, 34-52 with discussion from others through p. 83, and further important information in *Zetetic Scholar* 1982, 10, 43-81 and *Skeptical Inquirer* 1983, 7(3), 77-82.

Dean, G. (1983). Shortwave radio propogation: non-correlation with planetary positions. *Correlation*, 3(1), 4-37. A critical analysis of the forecasts of John H. Nelson.

Dean, G. (1984a). More on astrology, paradigms, and research (letter and responses). *Correlation* 4(1), 35-37.

Dean, G. (1984b). More on how to approach and research astrology (letter). *Correlation* 4(2), 33-36.

Dean, G. (1984c). Astrology vs. recent findings about human ovulation (letter). *Correlation*, 4, 2, 39-41.

Dean, G. (1986a). Can astrology predict E and N? Part 3: Discussion and further research. *Correlation*, 6(2), 7-52. When ranked in terms of validity and reliability against palmistry, graphology, and orthodox psychological tests, astrology comes last. A detailed survey of the evidence with 110 references.

Dean, G. (1986b). Astrology as an alternative reality (letter). *Astrological Journal* 28(5), 276-277.

Dean, G. (1987). Does astrology need to be true? *Skeptical Inquirer* 9(2), 166-184 and 9(3), 257-273. In two parts with 100 references.

Dean, G. and Mather, A. (1985). Superprize results. *FAA Journal* (Australia), 15(3-4), 19-32 and 1986, 16(1), 65-72. In two parts with description of each entry.

Dean, Geoffrey and Mather, Arthur. (1977) *Recent Advances in Natal Astrology: A Critical Review 1900-1976*. Rockport, Mass.: Para Research.

Dobyns, Z. (1986). Personal perspective [on astrology] *Astro-Psychological Problems*, 4(3), 43-45.

Dutton, D. (1987). Stars for fun (letter), *Skeptical Inquirer*, 12(1), 109-110.

Elwell, D. (1986). Astrology: an alternative reality. *Astrological Journal* 28(4), 143-151.

Elwell, D. (1987a). Models of reality (letter). *Astrological Journal* 29(1), 37-39.

Elwell, D. (1987b). *Cosmic Loom: The New Science of Astrology*. London: Unwin Hyman. Provocative but contains no science despite the title, and offers little evidence in support of the "new science."

Ertel, S. (1987a). Further grading of eminence: planetary correlations with musicians, painters, writers. *Correlation*, 7(1), 4-17.

Ertel, S. (1987b). Good news and bad news from Gauquelin replication trials. Unpublished work based on a paper presented at the 6th London Astrological Research Conference, 20-22 November 1987. Brief details appear in *Correlation* 1987, 7(2), 4-5 with important corrections in the following issue (in press).

Ertel, S. (1988). Raising the hurdle for the athletes' mars effect: Association covaries with eminence. *Journal of Scientific Exploration*, 2(1), 53-82. Despite appearing later than the above, it was written first.

Ertel, S. (1989). Purifying Gauquelin's grain of gold: planetary effects defy physical interpretation. *Correlation 9(1), 5-23.*

Eysenck, H., & Nias, D.K.B. (1982) *Astrology: Science or Superstition?* New York: St. Martin's Press. 230 references.

Eysenck, H. (1985). *Decline and Fall of the Freudian Empire*. Harmondsworth: Viking.

Flugel, J.C. (1964). *A Hundred Years of Psychology*, 3rd edition. London: Duckworth, pp. 31-38. A most sympathetic but devastating critique of phrenology that applies equally to astrology.

Fullam, F.A. (1984). *Contemporary Belief in Astrology*. Master's thesis, Department of Sociology, University of Chicago, July 1984. Surveys public opinion polls for Western countries.

Gauquelin, M. and F. (1972). *Series C. Profession-Heredity Results of Series A & B.* Paris: LERRCP, 1972. Gives results based on hand calculations. Our Figure 4 shows results based on computer calculations where available from Gauquelin (1984), which average 0.86 x those based on hand calculations. Where not available, the results are based on hand calculations x 0.86. Because most effect sizes are less than .05, any uncertainty is less than the rounding error. Later heredity results are from Gauquelin, M. *New birth data series volume 2*. Planetary Heredity: A Reappraisal on 50,000 Subjects. Paris: LERRCP, 1984.

Gauquelin, M. and F. (1978). *Supplement to Series C. Diurnal Positions of Sun, Mercury, Uranus, Neptune, Pluto.* Paris: LERRCP. The note under Gauquelin M. and F. applies here also.

Gauquelin, M. (1979). *Dreams and Illusions of Astrology*. Buffalo, N.Y.: Prometheus.

Gauquelin, M. (1983). *The Truth About Astrology*. London: Basil Blackwell. Published in the USA as *Birthtimes*. New York: St. Martin's.

Gauquelin, M. (1984). Profession and heredity experiments: computer re-analysis and new investigations on the same material. *Correlation* 4(1), 8-24.

Gauquelin, M. (1988a). *Written in the Stars*. Wellingborough UK: Aquarian Press (Thorsons). A compilation of the best parts of *L'Influence des Astres* (1955) and *Les Hommes et les Astres* (1960), together with a brief update to 1988.

Gauquelin, M. (1988b). *Planetary Heredity*. San Diego, Calif.: ACS. A translation of *L'Hérédité Planétaire* (1966) with an update to 1988.

Gauquelin, M. (1989). Personal Communications. *See:* Gauquelin, M., *The Mars Effect and Sports Champions: A New Replication on 432 Famous Europeans.* Paris: LERRCP, 1979, 13. *2145 Physicians, Army Leaders, Top Executives.* Paris: LERRCP 1984, 8. *1540 Authors, Artists, Actors, Politicians, Journalists.* Paris: LERRCP, 1984, 9.

Hergenhahn, B. R. (1982). *An Introduction to Theories of Learning.* (2nd ed.) Englewood Cliffs, N.J.: Prentice-Hall.

Hovelmann, G.H., and Krippner, S. (1986). Charting the future of parapsychology. *Parapsychology Review*, 17, 6, 1-5.

Hungerford, E. (1930). Poe and phrenology. *American Literature*, 2, 209-31.

Kelly, I.W. (1979). Astrology and science: a critical examination. *Psychological Reports* 44, 1231-1240.

Kelly, I. W. (1980). The scientific case against astrology. *Mercury* 9(6), 135-142.

Kelly, I.W. (1982). Astrology, cosmobiology, and humanistic astrology. In P. Grim (Ed). *Philosophy of Science and the Occult.* Albany, New York: State University of New York Press.

Kelly, I.W., and Krutzen, R. (1983). Humanistic astrology: a critique. *Skeptical Inquirer*, 8(1), 62-73.

Kelly, I.W., Culver, R., and Loptson, P. (1989). Arguments of the Astrologers. In S.K. Biswas, D.C.V. Malik and C.V. Vishveshwara (Eds.). *Cosmic Perspectives.* New York: Cambridge University Press.

Lynes, B. (1987). Trust Betrayed. In Mann (1987:173-185). The title refers to the lack of trust in mundane astrology (essentially the prediction of world events) shown by politicians, economists and journalists. Lynes quotes dramatic but uncontrolled (and therefore meaningless) anecdotal cases to demonstrate supposed correlations between the heavens and world events, and claims that "What we lose by suppressing the knowledge that astrology works cannot be measured."

Mann, A.T. (Ed). (1987). *The Future of Astrology.* London: Unwin Hyman.

Martens, R., Kelly, I.W., and Saklofske, D.H. (1988). Lunar phase and birthrate: a 50-year critical review. *Psychological Reports*, 63, 923-934.

Mather, A. (1979). Response to reviews of Dean and Mather 1977. *Zetetic Scholar*, 3-4, 94-96.

Müller, A. (1986). Lasst sich der Gauquelin-Effekt bestätigen? Untersuchungsergebnisse mit einer Stichprobe von 1288 hervorragenden Ärzten. *Zeitschrift für Parapsychologie und Grenzgebiete der Psychologie* 28(1-2), 87-103. A successful replication of the Gauquelin effect using 1288 eminent physicians. For a shortened version in English *see:* Can the Gauquelin Effect be confirmed? Results with a sample of 1288 eminent physicians. *NCGR Research Journal* (formerly *Astro-Psychological Problems*), Autumn 1989, 17-20.

New Scientist (1985). Questions of priorities. 21 February: 14.

Nias, P.K.B. and Dean, G.A. (1986). Astrology and Parapsychology. In Modgil, S., & Modgil, C. (Eds.), *Hans Eysenck: Consensus and Controversy.* London: Falmer, 357-371, 374.

Niehenke, P. (1984). The validity of astrological aspects: an empirical inquiry. *Astro-Psychological Problems* 1984, 2(3), 10-15. A summary of part of the work described in Niehenke (1987).

Niehenke, P. (1987). *Kritische Astrologie: Zur erkenntnistheoretischen und empirisch-psychologischen Prüfung ihres Anspruchs.* Freiburg: Aurum. 244 references.

Nielsen, K. (1985). *Philosophy and Atheism.* Buffalo, New York: Prometheus.

Noonan, G. C. (1976). Letter. *Journal of Geocosmic Research*, 2(1), 6.

Rotton, J. & Kelly, I.W. (1985). Much ado about the full moon: a meta-analysis of lunar-lunacy research. *Psychological Bulletin*, 97, 286-306.

Rudhyar, D. (1936/1970). *The Astrology of Personality*. Garden City, N.Y.: Doubleday (originally published, 1936).

Rudhyar, D. (1975). *From Humanistic to Transpersonal Astrology*. Berkeley, Calif.: Seed.

Rudhyar, D. (1977). The birth chart as a celestial message of the universal whole to an individual part. *Review Monthly*, May, 32-34.

Rudhyar, D. (1979a). Personal communication to Dean and Mather, October.

Rudhyar, D. (1979b). Review of Dean and Mather's 'Recent Advances in Natal Astrology.' *Zetetic Scholar* 3&4, 83-85.

Rudhyar, D. (1980). Personal communication to Dean and Mather, January.

Saklofske, D.H., and Eysenck, S. (1988) *Individual Differences in Childhood and Adolescence*. London: Hodder.

Schneider-Gauquelin, F. (1988). Attempts at replicating Dr. Fr. Stark's planetary results. *Astro-Psychological Problems* 6(1), 19-25.

Severn, J.M. (1913). *Popular phrenology*. London: Rider.

Startup, M.J. (1984). *The Validity of Astrological Theory as Applied to Personality, with Special Reference to the Angular Separation between Planets*. Ph.D. thesis, Goldsmiths' College. London University, 1984. University Microfilms international ref. 8500, 985. For the serious researcher this is the best available review of methodological problems and how to solve them. 350 references.

Szanto, G. (1985). *The Marriage of Heaven and Earth: The Philosophy of Astrology*. London: Arkana (Routledge & Kegan Paul).

Transit (1983). Unsigned cartoon. *Transit* (magazine of the Astrological Association) 40, 23.

Urania Trust (1988). *The Astrological Journal*, 30(4) Ad inside back cover.

Wallace, A.R. (1898). *The Wonderful Century: Its Successes and its Failures*.

Watson, A. (1988). Trouble with numbers. *New Scientist* 21 January, 66-67. He notes that sun-sign columns carry no warnings like "Danger: simple calculation can induce ridicule." The exact quote for Libra and gynecologists is on page 107 (of *Cosmopolitan*, Jan. 1988) and reads "February . . . the 22nd sees you keeping a rendezvous with your gynecologist." No if's or maybe's here.

Watson, H.C. (1836). *Statistics of Phrenology, being a Sketch of the Progress and Present State of that Science in the British Isles*. London: Longmans, 1836. Additional socio-economic data are from Cook, C. and Stevenson, J., *Longman Atlas of Modern British History: A visual guide to British society and politics 1700-1970*. London: Longman, 1978.

Suggested Readings

As noted in the general introduction, this is not intended as anything like a complete bibliography. The works mentioned are generally standard sources, for the most part easily available, which may be of help at the next stage of the reader's research.

General introductions to traditional astrology:
Davidson, Ronald C. *Astrology*. New York: Bell, 1963.
Manolesco, Sir John. *Scientific Astrology*. New York: Pinnacle Books, 1973.
Weingarten, Henry. *A Modern Introduction to Astrology*. New York: ASI, 1974.
West, John Anthony and Toonder, Jan Gerhard. *The Case for Astrology*. Baltimore: Penguin Press, 1973.

Spiritual Astrology:
Rudhyar, Dane. *Galactic Dimensions of Astrology*. New York: ASI, 1980. (Original title: *The Sun Is Also A Star*.) Person-centered Astrology. New York: ASI, 1980.

Gauquelin's hypotheses:
Gauquelin, Michel. *Cosmic Influences on Human Behavior*. New York: ASI, 1978.
———. *Dreams and Illusions of Astrology*. Buffalo, N.Y.: Prometheus Books, 1979.
———. *Scientific Basis of Astrology: Myth or Reality?* New York: Stein & Day, 1970.
———. *The Truth About Astrology*. London: Basil Blackwell, 1983.
———. *Written in the Stars*. Wellingborough, U.K.: Aquarian Press (Thorsons), 1988.
———. *Planetary Heredity*. San Diego, Ca: ACS, 1988.
———. "Is There a Mars Effect?" *Journal of Scientific Exploration* 2 (1988): 29-51.
Krips, Henry. "Astrology: Fad, Fiction, or Forecast?" *Erknenntnis* 14 (1979): 373-392. A review of Gauquelin's work.

There has been a lengthy and extremely acrimonious controversy over attempts to further test Gauquelin's "Mars Effect". Very briefly, the history is this: a first attempt at replication, with essentially positive results, was run by the Belgian Committee for Scientific Investigation of Alleged Paranormal Phenomena (Comité Para). See *Nouvelles Breves* 43 (1976), 327-343. A second replication attempt, using Gauquelin's data but attempting to better control for expected frequencies of Mars, was run by the American Committee for the Scientific Investigation of Claims of the Paranormal (CSICOP). CSICOP claimed the results were negative, but this was disputed by the Gauquelins. See esp. Paul Kurtz, George O. Abell, and Martin Zelen, "Is There a Mars Effect?", *The Humanist*, Nov.-Dec., 1977, and accompanying articles.

A third replication attempt, again run by Kurtz, Abell, and Zelen for CSICOP, used a sample of U.S. sports champions. Here again CSICOP reported results as negative, but once again this was disputed by the Gauquelins. See esp. Paul Kurtz, Martin Zelen, and George O. Abell, "Results of the U.S. Test of the 'Mars Effect' are Negative," *The Skeptical Inquirer*, Winter 1979-80, and accompanying articles.

In October 1981 astronomer Dennis Rawlins, former Fellow and Executive Council member of the CSICOP, published a piece in *Fate* magazine in which he alleged manipulation of the data in CSICOP's two replication attempts and a coverup of positive results favoring the Gauquelins. See "sTarbaby," *Fate*, vol. 34, No. 10 (October 1981), 67-98. Further replies to Rawlins appear in *The Skeptical Inquirer*, winter 1981, and in pieces available by mail from CSIOP.

A fairly dispassionate discussion of the entire controversy—with a charge that both sides have indulged in post-hoc reinterpretation of statistical data— is Patrick Curry's "Research on the Mars Effect," *Zetetic Scholar*, No. 9, March 1982, 33-83. See also Marcello Truzzi, "Personal Reflections on the Mars Effect Controversy," *Zetetic Scholar*, no. 10, December 1982.

Critical works:
George O. Abell and Bennett Greenspan. "The Moon and the Maternity Ward," from *The Skeptical Inquirer*, reprinted in Kendrick Frazier, ed. *Paranormal Borderlands of Science*. Buffalo, N.Y.: Prometheus Books, 1981.
Bok, Bart J., and Jerome, Lawrence E. *Objections to Astrology*. Buffalo, N.Y.: Prometheus Books, 1975.
Shawn Carlson, "A double-blind test of astrology," *Nature*, vol. 318, December 5, 1985, 419-425. An attempt at a scientific test (inconclusive).
R. B. Culver and P. A. Ianna, *The Gemini Syndrome: A Scientific Evaluation of Astrology*. Buffalo, New York: Prometheus Books, 1984.
Dean, Geoffrey, and Mather, Arthur. *Recent Advances in Natal Astrology: A Critical Review 1900-1976*. Rockport, Mass.: Para Research, 1977.
Eysenck, H. and Nias, D. K. B. *Astrology: Science or Superstition?* New York: St. Martin's Press, 1982.
Goldberg, Steven. "Is Astrology Science?," *The Humanist*, March/April 1979, pp. 9-16.
Jerome, Lawrence E. *Astrology Disproved*. Buffalo, N.Y.: Prometheus Books, 1977.

Kelly, Ivan. "Astrology, Cosmobiology, and Humanistic Astrology," in Patrick Grim, ed., *Philosophy of Science and the Occult*, first edition. Albany, N.Y.: SUNY Press, 1982.

Thagard, Paul. "Resemblance, Correlation, and Pseudoscience." In *Science, Pseudoscience, and Society*, edited by Marsha Hanen, Margaret Osler, and Robert Weyant. Ontario: Calgary Institute for the Humanities and Wilfrid Laurier University Press, 1980, pp. 1-9.

———. "Why Astrology is a Pseudoscience." In *Introductory Readings in the Philosophy of Science*, edited by E. D. Klemke, Robert Hollinger, and A. David Kline. Buffalo, N.Y.: Prometheus Books, 1980, pp. 66-75.

SCIENCE or PSEUDOSCIENCE?
THE PROBLEM of DEMARCATION

Introduction

Parapsychologists are quite unanimous in upholding their discipline as a science. Advocates of the existence of flying saucers often present studies that aim at being scientific, and many characterize themselves as "UFOlogists" in that spirit. There are those who claim that Eastern Mysticism parallels contemporary science, and some astrologers, though by no means all, think of astrology as a science.

Those antagonistic to the pursuits these represent routinely condemn them as "mere pseudoscience." The critics' use of the term "pseudoscience" suggests not that parapsychology, "UFOlogy," astrology, and what we will later call "quantum mysticism" are merely unpopular science, or even bad science, but that they are something quite different—something that does not qualify as science at all, popular or unpopular, good or bad. Those who reject these as "pseudoscience" hold them to be unscientific pursuits: poor facsimiles or corrupt imitations of science proper. Fool's gold—iron pyrite—is not gold at all. On this analogy, "pseudoscience" is intended as a label for something like "fool's science."

This immediately raises a fundamental question in philosophy of science. What is science, what is pseudoscience, and what is the essential difference between them? How are we to test astronomy and astrology, parapsychology and psychology, or particular claims or theories or modes of procedure in these areas in order to tell whether they qualify as scientific or not?

In general, philosophers have not wanted to say that those areas of inquiry commonly labeled "pseudoscientific" are merely unpopular areas of investigation, or that the line between "science" and "pseudoscience" is a line drawn for a variety of historical, sociological, and broadly political reasons but without any firm rational or philosophical justification. Surely science and pseudoscience must be fundamentally different—or so it seems. But where precisely does the difference lie? This, following Karl Popper, is what is called the "problem of demarcation." It is this question that we first raised in discussing the case of astrology, which we will address in general terms in this section, and which will linger in one form or another throughout the volume.

CLARK GLYMOUR and DOUGLAS STALKER start off this section with a humorous piece satirizing the popular and lucrative forms of pseudoscience they see all around us. They suggest six dubious principles for constructing one's own pseudoscience, present their own bogus "Alphabetology" and "Fascinating Rhythms" in some detail as illustrations, and finish by offering the reader a chance to snap up a franchise in a questionable enterprise called "Peruvian Pick-Up Sticks." This entertaining piece does not pretend to be a complete philosophical treatment of the problem of demarcation. But it does quite effectively raise the central issue.

The next selection is KARL POPPER'S classic formulation of the problem of demarcation. Popper asks, "When should a theory be ranked as scientific?" and distinguishes this question from "When is a theory true?" or "When is a theory acceptable?" Popper rejects the suggestions that those theories are scientific that appeal to empirical evidence or that are sufficiently verified or confirmed. Instead he proposes a number of requirements that can be summarized by saying that "The criterion of the scientific status of a theory is its falsifiability, or refutability, or testability." For Popper a theory is scientific only if it puts itself at empirical risk, and confirming evidence counts for a theory only if it results from an attempt to refute the theory. A theory that is not refutable by any conceivable event is non-scientific. On the basis of this falsifiability criterion of demarcation Popper condemns as pseudoscientific not only astrology but also Marxist theories of history and Freudian and Adlerian psychology.

DANIEL ROTHBART argues that Popper's solution to the problem of demarcation is too weak; that Popper's criterion of falsifiability fails to exclude too many clear cases of pseudoscience. On the basis of a discussion of criteria for criteria of demarcation that is intriguing in its own right, Rothbart proposes an alternative to Popper's criterion that is more historically oriented. Whether a theory is to be considered scientific or not, Rothbart maintains, may depend on what other theories it has as competitors at a particular time. In order to be scientific, he proposes, a theory must match the empirical and explanatory success of its rivals and must clash with its rivals by yielding inconsistent test implications. On this criterion astrology may have been a science in the context of ancient Greece, though it is not a science now because its competitors have changed so drastically. In his final section Rothbart pursues some of the implications of this approach and attempts to counter objections.

A more familiar historical approach is that of THOMAS S. KUHN. Kuhn here accuses Popper of taking as characteristic of science certain instances of scientific revolution that are in fact extraordinary and exceptional in the history of science. "Normal science" consists not of dramatic attempts to falsify major theories but of problem-solving within a context of assumed and unquestioned theory. What is standardly tested in normal science is not those theories assumed as background but the competence of the individual investigator. For Kuhn it is normal science that holds the most promise of distinguishing science from other pursuits, and thus "if a demarcation criterion exists . . . it may lie just in that part of science which Sir Karl ignores." Kuhn agrees with Popper that astrology is an instance of pseudoscience, but for quite different reasons. Astrology does not fail to qualify as a science for lack of falsifiability; the history of astrology "records many predictions that categorically failed." Astrology fails to qualify as a science, Kuhn maintains, because it lacks the puzzles and problem-solving characteristics of normal science, and thus "could not have become a science even if the stars had, in fact, controlled human destiny."

IMRE LAKATOS distinguishes two strands in Popper's thought, which he terms 'naive falsificationism' and 'sophisticated falsificationism'. Kuhn is right, Lakatos argues, in objecting to naive falsificationism—scientific growth shows a greater continuity and some scientific theories have a greater tenacity in the face of falsification than such a view would imply. But Kuhn is wrong, Lakatos argues, to think that by disposing of naive falsificationism he has succeeded in discrediting all forms of falsificationism or the Popperian view in general. Kuhn's own view, moreover, "would vindicate, no doubt, unintentionally, the basic political credo of contemporary religious maniacs":

> For Popper scientific change is rational or at least rationally reconstructible and falls in the realm of *logic of discovery*. For Kuhn scientific change—from one 'paradigm' to another—is a mystical conversion which cannot be governed by rules of reason and which falls totally within the realm of the (*social*) *psychology of discovery*. Scientific change is a kind of religious change.

"Thus *in Kuhn's view*," Lakatos emphatically states, "*scientific revolution is irrational, a matter of mob psychology*." Here Lakatos attempts to outline a Popperian 'sophisticated falsificationism' which will escape Kuhn's criticisms and yet avoid this kind of irrationalist consequence.

In "Kuhn, Popper, and the Normative Problem of Demarcation," ROBERT FELEPPA offers an importantly different evaluation and resolution of the debate between Popper and Kuhn. Feleppa criticizes Kuhn's treatment of Popper both on specific points of history and interpretation and on the more central ground that Kuhn has missed the major point of Popper's work. For Popper's attempt is not merely to draw a line between science and pseudoscience that is historically accurate as a description, but to *justify* the praise of some things as science and the condemnation of others as pseudo-science. What Kuhn has missed, Feleppa suggests, is precisely this *normative* problem of demarcation. Feleppa also offers a "reconstruction" of Kuhn's work, however, which seems better equipped than Kuhn's original to handle the tasks of the normative problem. Here Feleppa introduces a pragmatic notion of justification involving "reflective equilibrium," borrowing from the work of Carl Hempel, John Rawls, and Nelson Goodman. A similar approach to Popper's work, he suggests, allows us to see the relative strengths and weaknesses of a Kuhnian and Popperian approach to the normative problem.

ROGER COOTER offers a quite different perspective on science and pseudoscience. Cooter presents in some detail the social history of nine-teenth-century phrenology as an illustration of his more general claim that the use of the term "pseudoscience" has played an "ideologically conserva-tive and morally prescriptive social role" in the interests of an established capitalist order. Here Cooter explicitly adopts a relativist position concern-ing scientific truth, viewing those things that have been labeled "science" and "pseudoscience" as competing social institutions rather than as "cor-rect science" and "false science." In the end Cooter proposes that the term "pseudoscience" be replaced by "unorthodox science" or "non-establishment science" so as to more adequately represent "relative differences in scien-tific social perception rather than categorical methodological opposites or matters of truth and falsehood."

Cooter's piece is clearly in conflict with much that precedes it. The gen-eral attitude he conveys is diametrically opposed to that evident in Glymour and Stalker's opening piece, and his adoption of a relativist position clearly amounts to an abandonment of the Popperian rationalism for which Lakatos argues and of the normative problem of demarcation emphasized by Feleppa. Is Cooter's relativism an acceptable position? Does Cooter argue effectively for an abandonment of the attempt to construct and defend a viable criterion of demarcation?

Consider also a particular bone of contention that is bound to arise be-tween Popper and Cooter. Cooter's is a historical argument, in the general tradition of Marxism, for abandoning any attempt at demarcation in Popper's sense. But one of the things that Popper explicitly rejects as pseudoscience is Marxist theory of history.

In the piece included earlier, Lakatos claimed that

> The clash between Popper and Kuhn is not about a mere technical point in epistemology. It concerns our central intellectual values, and has implica-tions not only for theoretical physics but also for the underdeveloped social sciences and even for moral and political philosophy.

The final piece in the section, by ROBERT L. BAUM, is a general reflec-tion on Popper, Kuhn, and Lakatos in very much this spirit.

The selections of this section as a whole, and the disagreements they exhibit, will undoubtedly leave the reader with his or her own set of questions. What is the role, or the limit, of historical considerations in dealing with the issue of demarcation? Has one criterion of demarcation in particular, or one general approach, been shown preferable to its rivals? What is it to confirm a theory, and what types of theory are confirmable? Is demarcation in Popper's sense possible at all?

Our questions started with an examination of astrology in the previous section. In the following sections we will pursue these general questions further by examining two other controversial topics in more detail: parapsychology and quantum mysticism.

Winning through Pseudoscience

Clark Glymour and Douglas Stalker

Like you, we used to think that enterprise, diligence, and a positive mental attitude would make sure we got ahead, into the big bucks. Like you, perhaps, we struggled along on sixty grand a year, worrying about cash flow, complaining about taxes, never making the big bucks. Then we got wise, and after we got wise—well, we got rich. Now we're going to share our methods because, frankly, we can afford to.

To get to the big bucks, you need a gadget. There are lots of gadgets, but brains and guts won't get you most of them. A rich uncle is a gadget, especially if he dies and leaves it all to you, but you've either got a rich uncle or you haven't, and there's not much you can do about it either way. But there's one gadget that's available to anyone with a high school diploma, anyone who is not afraid to use his noggin. Erich von Däniken, Sydney Omarr, Sybil Leek, Bernard Gittelson, Immanuel Velikovsky, and Werner Erhard are doing fine. They have different gadgets, but they all have the same *kind* of gadget: pseudoscience. It sells. People just love to believe, and our research shows that eighty percent of them will believe things a gorilla wouldn't. They'll cheerfully empty their wallets to anyone with a twinkle on his tongue and a pseudoscience in his pocket. Astrology, biorhythms, ESP, numerology, astral projection, scientology, UFOlogy, pyramid power, psychic surgeons, Atlantis real estate—they're all good business. Even Bigfoot turns a buck.

We woke up to the fact that in America pseudoscience is the gadget to give you a leg up. But, you may say, virtually every imaginable pseudoscience already has its promoter, the market is cornered, there's no way for a new boy to break in. Wrong. You break in with your *own* pseudoscience, and we'll show you how. Just to make it easy, we are going to divulge the very principles behind it all. We'll even build two of these money-making pseudosciences right here, and show you how well they work. If you can't do better by yourself, we'll sell you a franchise in our very own Peruvian Pick-Up Sticks. Once you have your own pseudoscience, or even a franchise (with your own territory, of course), the possibilities for turning it into dollars are endless. To set you off in the right direction, we'll give you some tips, taken from the masters.

Fundamental Principles of Pseudoscience Construction

Building your very own pseudoscience, and even defending it against skeptics, is very easy if you keep a few simple principles in mind.

Principle 1: A coincidence in the hand is worth two in the bush. Start your pseudoscience from a few odd coincidences, then work backward to a preposterous theory that fits them. Some would-be pseudoscience builders make the serious mistake of thinking they must start with firm, clear assumptions and see if they can then obtain amazing predictions from these assumptions. They seldom succeed. Start with the coincidences you want, it's safer and easier. Coincidences are nearly everywhere. But if you have difficulty locating genuine coincidences that you like, start with any event or state of affairs—for example, the fact that you are wearing green socks today. Now trace back the history of the fact far enough, and—hey! presto!—it becomes enormously improbable, an amazing coincidence, that the thing ever happened. After all, what was the chance, eight million years ago, that the universe would evolve so that you are wearing green socks today? Dinky, right? With this technique, you can take any plain-looking fact and turn it into something really startling.

Principle 2: A purpose to everything and everything to its purpose. By faithfully applying Principle 1, you end up with a number of facts that cry out for an explanation. The easiest way to explain those amazing coincidences you have just manufactured is to say that they are no accident at all. They happened on purpose. Whose purpose, what purpose? Yours, dummy. Those green socks signify great things; they portend and presage, they are part of a plan, a cosmic aim or will, some big intention or resolution. Just remember: things don't just happen, they happen on purpose, and it's always your purpose. Pick one and see, but do your choosing after you've read Principle 3, which concerns making the best possible choice in these matters.

Principle 3: The taller the story, the harder it falls. Let your imagination run free when inventing explanations for those amazing coincidences, and give the cold theory a little personal touch, some poignancy, human or inhuman interest. There are lots of models. Oppressed dwarves driven to living inside a hollow earth, daring space travelers dropping human prototypes to rule over the apes (it helps if your theory contradicts a theory nobody much likes anyway), planets that go bump in the night. You can do better: chickens didn't exist on earth before the year 2000 B.C. That's because they are Martians in disguise, and they can influence the minds of people who eat egg yolks. See, it's easy.

Principle 4: Even physics isn't all that precise. But, you might say, chickens did exist before 2000 B.C. Well, so what? Don't get bogged down with facts. Nobody knows about things like what lived 4,000 years ago, or the laws of celestial mechanics or the theory of probability, except maybe a few stuffy professors that nobody reads anyway. (In case they do get in print somewhere, see Principle 6.) The plain truth about facts is this: if you need a fact, try a fiction. If you need something to be true, say it is.

Principle 5: Science is numbers and gauges. Or so think the unwashed, which means that your pseudoscience will have better sales potential if it makes use of a mysterious device, or a lot of calculations (but *simple* calculations). You have to watch out with the gadgets or the FDA will be after you, but the calculations are safe as pie. They should have some method to them, but not much. A few additions and subtractions, perhaps, and then one long division. The great models are astrology and biorhythms, and, we must say, our own Fascinating Rhythm, which is described below.

Principle 6: Saying no to nit-pickers. The world, alas, is filled with skeptics and troublemakers, some of whom may complain that your theoretical claims contradict well-known scientific results, or are unscientifically vague, or that you lack evidence for your pseudoscience. None of these individuals understands the free enterprise system. Don't be intimidated: follow our advice, show a little grit and you'll reduce them to quivering ectoplasm. Here are five sure-fire strategies.

(a) Tell them they don't understand (especially if they offer a counterexample). When confronted with counterexamples and cavils, make a distinction, any distinction. If really hard pressed, keep changing your claims (but use the same words as much as possible) while insisting that you are only saying what you meant all along.

(b) If the skeptics complain about a lack of evidence, just remember that, so far as the public is concerned, one explanation is worth a thousand tests. Remind them of all of the amazing coincidences your theory explains. How could that not be evidence?

(c) Quote a biggie at them. Emerson is best, that stuff about a foolish consistency being the hobgoblin of little minds (maybe leave out the "foolish").

(d) Wax philosophical. Talk about alternative conceptual schemes, evolving criteria of evidence, methodological opacity, the jiggle between theory and evidence, the necessity to believe in order to understand, the ever-present incommensurability.

(e) Counterattack! Point out that you and your work are victims of the narrow scientific establishment, just like Galileo and his work. Mention narrow-minded expertise, missing the forest for the trees, government plots, and the like.

Two Pseudosciences from Scratch

As we promised, we will unveil two money-makers right before your very eyes. First off, we explain Fascinating Rhythm, a true analog to biorhythms. It has some distinct advantages, including confirmation, glitter, and advertising appeal (*and* you can dance to it). With FR-3, short for Fascinating Rhythm, you can now chart the syncopated beat of our lives. It's computed in the very same way as that phony bio stuff. You start with your date of birth, and then chart through to the day you want. Here's how.

There are three important cycles in the life of each and every human being. The Mambo cycle runs its full course in 145 days, half on the upswing, half on the downbeat. The Mambo cycle controls your intellectual life, your powers of thinking and deciding. Next, there is the Samba cycle, and it takes a full course in 93 days. Once again, half on your good side, half on your bad. The Samba cycle governs your physical life, those nerves and muscles, that strength and endurance. Last but not least, we have our personal favorite, the Rhumba. Can you feel the beat? The Rhumba cycle goes a mere 48 days. Half up, half down. The Rhumba, as you might guess, regulates your emotional life; it's the meaning behind your fluctuating moods.

To see how well FR-3 works, and to gain a fuller understanding of its powers, we will chart the rhythms for two important people and important events in their lives. So let's do some real science with Wilt Chamberlain and Janis Joplin. On March 2, 1962 Wilt, you may recall, scored 100 points in one game of basketball. Consider what his FR-3 chart tells us about that day. Wilt

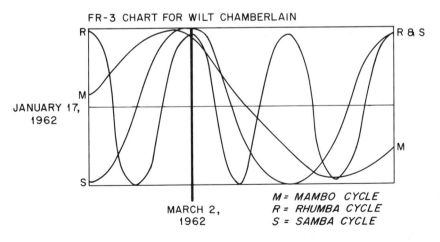

Figure 1.

Chamberlain was born on August 21, 1936. On that fated evening at Madison Square Garden he had been alive 9,324 days. Nine thousand three hundred twenty-four divided by 145 (the number of days in a Mambo cycle) gives 64 full Mambo cycles, with a remainder of 44 days. Nine thousand three hundred twenty-four divided by 93 gives 100 full Samba cycles, with 24 days left over. And divided by 48, 9,324 gives 194 full Rhumba cycles, with 12 days left over. So on March 2, 1962, Wilt the Stilt was 44 days into a Mambo cycle, 24 days into a Samba cycle, and 12 days into a Rhumba cycle. The chart of his rhythms (starting from January 17, 1962) appears in Figure 1. Even a child can see that FR-3 shows *why* Wilt had a great day. All three curves peak, peak, peak. But we know that you want more proof, and so we will draw up a chart for Janis Joplin, too. Recall that Janis was born on January 19, 1943. She died on October 4, 1970. It's easy enough to see what her chart looks like for her last day. She had then been alive 10,120 days. Dividing by the Mambo cycle (145 days), we find that she was 115 days into a new Mambo cycle. The Samba cycle is just as easy to calculate, and for Janis Joplin we find that she had gone through 108 full Samba cycles, which means that she was 76 days into a new Samba cycle. And, last but not least, Joplin had danced 210 full Rhumbas on that fatal day; she was just 40 days into her new Rhumba cycle. This is all plain when you glance at her chart in Figure 2. Once again, FR-3 explains the events. On October 4, 1970, all three of her cycles were close to their lowest, and most dangerous points. If you like, run a biorhythm for these same people and these same events. You will find what we found: FR-3 produces the best chart; it explains events that biorhythms can't. The only rhythms for you, it should be obvious, are the Fascinating ones. We wouldn't tap our toe to any other beat.

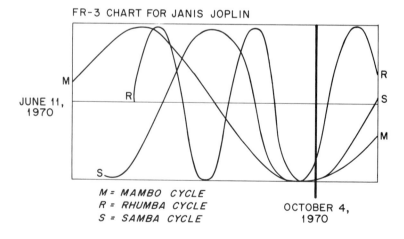

Figure 2.

Once is not enough. And so we describe for you a second, new and profound pseudoscience. It concerns the inner self. Since the beginning of philosophical reflection, the most challenging task has always been to know oneself, and nowadays people want to know themselves before they know you. We have discovered the one sure path to the true person, the way to locate the real them: Alphabetology. It's better than astrology and numerology put together, and easier too.

For centuries, men have noticed the metaphysical and theosophical importance of naming. After all, just look at the mystical and undeniable importance western religions place on baptizing, which concerns only the first name. The more important part of naming—the last name—has been overlooked. By carefully noting three things about your last name, you'll be ready to do a reading that reveals all about you, or anyone else for that matter. So write down your last name—print if you can, that's best for the process. Now you need to find three things in your name: the Vowel Signs, the Ascending Consonant, and the Syllable House. To find your Vowel Signs, locate the vowels in your name, then consult our handy chart:

A   sincere, strong, patient
E   lucky, fluid, engaging
I   proud, forceful, possible born leader
O   aggressive, impatient, fighting instinct
U   gloomy, scholarly, punctual
Y   unpredictable, skillful, natural executive

To find the Ascending Consonant, simply find the first consonant in your last name, and then consult the chart below:

Group 1:   b, c, d, f, g   benefits, advantages
Group 2:   h, j, k, l, m   conflicts, strife
Group 3:   n, p, q, r, s   difficulties, deviousness
Group 4:   t, v, w, x, z   resolutions, conclusions

Don't draw any sweeping conclusions yet. So far, we have only sketched the true person, chosen the paints for the portrait, as it were. A person is not just a collection of traits and happenings. No sir, these items go together, they combine, they dominate or recede. You know that a painting of a person is not just a list of colors. They must go together, and so must the Vowel Signs and Ascendant. How? You find out by finding your Syllable House. To do that, count up the number of syllables in your last name, and then consult our third chart:

1 syllable   commutative (combining independent of order)
2 syllables   distributive (diffusing more or less evenly or unevenly)
3 syllables   associative (relating by groups)

4 syllables   symmetric (having corresponding yet connecting points)
5 syllables   contrapositive (permutating major parts, probably not positive)

These houses, as you no doubt have noted, find their expression in principles that are somewhat like mathematical principles, but not just like them. This difference is important when it comes to dealing with real flesh-and-blood people. It means that our principles cannot be reduced to mechanical operations, principles whose application involves no human judgment. The human factor is ever present and irreducible here, as it is throughout our real lives and activities, in contrast to the fictional world scientists describe for us.

We put these principles into action by making up readings. We have two sample readings, each based on a fully drawn analysis of the names in question, the respective Signs, Houses, and Ascendants, as well as the skill of one of our most practiced readers.

Official Reading for: Nixon
Vowel Signs: i   proud, forceful, born leader
               o   aggressive, impatient, fighting instinct
Ascendant: n   difficulties, deviousness
House: distributive

Reading: Nixon is certainly a proud person, as his public appearances and writings clearly indicate. He is also forceful, aggressive, and impatient. These traits were shown time and time again in national and international politics. Nixon, more than most world figures, has a fighting instinct. After all, isn't he still fighting? His Ascending Consonant points to the Watergate affair, among many lesser problems. His Signs and Ascendant seem to distribute evenly, as can be seen even by the untutored eye. Nixon, devious? The reading tells all. So be it!

This reading seems so accurate it is hard to believe, but this is the kind of result you get from Alphabetology, a clearly superior brand of pseudoscience. Quick, accurate readings are its hallmark. For example, just glance at our second sample reading and see how accurate it is.

Official Reading for: Hitler
Vowel Signs: i   proud, forceful, born leader
               e   lucky, fluid, engaging
Ascendant: h   conflicts, strife
House: distributive

Reading: Although infamous, Hitler was admittedly proud, forceful, and engaging—a born leader with more than his share of luck to be at the right time to take over a mighty nation. Indeed, an amazingly lucky chain of events that brought Hitler to power, one only a very lucky man could chance on—from housepainter to Führer; think of it! His changing views and whims and fancies showed him to be fluid in his preferences, most often to his discredit and the dismay of others. One week you could be his friend, the next his enemy.

Certainly his life was marked with conflicts, so much so that his traits distributed in the direction of definite evil. An uneven total, regrettably. One can only wonder what he might have been if they had coalesced differently.

We would go on to show you more official readings, but this book is pretty high-class, and so the editor wants to save room for some more penetrating articles. Suffice it to say, try out a name, get some practice in before you don your gold chain and go to the bar tonight.

The Importance of Principle 6

The foregoing, we trust, illustrates for you most of the principles of pseudoscience construction. Naturally you will want to apply them in your own original way. Principle 6 is the great thing, however, and to give you a taste of how to use it we offer the following snippet of dialogue, taped while one of us was interviewed by a hostile reporter.

Reporter: Dr. Stalker, there are a lot of well-known scientists who say your Alphabetology is nothing but a sham. There's even a petition going round to try to prevent you from teaching it.
Stalker: What do you expect from people taught by rote? Orthodox scientists are totalitarians at heart, they can't stand nonconformity; as for my colleagues, they know that in the free market of ideas Alphabetology will survive because it is the fittest. That, of course, is why they will stop at nothing to prevent students from voting with their butts . . .
Reporter: Wouldn't Alphabetology forecast the same reading for both Jimmy Carter and Billy Carter? But they're not the same!
Stalker: What makes you think that Jimmy and Billy say "Carter" *exactly* the same way? No chance. You fail to understand the full ramifications of Alphabetology. You've failed to distinguish between long and short vowels, for one thing. A Group 1 Ascending Consonant can, when the Syllable House is distributive, go evenly or unevenly, especially when the Ascending Consonant precedes a short vowel. This can cause a polarity. For example, an A sign might shift to an O sign, or it might remain the same. And, you must remember, we are not omniscient. What mortal or mortal's theory is?
Reporter: What do you say to the criticism that none of your work on Alphabetology contains any properly controlled experiments?
Stalker: Well, what's likelier, that Alphabetology is true or that *just by chance* its principles correctly describe the character of such different people as Nixon, Ghandi, and Hitler? There are hundreds of people whose characters have been correctly deciphered on alphabetological principles: Julius La Rosa, Pablo Casals, T. G. McGonnigle, Irving Thalberg, Mary Ann Mobley, Randolph Scott, Jay Silverheels, Barney Newman, and Carl Andre, just to name a few. What do you think—that all of these analyses hold just by accident? Piffle.

We hope this excerpt helps you put Principle 6 into action. With the world filled with so many depressed and destructive types, you need to be adept with Principle 6. If you still feel uncertain about how to use the principle, we recommend a careful study of the *Playboy* interview with Erich von Däniken.

Cashing In

Once you have your pseudoscience, how is it to be turned into hard dollars? Write a book, and entitle it something like *My Struggles to Reveal the Scientific Basis of . . .*, or *Winning Through . . .* If you can't write a book, make a dingy device to go with your pseudoscience and market it. For example, for only $49.95 our own Peruvian Pick-Up Sticks comes with genuine Peruvian pick-up sticks and board, and an authentic Peruvian Protractor. As for outlets, first try mail-order magazines like *Argosy, Fate,* and *Popular Mechanics.* And, don't forget, your local head shop will be glad to take a few on consignment. If you're full of energy, start a weekend workshop on your pseudoscience. Advertise a little and the post-teenyboppers will flock around you, just as though you were Travis McGee. Avoid sex at first—remember, this is science—but give just the hint of Future Sex. That'll bring 'em back for the next workshop. Before long, you'll be on the radio talk shows, and even daytime TV, along with the accordion players and Tupperware saleswomen.

If you give it an honest try and you fail, either because of difficulties with your pseudoscience, or because of marketing difficulties, then do the simple thing: take out a franchise with us. Our Peruvian Pick-up Sticks has been market tested in Portland, Oregon with astounding success. Right now, we're selling franchises faster than we can draw up the proper papers. But, frankly, we want even more money than we have now, and that means you have a chance to buy your very own franchise. It comes with a pseudoscience starter kit that enables you, in one short weekend of occasional practice, to turn a dollar on Monday. On Tuesday, the word will be out, your flag will be flying, and you may be on your way to untold millions.

Your starter kit comes complete with a magnificent tall tale about the cosmic significance of Peru and its sticks. To quote only a small portion of the spiel (written by no less an expert than Spiro Agnew's nephew):

> In deepest, darkest Peru, the naïve natives find themselves puzzled, even placed in a pickle, by certain massive indentations in the earth. Unsound scholars have speculated that these superficial scrawls were made to guide aliens to safe berth in any storm. The truth is now revealed. These markings, these scrawls, are but reflections of the king's castings in the game of fate, in the game of life, in the game of death known to us of late as Peruvian Pick-up Sticks.

Pretty good, huh? It works a lot like mood music by Montavani, sets the atmosphere, gets the good vibes going and the thinking to a minimum.

Your starter kit also comes with a genuine Peruvian Pick-up Sticks board, which is made of laminated Beau d'Arc wood. The board measures a full 12″ by 12″. In each corner of the board is a symbol signifying the essence and concrescence of human aspiration and hominoid degradation. Up in the left corner, there is a bird, signifying power, freedom, and imagination. Over in the upper right corner, you will find a squirrel, which signifies timorousness,

PERUVIAN PICK-UP STICKS BOARD

Figure 3.

deep secrets, and an interest in nuts. Down in the lower right corner is a llama, signifying tenacity, patience, duration, and valuable hair. And last, over in the lower left corner, you can see a pig, which signifies solidity, firmness, character, and a deep attachment to the present. In addition to these figures (drawn in real enamel), the board contains a groove, indeed the Great Groove, which is a direct representation of the actual huge and great groove that scars the surface of Peru. The board is illustrated in Figure 3. In real life, and for longer wear, it is covered with a tough vinyl coating. We'll even throw in a nice-looking Naughahide carrying case, and put your initials on the flap.

Besides the board, Peruvian Pick-up Sticks requires three sticks, preferably made of thorns from the Beau d'Arc tree (although other materials may, in a pinch, be substituted). Frankly, we're in a pinch, what with these franchises selling like hotcakes. So your three sticks are made of solid pine, a good wood in its own right. Two of these sticks are called nebishes and may be of any color, though puce is our standard. The third is thicker than the other two, and it *must*—for obvious reasons—be black. Called "the Wong," it is the stick of premier importance. The three sticks are cast on the board in the manner of the children's game, pick-up sticks. The cast must be made by the person whose fate is to be read, but the reader of the cast may be the caster or some other person of suitable skill. In any case, an accurate reading requires full and, in these times of inflation, somewhat expensive training. To be sure, we have workshops all over the country, especially in California, at which we offer proper training all in one short weekend. Besides, we usually hold them at the Holiday Inn closest to the local airport, so you can get in some dandy socializing along the way.

So, you're no doubt asking, how do you play this game of fate, this game of life, this game of death known to us as Peruvian Pick-up Sticks? It's easy. A cast is made, and then you take out your genuine Peruvian Protractor, which also comes in your starter kit. Just as life is a bunch of angles, so is this game. You find three angles: the angles between each of the nebish sticks and the Great Groove, and the angle that indicates the lie of the Wong (a misnomer

A TYPICAL CAST

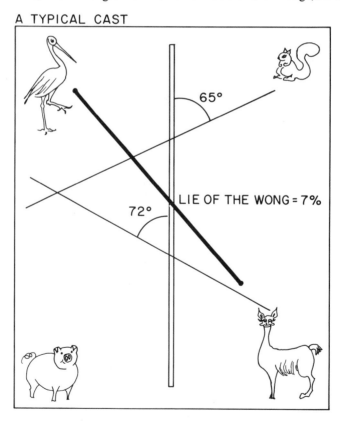

Figure 4.

to be sure, for the Wong does not lie). A typical cast is shown in Figure 4 complete with the angles indicated. In this case, the bird–llama line, which lies closest to the lie of the Wong, is said to *dominate* the cast; the domination is ameliorated, of course, by how far the Wong lies from the dominant line. Having ascertained these angles and the domination, the reader notes (1) the smallest nebish angle; (2) the largest nebish angle; (3) the sum of the nebish angles, and (4) the difference of the nebish angles. The first—the smallest nebish angle—concerns weakness and decay, the uncertainty of the future; the second concerns wisdom, power, and love; the third concerns material success and creativity; and the fourth is a measure of erotic magnetism, spirituality, and tragic prospects. Of course these indicators, these angles, cannot be interpreted mechanically; a skilled reader must have a thorough mastery of the ancient Peruvian principles, and of the many models left to us by the vanished master. These principles and models have been collected by us in a volume forthcoming from a distinguished university press and titled, in faithful translation from the Mayan, *My Thing*. Until the big-press edition hits the stands, we are including our vanity edition in your starter kit. Although not as handsome as the forthcoming edition, it will get you going. After all, why wait, the money is out there, so shouldn't you be?

It would be unprofessional of us to reveal the true reading of the cast we have illustrated. Some of our clients are international celebrities of one sort or another, and they have a right to their privacy, too. We can, however, tell you, merely as a hint, that the cast illustrated was thrown by a distinguished former monarch, that it is dominated by power and imagination, developed patiently and by an extreme tenacity of character (the Wong lying nearly on the llama); that, although the Wong may indicate the fixed character of the caster, the lay of the nebish sticks determines how this character will meet with circumstance. In the present case, the lay of the nebish sticks suggests, most strongly, weakness and decay, not to mention uncertainty. It should be noted that only a month after making the cast we have illustrated the unfortunate monarch was required by his countrymen to take an extended vacation in Morocco.

There you have it. And you can really have it, a full starter kit for only $49.95. Think of it: your own business at last, being your own boss; you'll be celebrated, and toasted, but most important, you'll be getting your piece of the big cake. By invention or by franchise, pseudoscience is your way to riches. We can think of no better envoi than the one L. Ron Hubbard (one of the great pseudoscientists of our time) used to end his classic best seller, *Dianetics: The Modern Science of Mental Health*. After 500 pages on topics from prenatal experiences to using your engrams, Ron put it all in a nutshell when he wrote: "Good hunting!"

Science: Conjectures and Refutations

Karl R. Popper

> Mr. Turnbull had predicted evil consequences, . . . and was now doing the best in his power to bring about the verification of his own prophecies.
> —Anthony Trollope

I

When I received the list of participants in this course and realized that I had been asked to speak to philosophical colleagues I thought, after some hesitation and consultation, that you would probably prefer me to speak about those problems which interest me most, and about those developments with which I am most intimately acquainted. I therefore decided to do what I have never done before: to give you a report on my own work in the philosophy of science, since the Autumn of 1919 when I first began to grapple with the problem, "*When should a theory be ranked as scientific?*" or "*Is there a criterion for the scientific character or status of a theory?*"

The problem which troubled me at the time was neither, "When is a theory true?" nor, "When is a theory acceptable?" My problem was different. *I wished to distinguish between science and pseudo-science*; knowing very well that science often errs, and that pseudo-science may happen to stumble on the truth.

I knew, of course, the most widely accepted answer to my problem: that science is distinguished from pseudo-science—or from "metaphysics"—by its *empirical method*, which is essentially *inductive*, proceeding from observation or experiment. But this did not satisfy me. On the contrary, I often formu-

This excerpt consists of the first two sections of a lecture given at Peterhouse, Cambridge, in Summer 1953, as part of a course on developments and trends in contemporary British philosophy, organized by the British Council. The whole of the original was originally published under the title "Philosophy of Science: a Personal Report" in *British Philosophy in Mid-Century*, ed. C. A. Mace, 1957. © 1963, 1965 by Karl R. Popper. Excerpted with permission from "Science: Conjectures and Refutations," *Conjectures and Refutations* (New York: Harper & Row, 1968), Ch. 1.

lated my problem as one of distinguishing between a genuinely empirical method and a non-empirical or even a pseudo-empirical method—that is to say, a method which, although it appeals to observation and experiement, nevertheless does not come up to scientific standards. The latter method may be exemplified by astrology, with its stupendous mass of empirical evidence based on observation—on horoscopes and on biographies.

But as it was not the example of astrology which led me to my problem I should perhaps briefly describe the atmosphere in which my problem arose and the examples by which it was stimulated. After the collapse of the Austrian Empire there had been a revolution in Austria: the air was full of revolutionary slogans and ideas, and new and often wild theories. Among the theories which interested me Einstein's theory of relativity was no doubt by far the most important. Three others were Marx's theory of history, Freud's psycho-analysis, and Alfred Adler's so-called "individual psychology."

There was a lot of popular nonsense talked about these theories, and especially about relativity (as still happens even today), but I was fortunate in those who introduced me to the study of this theory. We all—the small circle of students to which I belonged—were thrilled with the result of Eddington's eclipse observations which in 1919 brought the first important confirmation of Einstein's theory of gravitation. It was a great experience for us, and one which had a lasting influence on my intellectual development.

The three other theories I have mentioned were also widely discussed among students at that time. I myself happened to come into personal contact with Alfred Adler, and even to co-operate with him in his social work among the children and young people in the working-class districts of Vienna where he had established social guidance clinics.

It was during the summer of 1919 that I began to feel more and more dissatisfied with these three theories—the Marxist theory of history, psycho-analysis, and individual psychology; and I began to feel dubious about their claims to scientific status. My problem perhaps first took the simple form, "What is wrong with Marxism, psycho-analysis, and individual psychology? Why are they so different from physical theories, from Newton's theory, and especially from the theory of relativity?"

To make this contrast clear I should explain that few of us at the time would have said that we believed in the *truth* of Einstein's theory of gravitation. This shows that it was not my doubting the *truth* of those other three theories which bothered me, but something else. Yet neither was it that I merely felt mathematical physics to be more *exact* than the sociological or psychological type of theory. Thus what worried me was neither the problem of truth, at that stage at least, nor the problem of exactness or measurability. It was rather that I felt that these other three theories, though posing as science, had in fact more in common with primitive myths than with science; that they resembled astrology rather than astronomy.

I found that those of my friends who were admirers of Marx, Freud, and Adler, were impressed by a number of points common to these theories, and especially by their apparent *explanatory power*. These theories appeared to be able to explain practically everything that happened within the fields to which they referred. The study of any of them seemed to have the effect of an intellectual conversion or revelation, opening your eyes to a new truth hidden from those not yet initiated. Once your eyes were thus opened you saw confirming instances everywhere: the world was full of *verifications* of the theory. Whatever happened always confirmed it. Thus its truth appeared manifest; and unbelievers were clearly people who did not want to see the manifest truth; who refused to see it, either because it was against their class interest, or because of their repressions which were still "un-analysed" and crying aloud for treatment.

The most characteristic element in this situation seemed to me the incessant stream of confirmations, of observations which "verified" the theories in question; and this point was constantly emphasized by their adherents. A Marxist could not open a newspaper without finding on every page confirming evidence for his interpretation of history; not only in the news, but also in its presentation—which revealed the class bias of the paper—and especially of course in what the paper did *not* say. The Freudian analysts emphasized that their theories were constantly verified by their "clinical observations." As for Adler, I was much impressed by a personal experience. Once, in 1919, I reported to him a case which to me did not seem particularly Adlerian, but which he found no difficulty in analysing in terms of his theory of inferiority feelings, although he had not even seen the child. Slightly shocked, I asked him how he could be so sure. "Because of my thousandfold experience," he replied; whereupon I could not help saying: "And with this new case, I suppose, your experience has become thousand-and-one-fold."

What I had in mind was that his previous observations may not have been much sounder than this new one; that each in its turn had been interpreted in the light of "previous experience," and at the same time counted as additional confirmation. What, I asked myself, did it confirm? No more than that a case could be interpreted in the light of the theory. But this meant very little, I reflected, since every conceivable case could be interpreted in the light of Adler's theory, or equally of Freud's. I may illustrate this by two very different examples of human behaviour: that of a man who pushes a child into the water with the intention of drowning it; and that of a man who sacrifices his life in an attempt to save the child. Each of these two cases can be explained with equal ease in Freudian and in Adlerian terms. According to Freud the first man suffered from repression (say, of some component of his Oedipus complex), while the second man had achieved sublimation. According to Adler the first man suffered from feelings of inferiority (producing perhaps the need to prove to himself that he dared to commit some crime), and so did the second man (whose need was to prove to himself that he dared to rescue the child). I could not think of any human behaviour which could not be interpreted in terms of either theory. It was precisely this fact—that they always fitted, that they were always confirmed—which in the eyes of their admirers constituted the strongest argument in favour of these theories. It began to dawn on me that this apparent strength was in fact their weakness.

With Einstein's theory the situation was strikingly different. Take one typical instance—Einstein's prediction, just then confirmed by the findings of Eddington's expedition. Einstein's gravitational theory had led to the result that light must be attracted by heavy bodies (such as the sun), precisely as material bodies were attracted. As a consequence it could be calculated that light from a distant fixed star whose apparent position was close to the sun would reach the earth from such a direction that the star would seem to be slightly shifted away from the sun; or, in other words, that stars close to the sun would look as if they had moved a little away from the sun, and from one another. This is a thing which cannot normally be observed since such stars are rendered invisible in daytime by the sun's overwhelming brightness; but during an eclipse it is possible to take photographs of them. If the same constellation is photographed at night one can measure the distances on the two photographs, and check the predicted effect.

Now the impressive thing about this case is the *risk* involved in a prediction of this kind. If observation shows that the predicted effect is definitely absent, then the theory is simply refuted. The theory is *incompatible with certain possible results of observation*—in fact with results which everybody before Einstein would have expected.[1] This is quite different from the situation I have previously described, when it turned out that the theories in question were compatible with the most divergent human behaviour, so that it was practically impossible to describe any human behaviour that might not be claimed to be a verification of these theories.

These considerations led me in the winter of 1919-20 to conclusions which I may now reformulate as follows.

(1) It is easy to obtain confirmations, or verifications, for nearly every theory—if we look for confirmations.

(2) Confirmations should count only if they are the result of *risky predictions*; that is to say, if, unenlightened by the theory in question, we should have expected an event which was incompatible with the theory—an event which would have refuted the theory.

(3) Every "good" scientific theory is a prohibition: it forbids certain things to happen. The more a theory forbids, the better it is.

(4) A theory which is not refutable by any conceivable event is non-scientific. Irrefutability is not a virtue of a theory (as people often think) but a vice.

(5) Every genuine *test* of a theory is an attempt to falsify it, or to refute it. Testability is falsifiability; but there are degrees of testability: some theories are more testable, more exposed to refutation, than others; they take, as it were, greater risks.

(6) Confirming evidence should not count *except when it is the result of a genuine test of the theory*; and this means that it can be presented as a serious but unsuccessful attempt to falsify the theory. (I now speak in such cases of "corroborating evidence.")

(7) Some genuinely testable theories, when found to be false, are still upheld by their admirers—for example by introducing *ad hoc* some auxiliary assumption, or by re-interpreting the theory *ad hoc* in such a way that it escapes refutation. Such a procedure is always possible, but it rescues the theory from refutation only at the price of destroying, or at least lowering, its scientific status. (I later described such a rescuing operation as a "*conventionalist twist*" or a "*conventionalist stratagem*.")

One can sum up all this by saying that *the criterion of the scientific status of a theory is its falsifiability, or refutability, or testability.*

II

I may perhaps exemplify this with the help of the various theories so far mentioned. Einstein's theory of gravitation clearly satisfied the criterion of falsifiability. Even if our measuring instruments at the time did not allow us to pronounce on the results of the tests with complete assurance, there was clearly a possibility of refuting the theory.

Astrology did not pass the test. Astrologers were greatly impressed, and misled, by what they believed to be confirming evidence—so much so that they were quite unimpressed by any unfavourable evidence. Moreover, by making their interpretations and prophecies sufficiently vague they were able to explain away anything that might have been a refutation of the theory had the theory and the prophecies been more precise. In order to escape falsification they destroyed the testability of their theory. It is a typical soothsayer's trick to predict things so vaguely that the predictions can hardly fail: that they become irrefutable.

The Marxist theory of history, in spite of the serious efforts of some of its founders and followers, ultimately adopted this soothsaying practice. In some of its earlier formulations (for example in Marx's analysis of the character of the "coming social revolution") their predictions were testable, and in fact falsified. [2] Yet instead of accepting the refutations the followers of Marx re-interpreted both the theory and the evidence in order to make them agree. In this way they rescued the theory from refutation; but they did so at the price of adopting a device which made it irrefutable. They thus gave a "conventionalist twist" to the theory; and by this stratagem they destroyed its much advertised claim to scientific status.

The two psycho-analytic theories were in a different class. They were simply non-testable, irrefutable. There was no conceivable human behaviour which could contradict them. This does not mean that Freud and Adler were not seeing certain things correctly: I personally do not doubt that much of what they say is of considerable importance, and may well play its part one day in a psychological science which is testable. But it does mean that those "clinical observations" which analysts naively believe confirm their theory cannot do this any more than the daily confirmations which astrologers find in their practice.[3] And as for Freud's epic of the Ego, the Super-ego, and the Id, no substantially stronger claim to scientific status can be made for it than for Homer's collected stories from Olympus. These theories describe some facts, but in the manner of myths. They contain most interesting psychological suggestions, but not in a testable form.

At the same time I realized that such myths may be developed, and become testable; that historically speaking all—or very nearly all—scientific theories originate from myths, and that a myth may contain important anticipations of scientific theories. Examples are Empedocles' theory of evolution by trial and error, or Parmenides' myth of the unchanging block universe in which nothing ever happens and which, if we add another dimension, becomes Einstein's block universe (in which, too, nothing ever happens, since everything is, four-dimensionally speaking, determined and laid down from the beginning). I thus felt that if a theory is found to be non-scientific, or "metaphysical" (as we might say), it is not thereby found to be unimportant, or insignificant, or "meaningless," or "nonsensical."[4] But it cannot claim to be backed by empirical evidence in the scientific sense—although it may easily be, in some genetic sense, the "result of observation."

(There were a great many other theories of this pre-scientific or pseudo-scientific character, some of them, unfortunately, as influential as the Marxist interpretation of history; for example, the racialist interpretation of history—another of those impressive and all-explanatory theories which act upon weak minds like revelations.)

Thus the problem which I tried to solve by proposing the criterion of falsifiability was neither a problem of meaningfulness or significance, nor a problem of truth or acceptability. It was the problem of drawing a line (as well as this can be done) between the statements, or systems of statements, of the empirical sciences, and all other statements—whether they are of a religious or of a metaphysical character, or simply pseudo-scientific. Years later—it must have been in 1928 or 1929—I called this first problem of mine the *"problem of demarcation."* The criterion of falsifiability is a solution to this problem of demarcation, for it says that statements or systems of statements, in order to be ranked as scientific, must be capable of conflicting with possible, or conceivable, observations.

Notes

1. This is a slight oversimplification, for about half of the Einstein effect may be derived from the classical theory, provided we assume a ballistic theory of light.

2. See, for example, my *Open Society and Its Enemies*, ch. 15, section iii, and notes 13-14.

3. "Clinical observations," like all other observations, are *interpretations in the light of theories*; and for this reason alone they are apt to seem to support those theories in the light of which they were interpreted. But real support can be obtained only from observations undertaken as tests (by "attempted refutations"); and for this purpose *criteria of refutation* have to be laid down beforehand: it must be agreed which observable situations, if actually observed, mean that the theory is refuted. But what kind of clinical responses would refute to the satisfaction of the analyst not merely a particular analytic diagnosis but psycho-analysis itself? And have such criteria ever been discussed or agreed upon by analysts? Is there not, on the contrary, a whole family of analytic concepts, such as "ambivalence" (I do not suggest that there is no such thing as ambivalence), which would make it difficult, if not impossible, to agree upon such criteria? Moreover, how much headway has been made in investigating the question of the extent to which the (conscious or unconscious) expectations and theories held by the analyst influence the "clinical responses" of the patient? (To say nothing about the conscious attempts to influence the patient by proposing interpretations to him, etc.) Years ago I introduced the term "*Oedipus effect*" to describe the influence of a theory or expectation or prediction *upon the event which it predicts* or describes: it will be remembered that the causal chain leading to Oedipus' parricide was started by the oracle's prediction of this event. This is a characteristic and recurrent theme of such myths, but one which seems to have failed to attract the interest of the analysts, perhaps not accidentally. (The problem of confirmatory dreams suggested by the analyst is discussed by Freud, for example in *Gesammelte Schriften*, III, 1925, where he says on p. 314: "If anybody asserts that most of the dreams which can be utilized in an analysis . . . owe their origin to the analyst's suggestion, then no objection can be made from the point of view of analytic theory. Yet there is nothing in this fact," he surprisingly adds, "which would detract from the reliability of our results.")

4. The case of astrology, nowadays a typical pseudo-science, may illustrate this point. It was attacked, by Aristotelians and other rationalists, down to Newton's day, for the wrong reason—for its now accepted assertion that the planets had an "influence" upon terrestrial ("sublunar") events. In fact Newton's theory of gravity, and especially the lunar theory of the tides, was historically speaking an offspring of astrological lore. Newton, it seems, was most reluctant to adopt a theory which came from the same stable as for example the theory that "influenza" epidemics are due to an astral "influence." And Galileo, no doubt for the same reason, actually rejected the lunar theory of the tides; and his misgivings about Kepler may easily be explained by his misgivings about astrology.

Demarcating Genuine Science
from Pseudoscience

DANIEL ROTHBART

Most of us believe that a scientific theory differs fundamentally from a pseudoscientific theory. Darwinism and the divine theory of creation differ not simply in their respective content but in the way they each explain the phenomena. Astronomy and astrology clash in a way that prompts us to classify the former as genuine science and the latter as pseudoscience. The abundance of paradigm cases of science and pseudoscience may suggest that the defining characteristics of genuine science are easily within our grasp, and that the dividing line between science and non-science is so obvious as to be unworthy of serious philosophical effort.

But in the opinion of many philosophers, no proposal to date satisfactorily demarcates genuine science from pseudoscience, for none successfully captures the common strands underlying the paradigm "scientific" theories. Each proposal seems to clash repeatedly with clear cases of either science or pseudoscience. To aggravate the problem, many theories such as Freudian psychoanalysis and Marxism are not uncontroversially classifiable as either science or pseudoscience, prompting heated debates between advocates and critics of the particular theory. In order to be successful, a demarcation criterion must enable us to distinguish the paradigm cases of genuine science from those of pseudoscience, as well as clearly to settle the classification of the many doubtful cases. [1]

The purpose of this paper is to propose and defend a demarcation criterion. After first showing why Karl Popper's famous demarcation criterion fails, I offer an alternative criterion that is far more historically oriented than Popper's. According to the alternative criterion, a theory's classification as science or non-science rests significantly on rival theories under consideration at a particular time.

Problems with earlier drafts of this paper were avoided thanks to discussions with Professors Allen Blue, James Fletcher, Patrick Grim, Emmett Holman, Robert Pielke, and Donald Ross.

Popper on Demarcation

In his well-known attempt to solve the demarcation problem, Popper contends that the defining feature of science is its falsifiability, that is, its vulnerability to refutation by at least one possible event. A theory is scientific on this view if there exists some *possible* state of affairs whose actual occurrence would refute the theory. According to Popper, this potential for refutation reveals that the theory clashes with experience and so is an instance of empirical science. Newtonian mechanics is scientific since the occurrence of many possible events would falsify the theory, such as the failure of Halley's comet to return to earth around the calculated time of December 1758. Conversely, any theory immune to falsification by some possible event is by definition pseudoscientific on this view, since these theories fail to clash with experience, as illustrated by tautologous theories that are empirically empty (Popper, 1959, Chs. 1, 4).

But clearly, this criterion yields an impoverished view of science, and is far too weak to function as a defining standard. The criterion systematically classifies a wide range of pseudoscientific theories as instances of science (Agassi, 1964, p. 198). To satisfy Popper's falsifiability criterion, a theory need simply describe any one specific state of affairs by implying just one observation statement, or basic statement.[2] The basic statement need not actually be false, but were it false then the theory *would* be refuted. So on this view any theory is scientific if it yields *any observation statement* whatsoever, so long as the theory would be refuted if that statement happened to be false. It does not matter whether the statement is true or false, or whether the statement describes a past event (as in explanations) or a future event (as in predictions). But these conditions for science are far too weak, for they open the gates to a seemingly endless series of pseudoscientific theories that are falsifiable. Clearly, very many pseudoscientific theories have at least *some* contact with the world by describing at least *one* possible event. Velikovsky's theories make explicit reference to planetary orbits and specific terrestrial events; the theory describing dowsing rods to find underground water yields empirical predictions; the theory of divine creation is not only falsifiable by many possible events but has in fact been repeatedly falsified, as have witchcraft and voodoo (Gardner, 1952). Astrology as well has been empirically falsified, which reveals its falsifiable character (Jerome, 1975).[3] We can easily conjure up bizarre or ridiculous theories that satisfy the falsifiability standard by describing *some* possible event. A full account of an *Alice in Wonderland* universe purportedly describes certain specific events and therefore is falsifiable. So it appears that a successful demarcation criterion must provide for more restrictive standards for science than Popper recommends.

Criteria for a Demarcation Criterion

How might we construct a more plausible criterion of demarcation? Although Popper's criterion fails, he suggests a rather promising approach to solving the demarcation problem, by arguing that the demarcation between science and pseudoscience is to be drawn in terms of methodological characteristics of science, or on the basis of our theory of scientific knowledge (Musgrave, 1968, pp. 80–82). This much I will borrow from Popper since my theory, like his, will rely on a sketch of proper scientific methodology. By drawing the demarcation line in terms of certain methodological rules, I am assuming that genuine science by definition satisfies these methodological standards whereas pseudoscience by definition does not, that Ptolemaic astronomy, for example, meets certain requirements for rational science that superstition lacks. But exactly what type of methodological requirements are relevant in settling the demarcation issue, and what type of requirements can be dismissed as irrelevant?

At this point it may be helpful to distinguish between two types of methodological standards: (1) those testing standards used during and after the experimental test, such as repeatability of experimental results, versus (2) those testing standards used before any testing to determine which theory should be selected for testing. The latter pre-testing standards are indispensable, since scientists are, in principle, barraged by an infinite number of possible theories, which must be ordered by weighing strengths and weaknesses. In selecting a hypothesis for testing, scientists try to maximize certain methodological virtues, which usually suggest future success, and minimize methodological vices, which indicate possible failings. A rational procedure is needed for this a priori selection, that is, the selection that logically precedes testing (Hanson, 1972, Ch. 4).

Within the category of pre-testing requirements, a further distinction can be made between two types of pre-testing standards. The first type includes only those standards that *must* be fulfilled in order for a hypothesis to be rationally selected for testing. These can be called necessary requirements for a priori selection, as illustrated by the condition that, trivially, each hypothesis must produce test statements purportedly describing some future event. Any hypothesis that fails to yield such statements is immune to experimentation and clearly must not—and in a sense cannot—be selected for testing. Let us use the term "eligibility requirements" for all necessary pre-testing requirements, since they indicate whether a hypothesis is an eligible candidate for a priori selection, or whether it is viable for testing by displaying enough promise to be seriously considered for testing.

The second group exhausts all other pre-testing requirements, and consists of all pre-testing standards that, however important, are not strictly necessary for a priori selection for testing. For example, some philosophers maintain that one standard for a priori selection is the theory's concordance with the accepted background theories at a particular time. But this condition does not seem to be necessary for testing since other virtues of a theory may, according to these philosophers, counterbalance a lack of concordance with background knowledge (Schaffner, 1970, pp. 320–321). Simplicity as well has been widely heralded as a standard for a theory's selection, but again not a necessary requirement, since other factors, such as explanatory power, may warrant the theory's selection despite a lack of simplicity.

My purpose in distinguishing between necessary pre-testing requirements (eligibility requirements) and non-necessary pre-testing requirements is to suggest that it is the former that serve as the basis for a demarcation criterion. I thus propose that the demarcation between science and non-science is the satisfaction of all eligibility requirements (all necessary requirements) from our methodology for "rational science."[4] On this view the defining attribute of genuine science is its candidacy as a viable hypothesis for testing or its plausibility to be selected for experimentation—in a word, its *testworthiness*. Scientific theories must exhibit enough merit and promise to render them serious contenders for a priori selection. Although a scientific theory need not be the best testing candidate, it must be a viable candidate by satisfying all eligibility requirements. Conversely, the defining feature of non-science is its implausibility for testing, that is, its failure to meet at least one of the eligibility requirements. Witchcraft, magic, and voodoo are unscientific because they are *unworthy* of experimentation.[5]

Note that my proposal is not by itself a demarcation criterion, but is instead a requirement or adequacy condition for every plausible demarcation criterion, stipulating that the dividing line between science and non-science is the fulfillment of the eligibility requirements, whatever they happen to be. A demarcation criterion can be explicitly suggested only after we agree on the particular eligibility requirements, a problem that I address below. Thus my proposal might be described as a metacriterion for every demarcation criterion.

One argument in behalf of this metacriterion is that it complies with certain intuitions about what distinguishes genuinely scientific systems. The defining feature of science does not seem to be experimental success, for most clear cases of genuine science have been experimentally falsified. It seems instead to be some set of characteristics that theories share prior to testing that makes them scientific. Other theories, such as magic, seem to be classified as unscientific not because of experimental failure but because of some a priori deficiency. Conforming to these intuitions, the metacriterion outlined above maintains that the defining features of science rest exclusively on certain pre-testing requirements—the eligibility requirements—and that the deficiency of all unscientific systems is their untestworthy character.

Distinguishing Science from Non-Science

As noted above, the metacriterion cannot by itself settle the demarcation question without some specification of particular eligibility requirements. My immediate task is to suggest certain eligibility requirements which, in conjunction with the metacriterion, yield a satisfactory demarcation between genuine science and pseudoscience.

To determine the eligibility requirements, which are the pre-testing conditions every hypothesis must satisfy to be a viable testing candidate, consider what function an experimental test is intended to serve. One function that a test does *not* serve is to verify, in an absolute and final sense, the tested hypothesis. Hume and others have shown the impossibility of absolute verification in science. Another spurious function of a test is absolute falsification, which is the final and conclusive rejection of a hypothesis allegedly on empirical grounds. Pierre Duhem has shown that any hypothesis, however suspect, can be salvaged by "blaming" one of the linking auxiliary hypotheses (1962, Ch. 6). Absolute falsification in science is a myth. [6]

Rather than producing some final decision on a hypothesis, by either absolute verification or absolute falsification, the primary function of a test is to contribute to the preferential ranking for the hypothesis in relation to some rival theory. Testing provides empirical grounds for an epistemic ordering of rival hypotheses. Scientists usually perform tests in a context of a competitive struggle between some new hypothesis and some previously accepted background theory. Tests involving the theories of Darwin, Pasteur, and Mendel illustrate this competition in biology, as do tests from Galileo to Newton, from Faraday to Einstein, and from Rutherford to Heisenberg in physics. Tests can serve to justify the preference of one theory over competitors by revealing the empirical advantages of one theory with respect to the present state of scientific discussion (Popper, 1972, pp. 82–84). As a comparative matter, experimental success contributes favorably to the preferential ranking of the theory in relation to alternatives (Lakatos, 1970; Putnam, 1962). But no theory can be conclusively verified or can achieve a status of final success. Experimental refutation diminishes the theory's standing, but only in relation to rival theories at the time of testing. This is illustrated repeatedly in science by "crucial experiments," which lower the comparative standing of one theory in relation to its rival. But few theories have been falsified in isolation, since falsification usually requires some rival theory to fall back on (Lakatos, 1970, p. 179), and since the decision to abandon a theory usually involves a commitment to develop and examine an adequate alternative (Laudan, 1977, pp. 55–57).

Given this role for testing, we can expect the eligibility requirements for a priori selection to reflect the importance of alternative theories at a particular time. As a first eligibility requirement we might propose that the hypothesis selected for testing must account for all the phenomena that its rival background theory explains; that the hypothesis must encapsulate its rival's explanatory successes. If we have some theory on hand that solves an empirical problem, then any competing hypothesis must match this success in order to maintain its testworthiness. After Galilean mechanics explained the motion of falling bodies, any viable theory in the same domain had to account for such phenomena. If a hypothesis fails to match the explanatory successes of its rivals, the hypothesis shows no improvement, at least in relation to the known evidence, and thus fails to display enough promise to warrant testing. The goal that one theory should accumulate a rival's empirical successes is widely illustrated in science and defended by many philosophers and scientists (Post, 1971, pp. 228–235).[7]

The notion of explanation assumed here requires the following: (a) that the explanans in conjunction with statements of initial conditions imply the explanandum, (b) that the explanans include a lawlike generalization, and (c) that the explanans have some direct contact with the empirical phenomena.[8] Since our discussion concerns explanans with untested hypotheses, one condition that this notion of explanation does not include is that the explanans must be confirmed, corroborated, or true. Because of this exclusion, the notion of explanation assumed above more accurately defines a *potential* explanation as opposed to a *successful* explanation.

By grounding viability for testing in the theory's relationship to rival theories, however, this first eligibility requirement does *not* demand that a hypothesis must explain all known facts at a certain time. Few theories ever explain all the facts at a given time, for many theories fail to explain those events that their rivals could not explain (Feyerabend, 1975, pp. 55–60). For example, although mechanical theories of dynamics in the eighteenth century could not explain electromagnetic phenomena, scientists still tested many of these theories. A theory then may be allowed to fail empirically where its older rival theory failed, but still be viable for testing.

In addition to encapsulating the explanatory success of its rival, a viable theory must satisfy a second eligibility requirement. A viable theory must clash empirically with its rival by yielding test implications that are inconsistent with the rival theory. In Popper's terms, it must be susceptible to a severe test whose test statements clash with the given background knowledge accepted at the time of testing (1959, Ch. 10). This requirement is needed to ensure that a testworthy hypothesis has the potential to surpass empirically the background theories; to succeed where alternative theories fail. If a hypothesis fails to meet this requirement by yielding test implications all of which are consistent with the rival theory, then the hypothesis is simply "contained within" it and offers no new empirical information (Post, 1971, Ch. 3). Thus, the second eligibility requirement I propose is that any hypothesis viable for testing must be amenable to severe tests by implying test statements which background theories would tell us are probably false.

In addition to prohibiting hypotheses that are contained within background theories, this second eligibility requirement in effect rules out empirically empty hypotheses. Since empirically empty hypotheses are trivially true, they fail to clash empirically with any rival theory, as frequently illustrated by excessively vague hypotheses. Popper has shown that many "scientists" try to ward off refutations by proposing vague or unintelligible hypotheses, but at the cost of producing hypotheses that are trivally true in relation to accepted background knowledge as a result (1959, pp. 269–273). Consider, for example, the following rather simplistic hypothesis: (1) We all voluntarily act in ways that promote our own individual pleasure. This doctrine minimizes empirical conflict with background knowledge because of such a vague expression as "ways that promote our own pleasure." Some contend that an act promotes the agent's pleasure simply if the act is what the agent voluntarily performs. But on this reading (1) is reduced to utter triviality for it basically amounts to the following: (1′) We all voluntarily act in those ways in which we voluntarily act. Is it possible reasonably to deny that we voluntarily do the things that we voluntarily do? Since a falsifying instance to (1′) is impossible, it is immune to empirical refutation. It fails to clash with any hypothesis. (1′) is clearly a necessary truth, a tautology, rather than a contingent claim purporting to describe the world. This indicates that empirically empty hypotheses, such as (1′), are poor candidates to explain the world and should not be subject to testing, as Popper has convincingly shown.

I have proposed that two conditions determine which hypotheses are viable for testing: (a) a viable hypothesis must account for all phenomena explained by its rival theory, and (b) it must empirically clash with its rival. Both of these two conditions involve significant historical considerations, as both requirements appeal to the particular rival scientific theories at a certain time of discussion. Whether a theory meets these conditions depends on the particular rival theories at a particular time. In that sense the scientific status of a theory depends on its history.

Historical Dimensions of the Demarcation Criterion

According to the metacriterion given above, all scientific theories are by definition viable candidates for testing. Unscientific theories, on the other hand, are untestworthy. Given the two eligibility requirements proposed above, we can finally suggest a new demarcation criterion. Every scientific theory, by definition, must first match its rival's empirical successes by accounting for all phenomena that the rival theory explains, and second, must empirically clash with its rival by yielding inconsistent test implications.

As a dividing line between science and pseudoscience, this criterion relies heavily on the general competitive climate of discussion at a particular time, for the classification of a certain theory as scientific depends on how it stands in relation to its rivals at a given historical period. A scientific theory must compare favorably with its rivals' explanatory power as well as convey at least some conflicting empirical content in relation to its rivals. On this view science is essentially a relational property between a theory in question and some alternative. Ptolemaic cosmology is scientific in relation to Aristotelian cosmology since it equaled the explanatory successes of its predecessor and because it clashed empirically with Aristotle's theory concerning the planetary motions of retrogression. By incorporating the Copernican vision of a sun-dominated universe, Kepler proposed a theory that matched the explanatory power of Ptolemaic cosmology but that clashed with Ptolemy's theory concerning the orbital path of Mars. Kepler's theory is then scientific in relation to Ptolemy's. Galileo's dynamics could account for all the known phenomena explained by Aristotelian theory, but also implied many surprising results. Similarly, the special theory of relativity is scientific in relation to Newtonian theory of space and time. In addition to equaling the empirical successes of Newtonian theory, the theory of relativity predicts some surprising claims concerning the motion of high-speed subatomic particles.

If a theory fails to meet one of the eligibility requirements, the theory is not only untestworthy but also unscientific. One class of unscientific theories includes those that cannot explain some phenomenon that is successfully explained by a rival. In relation to today's conventionally accepted theories, unscientific hypotheses include the theory of divine creation, mythical beliefs of witchcraft and spiritual possession, and various forms of medical quackery (Gardner, 1952). Astrology is also unscientific in relation to contemporary rivals, for astrology fails to explain many known events, such as the identical fate of many victims of some human tragedy. But some of these unscientific theories conceivably could have reflected the most advanced state of knowledge at a certain period in history, and thus at that time deserved the label of science relative to their rivals. In ancient Greece astrology was a viable theory for testing in relation to alternatives, allowing us to identify astrology as science at least in that early historical period.

A second class of unscientific theories are by definition unworthy of testing because they yield no surprising results in relation to their rival. Since these theories are empirically contained within the rival background theory, they yield no potential for empirical improvement, as illustrated by empirically empty theories. Many critics charge that Freudian psychoanalysis, for example, is empirically empty because of its ability to explain any conceivable behavior (Popper, 1962, pp. 34–36) and because of its inability to clash with any theory.

The difference between the above demarcation criterion and Popper's is that the criterion proposed here imposes more stringent requirements for genuine science. The second eligibility requirement—that science must yield unexpected results—includes Popper's condition of falsifiability, since any hypothesis that yields surprising empirical consequences is falsifiable. But since falsifiability is an absolute property of a theory, Popper's criterion is thoroughly ahistorical, as is his classification of science and non-science. On my view science is primarily a relational property between a theory and some rival at a particular time.

Some critics with more formalist leanings may object that the historical character of this criterion yields counter-intuitive results. On this objection, the historical aspect of the criterion produces an unacceptable flexibility and uncertainty since the criterion allows for the possibility that some theories may be subject to a series of reclassifications in the future. The criterion permits Newtonian mechanics, for example, to be repeatedly reclassified: Newtonian mechanics is scientific in relation to its eighteenth-century rivals, but it is unscientific in relation to twentieth-century theories, but it may be scientific again in relation to some future theories, and so on. This intuitively unacceptable flexibility in the classification of science and non-science is the result of the view that science is primarily a relational concept between a theory and some rival at a particular time. The status of a theory should be immune to alteration by historical considerations, according to this objection.

But the force of this objection is at best rather weak, for our intuitions on this matter also seem to suggest that a historical demarcation criterion would be too restrictive by failing to allow for the reclassification of instances of science and non-science. Many domains in ancient Greece, for example, domains that today are called superstition, religion, magic, and the occult, were at that time clear cases of legitimate science (Crombie, 1953, pp. 5–7). But with the more stringent testing standards of modern science, these domains were removed from the rolls of science because they were immune to strict experimental appraisal (Crombie, 1953, p. 2). The advent of modern science brought with it a major reclassification of theories. The possibility of reclassifying theories in light of epistemological advances is an important intuitive feature for any demarcation criterion, one that an ahistorical criterion lacks. The point here is that our intuitive judgments on this issue are inconsistent—some intuitive considerations recommend a historical criterion whereas others recommend an ahistorical one. If this is the case, intuitions alone cannot determine which type of criterion, historical or ahistorical, is preferable. The objection that the criterion given above violates our intuitive leanings toward an ahistorical criterion is therefore unconvincing.

It should be noted that the proposed criterion does not define science on historical grounds alone, for some theories are classified as unscientific regardless of the historical climate at a given time. For example, a theory *t* that is unscientific at a given time because it fails to match the explanatory success of its rival will be unscientific relative to any other theory replacing the rival. Again, if *t* fails to account for some event that is explained by a rival theory, then *t* will fail to explain this event relative to any future theory replacing the particular rival. The reason is that all future theories that replace the rival theory will match the earlier explanatory success, following the criterion. This, in effect, prohibits *t* from every being worthy of testing. This ahistorical status of theories such as *t* is attributable to their failure to exhibit enough relative promise to merit testing, following the criterion. Because these theories are untestworthy, as opposed to some rivals, theories such as *t* will retain their status as pseudoscience. Moreover, theories that convey no empirical import, that is, theories that are immune to experimental testing, are always in violation of the requirement that all science must empirically clash with some rival. As shown above, such theories fail to clash empirically with any theory, which, according to the eligibility requirements given above, means that they can never be viable candidates for testing. The indictment against these theories is not that they are all false, but instead that they are always immune to empirical and "scientific" appraisal.

Admittedly, the proposed demarcation criterion will no doubt fail to accommodate at least some clear cases of science or non-science. This raises the perplexing issue of balancing advantages and disadvantages of any conceptual proposal. To what degree, for example, should our judgments about unobjectionable cases be sacrificed for conceptual clarity and simplicity? Although these problems obviously surpass the scope of this paper, I believe nevertheless that the criterion presented above avoids any major systematic conflict with our judgments about clear cases, and is thus an improvement over Popper's criterion which is subject to such conflicts. Moreover, because of its historical emphasis, the criterion is realistically grounded in the actual epistemic struggle between rival theories, competing for supremacy in a preferential ordering. On this criterion it is viability for testing, a matter that changes with history, that divides genuine science from pseudoscience.

Notes

1. The word "science" has many fundamentally different meanings in ordinary use. "Science" refers to a "rational" methodology for acquiring knowledge, a particular branch of knowledge, or systematic knowledge of the world in general. But none of these meanings captures the notion of science that concerns us here. For my purpose, "science" refers to certain systems of statements, pardigm cases being Newtonian mechanics and contemporary astrophysics as opposed to pseudoscientific systems such as witchcraft and voodoo.

2. Popper defines a basic statement as a singular existential statement "asserting that an observable event is occurring in a certain individual region of space and time" (1959, p. 103).

3. Even though Popper acknowledges the pseudoscientific nature of astrology (1962, pp. 34, 53, 188), astrology is falsifiable, as evidenced by the repetition of

falsifying instances. These falsifications arise from mass human tragedies like wars and earthquakes, since the victims suffer the same tragic fate together despite divergent astrological horoscopes. Other refuting instances occur when two people, born simultaneously with identical celestial influences, experience different fates (Jerome, 1975).

4. No philosopher, to my knowledge, has proposed this metacriterion, and some philosophers offer demarcation criteria that do not conform to the metacriterion. One eligibility requirement for Popper, for example, is that a scientific theory must convey a high degree of empirical content. But Popper does not use this requirement as a defining feature of science, although he clearly prescribes it as one standard for successful science.

5. My proposal restricts how we should define science, and in doing so, amounts to the following metacriterion, M, for any adequate demarcation criterion:

> M: Any adequate demarcation criterion defines a genuinely scientific theory as a system of statements satisfying all eligibility requirements from our accepted methodological theory of appraisal.

Despite this strict relativity of the notion of science to an accepted methodology, M is totally neutral with respect to the selection of particular eligibility requirements. But once such requirements are selected, the demarcation between science and non-science follows quite easily.

6. Some commentators have attributed to Popper the view that absolute falsification is the primary function of testing (Ayer, 1952, p. 32). But Popper asserts that conclusive falsification is unattainable, since we can always "transfer responsibility for the empirical refutation to one or more of the cooperating hypotheses" (1959, p. 50).

7. H. R. Post is one advocate of this aim in science and he cites a number of scientists who share this view, including Max Born, P. Duhem, and J. Clerk Maxwell (Post, 1971, p. 213). As William Whewell wrote,

> The principles which constituted the triumph of the preceding stages of the science, may appear to be subverted and ejected by the later discoveries, but in fact they are (so far as they were true) taken up into the subsequent doctrines and included in them. They continue to be an essential part of the science. The earlier truths are not expelled but absorbed, not contradicted but extended; and the history of each science, which may thus appear like a succession of revolutions, is in reality, a series of developments. (1857, p. 8, cited by Post, 1971)

8. The explanadum is, by definition, the sentence describing the phenomenon to be explained. The explanans is, by definition, the set of sentences used to account for the phenomenon.

References

Agassi, Joseph. "The Nature of Scientific Problems and Their Roots in Metaphysics." In *The Critical Approach to Science and Philosophy*, edited by Mario Bunge. London: Free Press, 1964.

Ayer, A. J. *Language, Truth and Logic*. New York: Dover, 1952.

Crombie, A. C. *Robert Grosseteste and the Origins of Experimental Science*. Oxford: Clarendon Press, 1953.

Duhem, Pierre. *The Aim and Structure of Physical Theory*. New York: Atheneum, 1962.

Feyerabend, Paul K. *Against Method*. London: Verso, 1975.

Gardner, Martin. *Fads & Fallacies*. New York: Dover, 1952.

Hanson, Norwood Russell. *Patterns of Discovery*. Cambridge: Cambridge University Press, 1972.

Jerome, Lawrence E. "Astrology—Magic or Science." *The Humanist* 35 (1975), No. 5: 10–16.

Lakatos, I. "Falsification and the Methodology of Scientific Research Programmes." In *Criticism and the Growth of Knowledge*, edited by Imre Lakatos and Alan Musgrave. New York: Cambridge University Press, 1970.

Laudan, Larry. *Progress and Its Problems*. Berkeley: University of California Press, 1977.

Musgrave, A. E. "On a Demarcation Dispute." In *Problems in the Philosophy of Science*, vol. 3 of *Proceedings of the International Colloquium in the Philosophy of Science*, edited by Imre Lakatos and Alan Musgrave. Amsterdam: North-Holland, 1968.

Popper, Karl R. *The Logic of Scientific Discovery*. Translated by Julius and Lan Freed. New York: Harper & Row, 1959.

Popper, Karl R. *Conjectures and Refutations*. New York: Harper & Row, 1962.

Popper, Karl R. *Objective Knowledge*. Oxford: Clarendon Press, 1972.

Post, H. R. "Correspondence, Invariance and Heuristics: In Praise of Conservative Induction." *Studies in History and Philosophy of Science* 2 (1971): 213–255.

Putnam, Hilary. *The Analytic and the Synthetic*. Minnesota Studies in the Philosophy of Science, edited by H. Feigl and G. Maxwell, vol. 3. Minneapolis: University of Minnesota Press, 1962.

Schaffner, Kenneth. *Outlines of a Logic of Comparative Theory Evaluation*. Minnesota Studies in the Philosophy of Science, edited by Roger H. Stuewer, vol. 5. Minneapolis: University of Minnesota Press, 1970.

Whewell, William. *History of the Inductive Sciences*. 3rd ed. London: 1857.

Logic of Discovery or Psychology of Research?

Thomas S. Kuhn

Among the most fundamental issues on which Sir Karl and I agree is our insistence that an analysis of the development of scientific knowledge must take account of the way science has actually been practiced. That being so, a few of his recurrent generalizations startle me. One of these provides the opening sentences of the first chapter of the *Logic of Scientific Discovery:* "A scientist," writes Sir Karl, "whether theorist or experimenter, puts forward statements, or systems of statements, and tests them step by step. In the field of the empirical sciences, more particularly, he constructs hypotheses, or systems of theories, and tests them against experience by observation and experiment." [1] The statement is virtually a cliché, yet in application it presents three problems. It is ambiguous in its failure to specify which of two sorts of "statements" or "theories" are being tested. That ambiguity can, it is true, be eliminated by reference to other passages in Sir Karl's writings, but the generalization that results is historically mistaken. Furthermore, the mistake proves important, for the unambiguous form of the description misses just that characteristic of scientific practice which most nearly distinguishes the sciences from other creative pursuits.

There is one sort of "statement" or "hypothesis" that scientists do repeatedly subject to systematic test. I have in mind statements of an individual's best guesses about the proper way to connect his own research problem with the corpus of accepted scientific knowledge. He may, for example, conjecture that a given chemical unknown contains the salt of a rare earth, that the obesity of his experimental rats is due to a specified component in their diet, or that a newly discovered spectral pattern is to be understood as an effect of nuclear spin. In each case, the next steps in his research are intended to try out or test the conjecture or hypothesis. If it passes enough or stringent enough tests, the scientist has made a discovery or has at least resolved the puzzle he had been set. If not, he must either abandon the puzzle entirely or attempt to solve it with the aid of some other hypothesis. Many research problems, though by no means all, take this form. Tests of this sort are a standard component of what I have elsewhere labelled "normal science" or "normal research," an enterprise which accounts for the overwhelming majority of the work done in basic science. In no usual sense, however, are such tests directed to current theory. On the contrary, when engaged with a normal research problem, the scientist must *premise* current theory as the rules of his game. His object is to solve a puzzle, preferably one at which others have failed, and current theory is required to define that puzzle and to guarantee that, given sufficient brilliance, it can be solved.[2] Of course the practitioner of such an enterprise must often test the conjectural puzzle solution that his ingenuity suggests. But only his personal conjecture is tested. If it fails the test, only his own ability not the corpus of current science is impugned. In short, though tests occur frequently in normal science, these tests are of a peculiar sort, for in the final analysis it is the individual scientist rather than current theory which is tested.

This is not, however, the sort of test Sir Karl has in mind. He is above all concerned with the procedures through which science grows, and he is convinced that "growth" occurs not primarily by accretion but by the revolutionary overthrow of an accepted theory and its replacement by a better one.[3] (The subsumption under "growth" of "repeated overthrow" is itself a linguistic oddity whose *raison d'être* may become more visible as we proceed.) Taking this view, the tests which Sir Karl emphasizes are those which were performed to explore the limitations of accepted theory or to subject a current theory to maximum strain. Among his favorite examples, all of them startling and destructive in their outcome, are Lavoisier's experiments on calcination, the eclipse expedition of 1919, and the recent experiments on parity conservation.[4] All, of course, are classic tests, but in using them to characterize scientific activity Sir Karl misses something terribly important about them. Episodes like these are very rare in the development of science. When they occur, they are generally called forth either by a prior crisis in the relevant field (Lavoisier's experiments or Lee and Yang's[5]) or by the existence of a theory which competes with the existing canons of research (Einstein's general relativity). These are, however, aspects of or occasions for what I have elsewhere called "extraordinary research," an enterprise in which scientists do display very many of the characteristics Sir Karl emphasizes, but one which, at least in the past, has arisen only intermittently and under quite special circumstances in any scientific speciality.[6]

I suggest then that Sir Karl has characterized the entire scientific enterprise in terms that apply only to its occasional revolutionary parts. His emphasis is natural and common: the exploits of a Copernicus or Einstein make better reading than those of a Brahe or Lorentz; Sir Karl would not be the first if he mistook what I call normal science for an intrinsically uninteresting enterprise. Nevertheless, neither science nor the development of knowledge is likely to be understood if research is viewed exclusively through the revolutions it occasionally produces. For example, though testing of basic commitments occurs only in extraordinary science, it is normal science that discloses both the points to test and the manner of testing. Or again, it is for the normal, not the extraordinary practice of science that professionals are trained; if they are nevertheless eminently successful in displacing and replacing the theories on which normal practice depends, that is an oddity which must be explained. Finally, and this is for now my main point, a careful look at the scientific enterprise suggests that it is normal science, in which Sir Karl's sort of testing does not occur, rather than extraordinary science which most nearly distinguishes science from other enterprises. If a demarcation criterion exists (we must not, I think, seek a sharp or decisive one), it may lie just in that part of science which Sir Karl ignores.

In one of his most evocative essays, Sir Karl traces the origin of "the tradition of critical discussion [which] represents the only practicable way of expanding our knowledge" to the Greek philosophers between Thales and Plato, the men who, as he sees it, encouraged critical discussion both between schools and within individual schools. [7] The accompanying description of Presocratic discourse is most apt, but what is described does not at all resemble science. Rather it is the tradition of claims, counter-claims, and debates over fundamentals which, except perhaps during the Middle Ages, have characterized philosophy and much of social science ever since. Already by the Hellenistic period mathematics, astronomy, statics and the geometric parts of optics had abandoned this mode of discourse in favour of puzzle solving. Other sciences, in increasing numbers, have undergone the same transition since. In a sense, to turn Sir Karl's view on its head, it is precisely the abandonment of critical discourse that marks the transition to a science. Once a field has made that transition, critical discourse recurs only at moments of crisis when the bases of the field are again in jeopardy. [8] Only when they must choose between competing theories do scientists behave like philosophers. That, I think, is why Sir Karl's brilliant description of the reasons for the choice between metaphysical systems so closely resembles my description of the reasons for choosing between scientific theories. [9] In neither choice, as I shall shortly try to show, can testing play a quite decisive role.

There is, however, good reason why testing has seemed to do so, and in exploring it Sir Karl's duck may at last become my rabbit. No puzzle-solving enterprise can exist unless its practitioners share criteria which, for that group and for that time, determine when a particular puzzle has been solved. The same criteria necessarily determine failure to achieve a solution, and anyone who chooses may view that failure as the failure of a theory to pass a test. Normally, as I have already insisted, it is not viewed that way. Only the practitioner is blamed, not his tools. But under the special circumstances which induce a crisis in the profession (e.g. gross failure, or repeated failure by the most brilliant professionals) the group's opinion may change. A failure that had previously been personal may then come to seem the failure of a theory under test. Thereafter, because the test arose from a puzzle and thus carried settled criteria of solution, it proves both more severe and harder to evade than the tests available within a tradition whose normal mode is critical discourse rather than puzzle solving.

In a sense, therefore, severity of test-criteria is simply one side of the coin whose other face is a puzzle-solving tradition. That is why Sir Karl's line of demarcation and my own so frequently coincide. That coincidence is, however, only in their *outcome*; the *process* of applying them is very different, and it isolates distinct aspects of the activity about which the decision—science or non-science—is to be made. Examining the vexing cases, for example, psychoanalysis or Marxist historiography, for which Sir Karl tells us his criterion was initially designed, [10] I concur that they cannot now properly be labelled "science." But I reach that conclusion by a route far surer and more direct than his. One brief example may suggest that of the two criteria, testing and puzzle solving, the latter is at once the less equivocal and the more fundamental.

To avoid irrelevant contemporary controversies, I consider astrology rather than, say, psychoanalysis. Astrology is Sir Karl's most frequently cited example of a "pseudo-science." [11] He says: "By making their interpretations and prophecies sufficiently vague they [astrologers] were able to explain away anything that might have been a refutation of the theory had the theory and the prophecies been more precise. In order to escape falsification they destroyed the testability of the theory." [12] Those generalizations catch something of the spirit of the astrological enterprise. But taken at all literally, as they must be if they are to provide a demarcation criterion, they are impossible to support. The history of astrology during the centuries when it was intellectually reputable records many predictions that categorically failed. [13] Not even astrology's most convinced and vehement exponents doubted the recurrence of such failures. Astrology cannot be barred from the sciences because of the form in which its predictions were cast.

Nor can it be barred because of the way its practitioners explained failure. Astrologers pointed out, for example, that, unlike general predictions about, say, an individual's propensities or a natural calamity, the forecast of an individual's future was an immensely complex task, demanding the utmost skill, and extremely sensitive to minor errors in relevant data. The configuration of the stars and eight planets was constantly changing; the astronomical tables used to compute the configuration at an individual's birth were notoriously imperfect; few men knew the instant of their birth with the requisite precision. [14] No wonder, then, that forecasts often failed. Only after astrology itself became implausible did these arguments come to seem question-begging. [15] Similar arguments are regularly used today when explaining, for example, failures in medicine or meteorology. In times of trouble they are also deployed in the exact sciences, fields like physics, chemistry, and astronomy. [16] There was nothing unscientific about the astrologer's explanation of failure.

Nevertheless, astrology was not a science. Instead it was a craft, one of the practical arts, with close resemblances to engineering, meteorology, and medicine as these fields were practised until little more than a century ago. The parallels to an older medicine and to contemporary psychoanalysis are, I think, particularly close. In each of these fields shared theory was adequate only to establish the plausibility of the discipline and to provide a rationale for the various craft-rules which governed practice. These rules had proved their use in the past, but no practitioner supposed they were sufficient to prevent recurrent failure. A more articulated theory and more powerful rules were desired, but it would have been absurd to abandon a plausible and badly needed discipline with a tradition of limited success simply because these desiderata were not yet at hand. In their absence, however, neither the astrologer nor the doctor could do research. Though they had rules to apply, they had no puzzles to solve and therefore no science to practise. [17]

Compare the situations of the astronomer and the astrologer. If an astronomer's prediction failed and his calculations checked, he could hope to set the situation right. Perhaps the data were at fault; old observations could be re-examined and new measurements made, tasks which posed a host of calculational and instrumental puzzles. Or perhaps theory needed adjustment, either by the manipulation of epicycles, eccentrics, equants, etc., or by more fundamental reforms of astronomical technique. For more than a millennium these were the theoretical and mathematical puzzles around which, together with their instrumental counterparts, the astronomical research tradition was constituted. The astrologer, by contrast, had no such puzzles. The occurrence of failures could be explained, but particular failures did not give rise to research puzzles, for no man, however skilled, could make use of them in a constructive attempt to revise the astrological tradition. There were too many possible sources of difficulty, most of them beyond the astrologer's knowledge, control, or responsibility. Individual failures were correspondingly uninformative, and they did not reflect on the competence of the prognosticator in the eyes of his professional compeers. [18] Though astronomy and astrology were regularly practised by the same people, including Ptolemy, Kepler, and Tycho Brahe, there was never an astrological equivalent of the puzzle-solving astronomical tradition. And without puzzles, able first to challenge and then to attest the ingenuity of the individual practioner, astrology could not have become a science even if the stars had, in fact, controlled human destiny.

In short, though astrologers made testable predictions and recognized that these predictions sometimes failed, they did not and could not engage in the sorts of activities that normally characterize all recognized sciences. Sir Karl is right to exclude astrology from the sciences, but his over-concentration on science's occasional revolutions prevents his seeing the surest reason for doing so.

That fact, in turn, may explain another oddity of Sir Karl's historiography. Though he repeatedly underlines the role of tests in the replacement of scientific theories, he is also constrained to recognize that many theories, for example the Ptolemaic, were replaced before they had in fact been tested. [19] On some occasions, at least, tests are not requisite to the revolutions through which science advances. But that is not true of puzzles. Though the theories Sir Karl cites had not been put to the test before their displacement, none of these was replaced before it had ceased adequately to support a puzzle-solving tradition. The state of astronomy was a scandal in the early sixteenth century. Most astronomers nevertheless felt that normal adjustments of a basically Ptolemaic model would set the situation right. In this sense the theory had not failed a test. But a few astronomers, Copernicus among them, felt that the difficulties must lie in the Ptolemaic approach itself rather than in the particular versions of Ptolemaic theory so far developed, and the results of that conviction are already recorded. The stituation is typical. [20] With or without tests, a puzzle-solving tradition can prepare the way for its own displacement. To rely on testing as the mark of a science is to miss what scientists mostly do and, with it, the most characteristic feature of their enterprise.

Notes

1. Popper [1959], p. 27.
2. For an extended discussion of normal science, the activity which practitioners are trained to carry on, see my [1962], pp. 23–42, and 135–42. It is important to notice that when I describe the scientist as a puzzle solver and Sir Karl describes him as a problem solver (e.g. in his [1963], pp. 67, 222), the similarity of our terms disguises a fundamental divergence. Sir Karl writes (the italics are his), "Admittedly, our expectations, and thus our theories, may precede, historically, even our problems. *Yet science starts only with problems.* Problems crop up especially when we are disappointed in our expectations, or when our theories involve us in difficulties, in contradictions." I use the term "puzzle" in order to emphasize that the difficulties which *ordinarily* confront even the very best scientists are, like crossword puzzles or chess puzzles, challenges only to his ingenuity. *He* is in difficulty, not current theory. My point is almost the converse of Sir Karl's.
3. Cf. Popper [1963], pp. 129, 215 and 221, for particularly forceful statements of this position.

4. For example, Popper [1963], p. 220.

5. For the work on calcination see, Guerlac [1961]. For the background of the parity experiments see, Hafner and Presswood [1965].

6. The point is argued at length in my [1962], pp. 52–97.

7. Popper [1963], chapter 5, especially pp. 148–52.

8. Though I was not then seeking a demarcation criterion, just these points are argued at length in my [1962], pp. 10–22 and 87–90.

9. Cf. Popper [1963], pp. 192–200, with my [1962], pp. 143–58.

10. Popper [1963], p. 34.

11. The index to Popper [1963] has eight entries under the heading "astrology as a typical pseudoscience."

12. Popper [1963], p. 37.

13. For examples see, Thorndike [1923–58], 5, pp. 225 ff.; 6, pp. 71, 101, 114.

14. For reiterated explanations of failure see, *ibid. x*, pp. 11 and 514ff.; 4, 368; 5, 279.

15. A perceptive account of some reasons for astrology's loss of plausibility is included in Stahlman [1956]. For an explanation of astrology's previous appeal see, Thorndike [1955].

16. Cf. my [1962], pp. 66–76.

17. This formulation suggests that Sir Karl's criterion of demarcation might be saved by a minor restatement entirely in keeping with his apparent intent. For a field to be a science its conclusions must be *logically derivable* from *shared premises*. On this view astrology is to be barred not because its forecasts were not testable but because only the most general and least testable ones could be derived from accepted theory. Since any field that did satisfy this condition *might* support a puzzle solving tradition, the suggestion is clearly helpful. It comes close to supplying a sufficient condition for a field's being a science. But in this form, at least, it is not even quite a sufficient condition, and it is surely not a necessary one. It would, for example, admit surveying and navigation as sciences, and it would bar taxonomy, historical geology, and the theory of evolution. The conclusions of a science may be both precise and binding without being fully derivable by logic from accepted premises. Cf. my [1962], pp. 35–51, and also the discussion in Section III, *below*.

18. This is not to suggest that astrologers did not criticize each other. On the contrary, like practioners of philosophy and some social sciences, they belonged to a variety of different schools, and the inter-school strife was sometimes bitter. But these debates ordinarily revolved about the *implausibility* of a particular theory employed by one or another school. Failures of individual predictions played very little role. Compare Thorndike [1923–58], 5, p. 233.

19. Cf. Popper [1963], p. 246.

20. Cf. my [1962], pp. 77–87.

References

Guerlac [1961]: *Lavoisier—The Crucial Year*, 1961.

Hafner and Presswood [1965]: "Strong Interference and Weak Interactions," *Science*, 149, pp. 503–10.

Kuhn [1962]: *The Structure of Scientific Revolutions*, 1962.

Popper [1959]: *Logic of Scientific Discovery*, 1959.

Popper [1963]: *Conjectures and Refutations*, 1963.

Stahlman [1956]: "Astrology in Colonial America: An Extended Query," *William and Mary Quarterly*, 13, pp. 551–63.

Thorndike [1923–58]: *A History of Magic and Experimental Science*, 8 vols, 1923–58.

Thorndike [1955]: "The True Place of Astrology in the History of Science," *Isis*, 46, pp. 273–8.

The Popperian versus the
Kuhnian Research Programme

IMRE LAKATOS

Science: Reason or Religion?

... Popper's distinction lies primarily in his having grasped the full implications of the collapse of the best-corroborated scientific theory of all times:
Newtonian mechanics and the Newtonian theory of gravitation. In his view
virtue lies not in caution in avoiding errors, but in ruthlessness in eliminating them. Boldness in conjectures on the one hand and austerity in refutations on the other: this is Popper's recipe. Intellectual honesty does not
consist in trying to entrench, or establish one's position by proving (or
'probabilifying') it—intellectual honesty consists rather in specifying precisely the conditions under which one is willing to give up one's position.
Committed Marxists and Freudians refuse to specify such conditions: this
is the hallmark of their intellectual dishonesty. *Belief* may be a regrettably
unavoidable biological weakness to be kept under the control of criticism:
but *commitment* is for Popper an outright crime.

Kuhn thinks otherwise. He too rejects the idea that science grows by
accumulation of eternal truths.[1] He too takes his main inspiration from
Einstein's overthrow of Newtonian physics. His main problem too is *scientific revolution*. But while according to Popper science is 'revolution in
permanence', and criticism the heart of the scientific enterprise, according
to Kuhn revolution is exceptional and, indeed, extra-scientific, and criticism is, in 'normal' times, anathema. Indeed for Kuhn the transition from
criticism to commitment marks the point where progress—and 'normal'

Excerpted with permission from Imre Lakatos, "Falsification and the Methodology of Scientific Research Programmes," in Lakatos, Imre and Musgrave, Alan, eds., *Criticism and the
Growth of Knowledge*, New York: Cambridge University Press, 1970, pp. 91-197. Slight changes
have been made in the style of bibliographical references.

science—begins. For him the idea that on 'refutation' one can demand the rejection, the elimination of a theory, is 'naive' falsificationism. Criticism of the dominant theory and proposals of new theories are only allowed in the rare moments of 'crisis'. This last Kuhnian thesis has been widely criticized[2] and I shall not discuss it. My concern is rather that Kuhn, having recognized the failure both of justificationism and falsificationism in providing rational accounts of scientific growth, seems now to fall back on irrationalism.

For Popper scientific change is rational or at least rationally reconstructible and falls in the realm of the *logic of discovery*. For Kuhn scientific change—from one 'paradigm' to another—is a mystical conversion which is not and cannot be governed by rules of reason and which falls totally within the realm of the (*social*) *psychology of discovery*. Scientific change is a kind of religious change.

The clash between Popper and Kuhn is not about a mere technical point in epistemology. It concerns our central intellectual values, and has implications not only for theoretical physics but also for the underdeveloped social sciences and even for moral and political philosophy. If even in science there is no other way of judging a theory but by assessing the number, faith and vocal energy of its supporters, then this must be even more so in the social sciences: truth lies in power. Thus Kuhn's position would vindicate, no doubt, unintentionally, the basic political *credo* of contemporary religious maniacs ('student revolutionaries').

In this paper I shall first show that in Popper's logic of scientific discovery two different positions are conflated. Kuhn understands only one of these, 'naive falsificationism' (I prefer the term 'naive methodological falsificaionalism'); I think that his criticism of it is correct, and I shall even strengthen it. But Kuhn does not understand a more sophisticated position the rationality of which is not based on 'naive' falsificationism. I shall try to explain—and further strengthen—this stronger Popperian position which, I think, may escape Kuhn's strictures and present scientific revolutions as constituting rational progress rather than as religious conversions.

Fallibilism Versus Falsificationism . . .

Sophisticated versus naive methodological falsificationism. Progressive and degenerating problemshifts.

Sophisticated falsificationism differs from naive falsificationism both in its rules of *acceptance* (or 'demarcation criterion') and its rules of *falsification* or elimination. For the naive falsificationist any theory which can be interpreted as experimentally falsifiable, is 'acceptable' or 'scientific'. For the sophisticated falsificationist a theory is 'acceptable' or 'scientific' only if it has corroborated excess empirical content over its predecessor (or rival), that is, only if it leads to the discovery of novel facts. This condition can be analysed into two clauses: that the new theory has excess empirical content (*'accept-ability'*) and that some of this excess content is verified (*'acceptability'*). The first clause can be checked instantly by *a priori* logical analysis; the second can be checked only empirically and this may take an indefinite time.

Again, for the naive falsificationist a theory is *falsified* by a ('fortified') 'observational' statement which conflicts with it (or rather, which he decides to interpret as conflicting with it). The sophisticated falsificationist regards a scientific theory T as falsified if and only if another theory T' has been proposed with the following characteristics: (1) T' has excess empirical content over T: that is, it predicts *novel* facts, that is, facts improbable in the light of, or even forbidden, by T;[3] (2) T' explains the previous success of T, that is, all the unrefuted content of T is contained (within the limits of observational error) in the content of T'; and (3) some of the excess content of T' is corroborated. . . . [4]

Let us take a series of theories, T_1, T_2, T_3, . . . where each subsequent theory results from adding auxiliary clauses to (or from semantical reinterpretations of) the previous theory in order to accommodate some anomaly, each theory having at least as much content as the unrefuted content of its predecessor. Let us say that such a series of theories is *theoretically progressive* (*or 'constitutes a theoretically progressive problemshift'*) if each new theory has some excess empirical content over its predecessor, that is, if it predicts some novel, hitherto unexpected fact. Let us say that a theoretically progressive series of theories is also *empirically progressive* (*or 'constitutes an empirically progressive problemshift'*) if some of this excess empirical content is also corroborated, that is, if each new theory leads us to the actual discovery of some *new fact*.[5] Finally, let us call a problemshift *progressive* if it is both theoretically and empirically progressive, and *degenerating* if it is not.[6] We *'accept'* problemshifts as 'scientific' only if they are at least theoretically progressive; if they are not, we *'reject'* them as 'pseudoscientific'. Progress is measured by the degree to which a problemshift is progressive, by the degree to which the series of theories leads us to the discovery of novel facts. We regard a theory in the series 'falsified' when it is superseded by a theory with higher corroborated content.

This demarcation between progressive and degenerating problemshifts sheds new light on the appraisal of *scientific — or, rather, progressive — explanations*. If we put forward a theory to resolve a contradiction between a previous theory and a counterexample in such a way that the new theory, instead of offering a content-increasing (scientific) *explanation*, only offers a content-decreasing (linguistic) *reinterpretation*, the contradiction is resolved in a merely semantical, unscientific way. *A given fact is explained scientifically only if a new fact is also explained with it.*[7]

Sophisticated falsificationism thus shifts the problem of how to appraise *theories* to the problem of how to appraise *series of theories*. Not an isolated *theory*, but only a series of theories can be said to be scientific or unscientific: to apply the term 'scientific' to one *single* theory is a category mistake.[8]

The time-honoured empirical criterion for a satisfactory theory was agreement with the observed facts. Our empirical criterion for a series of theories is that it should produce new facts. *The idea of growth and the concept of empirical character are soldered into one.*

This revised form of methodological falsificationism has many new features. First, it denies that 'in the case of a scientific theory, our decision depends upon the results of experiments. If these confirm the theory, we may accept it until we find a better one. If they contradict the theory, we reject it.'[9] It denies that 'what ultimately decides the fate of a theory is the result of a test, *i.e.* an agreement about basic statements'.[10] Contrary to naive falsificationism, *no experiment, experimental report, observation statement or well-corroborated low-level falsifying hypothesis alone can lead to falsification. There is no falsification before the emergence of a better theory.*[11] But then the distinctively negative character of naive falsificationism vanishes; criticism becomes more difficult, and also positive, constructive. But, of course, if falsification depends on the emergence of better theories, on the invention of theories which anticipate new facts, then falsification is *not* simply a relation between a theory and the empirical basis, but a multiple relation between competing theories, the original 'empirical basis', and the empirical growth resulting from the competition. Falsification can thus be said to have a *'historical character'*.[12] Moreover, some of the theories which bring about falsification are frequently proposed *after* the 'counterevidence'. This may sound paradoxical for people indoctrinated with naive falsificationism. Indeed, this epistemological theory of the relation between theory and experiment differs sharply from the epistemological theory of naive falsificationism. The very term 'counterevidence' has to be abandoned in the sense that no experimental result must be interpreted directly as 'counterevidence'. If we still want to retain this time-honoured term, we have to redefine it like this: 'counterevidence to T_1' is a corroborating instance to T_2 which is either inconsistent with or independent of T_1 (with the *proviso* that T_2 is a theory which satisfactorily explains the empirical success of T_1). This shows that *'crucial counter evidence'* — or *'crucial experiments'* — can be recognized as such among the scores of anomalies only *with hindsight*, in the light of some superseding theory.[13]

Thus the crucial element in falsification is whether the *new theory* offers any novel, excess information compared with its predecessor and whether some of this excess information is corroborated. Justificationists valued 'confirming' instances of a theory; naive falsificationists stressed 'refuting' instances; for the methodological falsificationists it is the — rather rare — corroborating instances of the *excess* information which are the crucial ones; these receive all the attention. We are no longer interested in the thousands of trivial verifying instances nor in the hundreds of readily available anomalies: the few crucial *excess-verifying instances* are decisive.[14] This consideration rehabilitates—and reinterprets—the old proverb: *Exemplum docet, exempla obscurant.*

'Falsification' in the sense of naive falsificationism (corroborated counter-evidence) is not a *sufficient* condition for eliminating a specific theory: in spite of hundreds of known anomalies we do not regard it as falsified (that is, eliminated) until we have a better one.[15] Nor is 'falsification' in the naive sense necessary for falsification in the sophisticated sense: a progressive problemshift does not have to be interspersed with 'refutations'. Science can grow without any 'refutations' leading the way. Naive falsificationists suggest a linear growth of science, in the sense that theories are followed by powerful refutations which eliminate them; these refutations in turn are followed by new theories.[16] It is perfectly *possible* that theories be put forward 'progressively' in such a rapid succession that the 'refutation' of the *n*-th appears only as the corroboration of the *n* + 1-th. The problem fever of science is raised by proliferation of rival theories rather than counterexamples or anomalies.

This shows that the slogan of *proliferation of theories* is much more important for sophisticated than for naive falsificationism. For the naive falsificationist science grows through repeated experimental overthrow of theories; new rival theories proposed before such 'overthrows' may speed up growth but are not absolutely necessary[17]; constant proliferation of theories is optional but not mandatory. For the sophisticated falsificationist proliferation of theories cannot wait until the accepted theories are 'refuted' (or until their protagonists get into a Kuhnian crisis of confidence).[18] While naive falsificationism stresses "the urgency of replacing a *falsified* hypothesis by a better one",[19] sophisticated falsificationism stresses the urgency of replacing *any* hypothesis by a better one. Falsification cannot "compel the theorist to search for a better theory",[20] simply because falsification cannot precede the better theory . . .

The Popperian Versus the Kuhnian Research Programme

Let us now sum up the Kuhn-Popper controversy.

We have shown that Kuhn is right in objecting to naive falsificationism, and also in stressing the *continuity* of scientific growth, the *tenacity* of some scientific theories. But Kuhn is wrong in thinking that by discarding naive falsificationism he has discarded thereby all brands of falsificationism. Kuhn objects to the entire Popperian research programme, and he excludes *any* possibility of a rational reconstruction of the growth of science. In a succinct comparison of Hume, Carnap and Popper, Watkins points out that the growth of science is inductive and irrational according to Hume, inductive and rational according to Carnap, non-inductive and rational according to Popper.[21] But Watkins's comparison can be extended by adding that it is non-inductive and irrational according to Kuhn. *In Kuhn's view there can be no logic, but only psychology of discovery.*[22] For instance, in Kuhn's conception, anomalies, inconsistencies *always* abound in science, but in 'normal' periods the dominant paradigm secures a pattern of growth which is eventually overthrown by a 'crisis'. There is no particular rational cause for the appearance of a Kuhnian 'crisis'. 'Crisis' is a psychological concept; it is a contagious panic. Then a new 'paradigm' emerges, incommensurable with its predecessor. There are no rational standards for their comparison. Each paradigm contains its own standards. The crisis sweeps away not only the old theories and rules but also the standards which made us respect them. The new paradigm brings a totally new rationality. There are no super-paradigmatic standards. The change is a bandwagon effect. Thus *in Kuhn's view scientific revolution is irrational, a matter for mob psychology.*

The reduction of philosophy of science to psychology of science did not start with Kuhn. An earlier wave of 'psychologism' followed the breakdown of justificationism. For many, justificationism represented the only possible form of rationality: the end of justificationism meant the end of rationality. The collapse of the thesis that scientific theories are provable, that the progress of science is cumulative, made justificationists panic. If 'to discover is to prove', but nothing is provable, then there can be no discoveries, only discovery-claims. Thus disappointed justificationists— exjustificationists—thought that the elaboration of rational standards was a hopeless enterprise and that all one can do is to study—and imitate—the Scientific Mind, as it is exemplified in famous scientists. After the collapse of Newtonian physics, Popper elaborated new, non-justificationist critical standards. Now some of those who had already learned of the collapse of justificationist rationality now learned, mostly by hearsay, of Popper's colourful slogans which suggested naive falsificationism. Finding them untenable, they identified the collapse of naive falsificationism with the end of rationality itself. The elaboration of rational standards was again regarded as a hopeless enterprise; the best one can do is to study, they thought once again, the Scientific Mind.[23] Critical philosophy was to be replaced by what Polanyi called a 'post-critical' philosophy. But the Kuhnian research programme contains a new feature: we have to study not the mind of the individual scientist but the mind of the Scientific Community. Individual psychology is now replaced by social psychology; imitation of the great scientists by submission to the collective wisdom of the community.

But Kuhn overlooked Popper's sophisticated falsificationism and the research programme he initiated. Popper replaced the central problem of classical rationality, *the old problem of foundations*, with *the new problem of fallible-critical growth*, and started to elaborate objective standards of this growth. In this paper I have tried to develop his programme a step further. I think this small development is sufficient to escape Kuhn's strictures . . .[24]

Notes

1. Indeed he introduces his [1962] by arguing against the 'development-by-accumulation' idea of scientific growth. But his intellectual debt is to Koyré rather than to Popper. Koyré showed that positivism gives bad guidance to the historian of science, for the history of physics can only be understood in the context of a succession of 'metaphysical' research programmes. Thus scientific changes are connected with vast cataclysmic metaphysical revolutions. Kuhn develops this message of Burtt and Koyré and the vast success of his book was partly due to his hard-hitting, direct criticism of justificationist historiography—which created a sensation among ordinary scientists and historians of science who Burtt's, Koyré's (or Popper's) message has not yet reached. But, unfortunately, his message had some authoritarian and irrationalist overtones.

2. Cf. e.g. Watkins's and Feyerabend's contributions in Lakatos, Imre and Musgrave, Alan, eds., *Criticism and the Growth of Knowledge, op. cit.*

3. I use 'prediction' in a wide sense that includes 'postdiction'.

4. *For a detailed discussion of these acceptance and rejection rules and for references to Popper's work*, cf. my "Changes in the Problem of Inductive Logic," in Lakatos (*ed.*): *The Problem of Inductive Logic*, 1968, pp. 315-417; pp. 375-90.

5. If I already know P_1: 'Swan A is white', $P\omega$ 'All swans are white' represents no progress, because it may only lead to the discovery of such further similar facts as P_2: 'Swan B is white'. So-called 'empirical generalizations' constitute no progress. A *new* fact must be improbable or even impossible in the light of previous knowledge . . .

6. The appropriateness of the term 'problemshift' for a series of theories rather than of problems may be questioned. I chose it partly because I have not found a more appropriate alternative—'theoryshift' sounds dreadful—partly because theories are always problematical, they never solve all the problems they have set out to solve. Anyway, in the second half of the paper, the more natural term 'research programme' will replace 'problemshifts' in the most relevant contexts.

7. Indeed, in the original manuscript of my "Changes in the Problem of Inductive Logic," *op. cit.*, I wrote: "A theory without excess corroboration has no excess explanatory power; *therefore, according to Popper, it does not represent growth and therefore it is not 'scientific': therefore, we should say, it has no explanatory power*" (p. 386). I cut out the italicized half of the sentence under pressure from my colleagues who thought it sounded too eccentric. I regret it now.

8. Popper's conflation of 'theories' and 'series of theories' prevented him from getting the basic ideas of sophisticated falsificationism across more successfully. His ambiguous usage led to such confusing formulations as 'Marxism [as the core of a series of theories or of a "research programme"] is irrefutable' and, at the same time, 'Marxism [as a particular conjunction of this core and some specified auxiliary hypotheses, initial conditions and a *ceteris paribus* clause] has been refuted.' (Cf. Popper, *Conjectures and Refutations*).

Of course, there is nothing wrong in saying that an isolated, single theory is 'scientific' if it represents an advance on its predecessor, as long as one clearly realizes that in this formulation we appraise the theory as the outcome of—and in the context of—a certain historical development.

9. Popper, *The Open Society and Its Enemies*, vol. II, 1945, p. 233. Popper's more sophisticated attitude surfaces in the remark that "concrete and practical consequences can be *more* directly tested by experiment" (*ibid.*, my italics).

10. Popper, *Logik der Forschung*, 1935, section 30 (expanded English edition: *The Logic of Scientific Discovery*, 1959.)

11. "In most cases we have, before falsifying a hypothesis, another one up our sleeves" (Popper, *The Logic of Scientific Discovery*, p. 87, footnote *1). But, as our argument shows, we *must* have one. Or, as Feyerabend put it: "The best criticism is provided by those theories which can replace the rivals they have removed" ("Reply to Criticism," in Cohen and Wartofsky (*eds.*): *Boston Studies in the Philosophy of Science*, II, pp. 223-61; p. 227. He notes that in *some* cases "alternatives will be quite indispensable for the purpose of refutation" (*ibid.*, p. 254). But according to our argument *refutation without an alternative shows nothing but the poverty of our imagination in providing a rescue hypothesis.*

12. Cf. my "Changes in the Problem of Inductive Logic," *op. cit.*, pp. 387 ff.

13. In the distorting mirror of naive falsificationsim, new theories which replace old refuted ones, are themselves born unrefuted. Therefore they do not believe that there is a relevant difference between anomalies and crucial counterevidence. For them, anomaly is a dishonest euphemism for counterevidence. But in actual history new theories are born refuted: they inherit many anomalies of the old theory. Moreover, frequently it is *only* the new theory which dramatically predicts that fact which will function as crucial counterevidence against its predecessor, while the 'old' anomalies may well stay on as 'new' anomalies.

All this will be still clearer when we introduce the idea of 'research programme'.

14. *Sophisticated falsificationism adumbrates a new theory of learning . . .*

15. It is clear that the theory T' may have excess corroborated empirical content over another theory T even if both T and T' are refuted. Empirical content has nothing to do with truth or falsity. Corroborated contents can also be compared irrespective of the refuted content. Thus we may see the rationality of the elimination of Newton's theory in favour of Einstein's, even though Einstein's theory may be said to have been born—like Newton's—'refuted'. We have only to remember that 'qualitative confirmation' is a euphemism for 'quantitative disconfirmation'. (Cf. my "Changes in the Problem of Inductive Logic," *op. cit.*, pp. 384-86.)

16. Cf. Popper, *Logik der Forschung*, 1934, section 85, p. 279 of *The Logic of Scientific Discovery*, 1959.

17. It is true that a certain type of *proliferation of rival theories* is allowed to play an accidental heuristic role in falsification. In many cases falsification heuristically "depends on [the condition] that sufficiently many and sufficiently different theories are offered" (Popper, "What is Dialectic?", *Mind*, N.S. 49, pp. 403-426; reprinted in Popper, *Conjectures and Refutations*, 1963, pp. 312-35.) For instance, we may have a theory T which is apparently unrefuted. But it may happen that a new theory T', inconsistent with T is proposed which equally fits the available facts: the differences are smaller than the range of observational error. In such cases the inconsistency prods us into improving our 'experimental techniques', and thus refining the 'empirical basis' so that either T or T' (or, incidentally, both) can be falsified: "We need [a] new theory in order to find out where the old theory was deficient" (Popper, *Conjectures and Refutations*, 1963, p. 246). But the role of this proliferation is *accidental* in the sense that, once the empirical basis is refined, the fight is between this refined empirical basis and the theory T under test; the rival theory T' acted only as a *catalyst* . . .

18. Also cf. Feyerabend, "Reply to Criticism," *op. cit.*, pp. 254-55.

19. Popper, *The Logic of Scientific Discovery*, p. 87, footnote *1.

20. Popper, *Logik der Forschung*, section 30.

21. Watkins, "Hume, Carnap and Popper", in Lakatos (*ed.*): *The Problem of Inductive Logic*, 1968, pp. 271-82.

22. Kuhn, "Logic of Discovery or Psychology of Research?", excerpted in this volume. But this position is already implicit in his *The Structure of Scientific Revolutions*, 1962.

23. Incidentally, just as some earlier ex-justificationists led the wave of sceptical irrationalism, so now some ex-falsificationists lead the *new* wave of sceptical irrationalism and anarchism. This is best exemplified in Feyerabend, "Against Method," *Minnesota Studies for the Philosophy of Science*, 4, 1970.

24. Indeed, as I had already mentioned, *my concept of a 'research programme' may be construed as an objective, 'third world' reconstruction of Kuhn's socio-psychological concept of 'paradigm':* thus the Kuhnian 'Gestalt-switch' can be performed without removing one's Popperian spectacles.

(I have not dealt with Kuhn's and Feyerabend's claim that theories cannot be eliminated on any *objective* grounds because of the 'incommensurability' of rival theories. Incommensurable theories are neither inconsistent with each other, nor comparable for content. But we can *make* them, by a dictionary, inconsistent and their content comparable. If we want to eliminate a programme, we need some methodological determination. This determination is the heart of methodological falsificationism; for instance, no result of statistical sampling is ever inconsistent with a statistical theory unless we *make them* inconsistent with the help of Popperian rejection rules . . .)

Kuhn, Popper, and the
Normative Problem of Demarcation

Robert Feleppa

Popper is not merely trying to distinguish science from pseudoscience. He is also making an important *evaluational* claim—a ranking of one as *better* than the other. Also, his demarcation criterion is meant to indicate scientific progress and improvement, and to establish science as *rational* in ways that pseudoscience is not. Ideally, adequate criteria of demarcation will prescriptively guide scientific behavior, rather than simply describe it, and will also sharpen our understanding of what it is to be rational.

Given all this, the task Popper sets is not then merely to identify criteria of demarcation, but to justify their evaluative force.

Popper's account of demarcation of course hinges centrally on his falsifiability criterion, and he has been widely criticized by Kuhn and others precisely for stressing this feature. In its place, Kuhn offers a cluster of criteria culled from what he takes to be a more sensitive reading of the historical record of science and from current scientific practice. But on the face of it, simply describing what scientists do seems to fall short of what is needed: the question of why they *ought* to act as they do seems unanswered. As many critics have pointed out, Kuhn's basis for inferring scientific and rational prescriptions on the basis of his historical descriptions is unclear: describing the genesis of a set of conventions does not alone serve to justify them.

How is the dispute between Popper and Kuhn to be resolved? Given that demarcation is fundamentally a normative issue, such resolution demands first a clearer understanding of the strategy each side employs in deriving prescriptive criteria from scientific practice. The debate goes much deeper than simply the question of whether one side or the other is correctly

An earlier draft of this paper was presented at an NEH Summer Seminar for College Teachers, the University of Maryland, August 1980. I am indebted to its director, Dudley Shapere, and to Ken Goosens, Dale Moberg, and Gene Laschyck for helpful discussion and comments.

characterizing scientific practice. The different conclusions they draw from substantially the same historical record suggest that they may well disagree on some basic points concerning the proper way to derive a methodological "ought" from a historical "is." But Kuhn and Popper both take justification roughly to involve a movement from the description of the practice to be justified to the development of justifying principles. This suggests a strong compatibility between their implicit views on justification and the "reflective equilibrium" account of justification expounded by Nelson Goodman and John Rawls. In what follows I hope to clarify the normative aspects of the Popper-Kuhn controversy, and to work toward a more fruitful comparison of their positions, by appealing to this notion.

In Defense of Popper

In an earlier essay in this collection Kuhn uses Popper's often-cited case of astrology to explicate Popper's central contention that essential to proper scientific procedure is the subjectability of theory to experimental tests, which are, in Kuhn's words, "performed to explore the limitations of accepted theory or to subject a current theory to maximum strain." A scientific theory is better insofar as it risks more, and a "theory" that risks nothing, that is, one so constructed as to be immune to falsification, is not really an empirical theory at all. Popper deems astrology a pseudoscience because astrological systems are too easily made immune to empirical failure, either by explaining away predictive failure by citing unaccounted or unknown variables or through the use of language so vague as to allow easy reinterpretation of predictions to make them agree with subsequent events. Thus "saving the appearances" works to the discredit of astrological systems; on Popper's view, they fail to meet what he intends as a central *necessary* condition for being genuinely scientific. According to a relatively recent formulation (1963) p.256:

> a system [1] is to be considered scientific only if it makes assertions which may clash with observations; and a system is, in fact, tested by attempts to refute it. Thus testability is the same as refutability, and can therefore likewise be taken as a criterion of demarcation.[2]

Kuhn's main objection to this, in the essay earlier in this volume, is that Popper's criterion is derived only from certain episodic—and atypical— fragments from the history of science, and thus presents a distorted account of scientific practice (keep in mind that ultimately this charge is leveled against Popper's justificatory account):

> a careful look at the scientific enterprise suggests that it is normal science, in which Sir Karl's sort of testing does not occur, rather than extraordinary science which most nearly distinguishes science from other enterprises. If a

141

demarcation criterion exists (we must not, I think, seek a sharp or decisive one), it may lie just in that part of science which Sir Karl ignores.

As he indicates in this passage, Kuhn looks instead to the practice of normal science in its periods of relatively crisis-free development in order to discover criteria of demarcation, contrasting astronomy with astrology in terms of the former's greater articulation and more powerful governing rules (as opposed to the latter's weaker "craft-rules").

Kuhn also questions Popper's account of the pseudoscientific character of astrology on the grounds that astrological practice actually quite often conforms to Popper's criterion. Although he agrees that Popper's focus on the untestability of astrology "catches something of the spirit of the astrological enterprise," Kuhn rejects Popper's criterion because "the history of astrology during the centuries when it was intellectually reputable records many predictions that categorically failed. Not even astrology's most convinced and vehement exponents doubted the recurrence of such failures."

But clearly Popper has a legitimate reply to Kuhn here on a number of counts. First, as I emphasized above, Popper should not be too quickly interpreted as offering a *sufficient* condition for demarcating science. The above passage indicates that it is to be taken only as a necessary condition, and hence Kuhn's insistence that a pseudoscience in historical fact is refutable on occasion does not speak directly to Popper's point.[3]

Also, even were this pertinent (Popper has on occasion suggested a stronger logical status for this criterion), why should the *occasional* adherence of astrologers to the falsifiability criterion serve to refute his contention? Is he not demanding unfailing, or at least general majority acceptance of this criterion within the relevant communities? And surely astrologers have been and still are guilty of the sort of exploitation of indeterminacy and vagueness that Popper decries.

This might lead us to ask why Kuhn cites the *earlier* history of astrology—when it was not only more reputable, as Kuhn remarks, but also far less easily distinguishable from astronomy in the eyes of its practitioners at the time as well as of subsequent historians. If Popper is to be refuted here, it would seem preferable to examine more recent astrological strains that are more easily distinguishable from what we take to be genuine science. Otherwise Popper could cite just this very intermixing of legitimate and metaphysical strands to explain why the astrologer-astronomer's behavior happened to conform more closely to his proposed standard in earlier times.

Moreover, Kuhn makes no mention of the replacement or revision of any astrological system as the result of empirical failure—and it is these sorts of consequences in which Popper is primarily interested: he wants more than the *admission* of error. His view is that scientists *learn* from their mistakes in the sense of using them to generate new theoretical products. (In this regard he could aptly cite Kuhn's remark that no astrologer "could make use of [research puzzles] in a constructive attempt to revise the astrological tradition.")

However, there is a deeper puzzle behind these various objections concerning just how we are to learn from the historical record what scientists, and rational persons generally, ought to be doing. The gist of Kuhn's criticisms seems simply to be that scientists do not behave in the way Popper claims they should (or that nonscientists do behave this way at times). He criticizes Popper for drawing on insufficient historical data and thus giving a distorted account of what they do *and* of what they ought to do. But should we not expect that a normative account will deviate from actual practice to some degree? Simply showing that Popper is appealing only to certain features of the historical record—which *exemplify* his chosen standard—does not prove that he is *distorting* things; we must first look more carefully at Popper's reasons for selecting as he does. Showing that one or the other account is "truer" to the historical record is beside the point where the primary question is one of justification: normative criteria are not the mere reflection of actual practice.

The cogency of Kuhn's objections is further clouded by the fact that it is not clear that Popper *has* ignored normal science. In specific rejoinder to the criticisms I have discussed, Popper (Lakatos and Musgrave, p. 51) points out that he noted the fundamental role of normal practice as a background for progressive change (through conjecture and refutation) as early as the writing of the opening paragraph of *The Logic of Scientific Discovery*:

> A scientist engaged in a piece of research . . . can attack his problem straight away. He can go at once to the heart of an organized structure. For a structure of scientific doctrines is already in existence; and with it, a generally accepted problem-situation. This is why he may leave it to others to fit his contribution into the framework of scientific knowledge.

Popper does not wish so much to discredit or ignore normal science as to uncover what it is that makes for *good* science, normal or abnormal. And if one considers that it is in reaction to some widely current forms of normal Freudian psychology and Marxist historiography that Popper often writes, it becomes clearer that simply citing deviations of normal practice from Popper's prescriptions is insufficient.

Popper gives a lot of attention to extraordinary science, but he does not thus mean to *ignore* normal practice. Instead, he intends to determine *good* normal practice in virtue of its *continuity* with extraordinary science. Adherence to the falsifiability criterion is intended to ensure that normal, nonrevolutionary practice never becomes so structured as to preclude the possibility of empirical refutation or theoretical revolution. Whatever lengths a scientist may go to protect a theory, Popper desires that these defenses never attain perfect impregnability to recalcitrant experience. So, for instance, Kuhn's remarks concerning the similar appeal made by meteorologists and astrologers to unknown and complex variables do not present any immediate threat tò Popper's demarcation of those fields: what is important for him is not *that* such a theory-saving strategy is used, but instead the degree to which it is used.

In Defense of Kuhn: A Reconstruction

To thus cite normal practice so uncritically in responding to Popper is to miss, then, his most central point. It is for this reason that Popper (Lakatos and Musgrave, p. 52) is so harshly critical of Kuhn in this regard:

> "Normal" science, in Kuhn's sense, exists. It is the activity of the non-revolutionary, or more precisely, the not-too-critical professional: of the student who accepts the ruling dogma of the day; who does not wish to challenge it; and who accepts a new revolutionary theory only if almost everybody else is ready to accept it—if it becomes fashionable by a kind of bandwagon effect. To resist a new fashion needs perhaps as much courage as was needed to bring it about.

Popper and others contend that Kuhn has offered nothing by way of an independent and critical standpoint for justifying what scientists do—and yet, in Popper's view, this is precisely what is needed if one is justifiably to judge science as progressing rationally or as improving over time. Kuhn, in answering charges that he has made science irrational, cites a changing but reasonably stable list of criteria by which to judge the genuine progress of science. Despite substantive theoretical disagreements, scientists generally agree that competing theories are to be judged in terms of their generality in scope, simplicity, accuracy, consistency, fruitfulness, and so on; and, furthermore, Kuhn contends that these criteria provide an objective measure of scientific progress (Lakatos and Musgrave, p. 264; Kuhn, 1977, pp. 321ff.). But merely citing the fact that professional agreement on these criteria spans disagreements over theoretical content seems insufficient to justify their normative status. It does not seem enough to say that these criteria are those used by members of the scientific community in rating theories and measuring progress—not if one is not willing simply to accept what scientists do as correct but is instead trying to *evaluate* what they do and to *justify* certain procedures and label others as "unscientific" (though they might comprise the common practice of certain scientists). Since people do not always do what they ought to do, a "justification" that consists of making norms conform to practice seems vacuous and circular (perhaps, Popper might contend, manifesting the "bandwagon" effect again at another level).

As Carl Hempel (1981) has recently noted the problem here is not necessarily that Kuhn's strategy is misguided—it is just that certain key justificatory steps are not made evident. Although he is generally in agreement with Kuhn's position (this has not always been the case), Hempel contends that Kuhn's arguments to establish the rationality of science do not go far enough. They show only that since scientists ought to behave as if their concern was to improve scientific knowledge, they ought then to adhere to their generally accepted criteria of theory-choice, since no other mode of behavior seems to serve this basic purpose as well. However, Hempel (p. 402) argues, aside from the fact that Kuhn has neither clarified what he takes the essential characteristics of science to be nor given an adequate account of what is to count as "improving scientific knowledge," we are left by this functional account with a glaring gap in normative reasoning:

> on Kuhn's view, scientific theory choice (and other facets of scientific behavior) are not affected by means of procedures that are deliberately adopted by the scientific community as a presumably optimal means for advancing scientific knowledge; and I would think that any kind of action, including scientific theory choice, can be called *rational* only if it can be causally linked to deliberation or reasoning aimed at achieving specific ends. Scientific theory choice as characterized by Kuhn would not be rational in this sense, but would rather be akin to what in anthropology are called *latently functional* behavior patterns, which serve a function they were never *chosen* to fulfill.

Hempel's contention is that the evaluation of the rationality of scientific theory-choice must be made in the light of general norms of rationality. And, in particular, that whatever the other virtues of the Kuhnian account, it cannot succeed in establishing the rationality of scientific practice without at least making explicit the connection between the adoption of choice-criteria and deliberation about how these criteria are to serve the basic purposes of science.

Hempel's criticism here is deep, but it nevertheless manifests a fundamental agreement with Kuhn's "historic-pragmatist" approach to science. For Hempel's criticism appeals to a norm drawn from our intuitive background of norms regarding rationality, and he does not claim to give it any more transcendental or *a priori* status than that. He agrees that norms of rational and scientific behavior are derived from our common practices wherein such norms operate. His only objection is that Kuhn does not seem to take into account all the pertinent norms in giving his justificatory account of science.

Hempel believes that the crucial connection between the evolution of choice-criteria and deliberation about practice is established in a pragmatist account of justification developed by Nelson Goodman in addressing the problem of induction in *Fact, Fiction, and Forecast*. Furthermore, he contends, Goodman's view can easily be adapted to the justificatory context that concerns Kuhn. Hempel argues that where the rules of choice can be made sufficiently explicit,[4] their justification can be achieved not by finding indubitable first principles, but by realizing with Goodman that justification (in any context) consists not in giving an *a priori* grounding of the practice in question, but rather in *codifying* that practice. In Goodman's view, we justify our inductive (and, for that matter, our deductive) inferences by realizing from the start that we *do* make *some* valid inferences, prior to any full realization of principles that differentiate valid from invalid inferences. Just as Aristotle did in the *Organon*, we work initially from a set of inferences that we take to be clearly valid and a set that we take to be clearly invalid, and we endeavor to find principles that will succeed in generating the valid ones and in ruling out the invalid ones: the main objective being to improve our ability to assess inductive or deductive validity through this explicit formulation of principles. "Justification," in this modified sense, is achieved by giving more precise formulation to the principles we already use to assess our practice, and by seeking gradually to attain a suitable fit between those rules and the practice. Rawls, in adapting this notion of justification to ethical contexts in *A Theory of Justice*, has called this intended balance between principles and practice "reflective equilibrium." The principles "govern" the practice in that they are, after codification, appealed to in order to make determinations of validity or invalidity, and so on, and in that they provide certain advantages over the uncodified practice such as enabling one to judge what before were indeterminate "borderline" cases. However, it is important to emphasize that on the Goodman-Rawls account these principles are not regarded as fixed Platonic absolutes derived from careful reasoned consideration of the flux of scientific practice (cf. my discussion of Popper below). Instead, these principles are always in an important way answerable to the intuitive, presystematic judgments from which they were derived. Where these "governing" principles conflict overtly with the stronger of these intuitive judgments (i.e., judgments made without recourse to the principles), the principles themselves may have to be modified.

We operate, then, in Goodman's (1973, p. 64) view, in a virtuous circle:

> The point is that rules and particular inferences alike are justified by being brought into agreement with each other. A rule is amended if it yields an inference we are unwilling to accept, an inference is rejected if it violates a rule we are unwilling to amend. The process of justification is the delicate one of making mutual adjustments between rules and accepted inferences; and in the agreement achieved lies the only justification needed for either.[6]

Since Kuhn is concerned with a problem of justification that is very similar in character to Goodman's—namely, one of hypothesis- and theory-acceptance in science—and since Kuhn and Goodman share a similar pragmatist orientation, Hempel believes that Goodman's view of justification can serve to supplement Kuhn's argument. This may be possible without forcing any significant modification in what Kuhn does say. We now see how a justification of a practice *can* be developed without vicious circularity: we find what sort of codification of principles will improve on the existing conventions, and realize that the justification we seek lies'in our achieving improvements without incurring serious violations of our intuitive practice of theory-evaluation, theory-choice, and so on. Although this sort of reconstrual of such a fundamental notion as "justification" may strike some as "having all the advantages of theft over honest toil," it is an appealing solution to a number of age-old normative problems—the main point being that we should come to realize that all the troubles we have had in actually nailing down the justification of induction, a mode of inference we are sure is legitimate and justifiable in many instances, stem not from lack of ingenuity but from a deeper problem: namely, a miscast *a prioristic* attitude toward the nature of our task. And the solution to the problem lies precisely in accepting the fact that our presystematic judgments are to some degree legitimate, and that justification is achieved by moving, via codification, from an already existing set of values to a better one. We do not simply infer what we ought to do from what we do, but instead develop more general principles, and a systematically more cohesive understanding, concerning what is taken to be an already partially viable process of valuation.

Popper and Reflective Equilibrium

Thus Kuhn's position might be saved from the attacks of Popper and others by an appeal to a justificatory notion of reflective equilibrium. But might not Popper's position also be characterized in terms of reflective equilibrium? If so, we would have a firm basis of comparison—something evidently lacking in the controversy as initially surveyed above.

It might seem that the core idea of reflective equilibrium justification—to get governing principles without determining absolute essences—is fundamentally at odds with what Popper is trying to do. Popper could easily be read as what Lakatos terms a "dogmatic falsificationist"—offering falsifiability as a basic, transcendent feature of empirical science to which all practice must conform. This might well seem the case given the attention Popper directs to his criterion, its centrality to his account, and the vigor (some would use a less complimentary expression) with which he has defended it.

But there are a number of passages in Popper's general discussion of falsifiability in *The Logic of Scientific Discovery* that reveal, I think, certain important affinities with Goodman's views on justification (and hence with the Hempel-Kuhn account). Whatever Popper's actual intentions, these affinities suggest a reconstruction of his position along reflective equilibrium lines. First, Popper strongly emphasizes that his criterion of demarcation must "be regarded as a proposal for an agreement or convention" among "parties having some purpose in common." Furthermore, he admits that his derivation of justificatory principles for science has "been guided, in the last analysis, by value judgments and predilections" (1959, p. 37). However, in light of these points he contends that he must then abandon the hope of giving any *a priori* grounding to his principles or of "representing them as the true or the essential aims of science"—they can be justified only through analyzing their logical consequences and determining "their fertility [and] their power to elucidate the problems of the theory of knowledge" (p. 38). Thus:

> My only reason for proposing my criterion of demarcation is that it is fruitful: that a great many points can be clarified and explained with its help. . . . It is only from the consequences of my definition of empirical science, and from the methodological decisions which depend upon this definition, that the scientist will be able to see how far it conforms to his intuitive idea of the goal of his endeavors. (p. 55)

In addition to giving this general account of his strategy, Popper attempts to give his account a clear structure, one that sets up a "first principle"—the falsifiability criterion—to generate methodological consequences that are to be examined in light of the intuitive judgments of scientists regarding both the basic aims of science and the presystematically adopted norms of theory- and hypothesis-acceptance. These "lower-order" norms are just the criteria of simplicity, scope, accuracy, and so on that are taken to function with some degree of legitimacy in science (p. 54).

Having thus outlined the strategy and structure of his program, Popper devotes subsequent sections of *The Logic of Scientific Discovery* to discussions of how the falsifiability criterion generates these standard criteria of adequacy. So, for instance, he argues that a theory that makes more precise predictions increases the number of experimental values that can clash with it, thus increasing its falsifiability, and is thus justifiably to be valued. Analogous arguments are offered for generality, simplicity, and so on.[7] In addition to these case judgments regarding the relative merits of legitimate theories, there are judgments of scientific legitimacy made throughout Popper's work that are to be measured by reference to the presystematic background. Hence the falsifiability criterion should not serve to rule out, say, quantum chromodynamics as nonempirical metaphysics.

Popper also contends that this criterion should provide an improved formulation of our already existing and evolving concept of science—that is, the set of value judgments commonly accepted regarding the basic aims and purposes of science. Among the values cited in Popper's work are the following: New theories should represent genuinely new discoveries that enable a deeper understanding about the world. Science should probe progressively deeper into the structure of the universe. Scientists should eschew dogmatist tendencies. Theories should have rich content and should reveal maximal "coherence," "compactness," and "organicity" in the states of affairs they describe. Theories should have heuristic power and should provide guidance and stimulation for our imagination and intuition.[8]

By insisting that theorists adhere to the falsifiability criterion, Popper hopes to assure compliance to these general background norms, while clarifying precisely what constitutes such compliance. Thus, for instance, he contends that the progressively greater depth of advancing theories is ensured by the reduction and correction of earlier, refutable theories by more comprehensive successors; and their "world-aboutness" is guaranteed by assuring that they are empirically refutable.

Under the present reconstruction, then, the falsifiability criterion is to be "tested" on the basis of (1) whether it generates rules of practice, and, derivatively, evaluations of theories or parts of theories as better or worse, genuinely empirical or metaphysical (or utterly pseudoscientific), that accord with our presystematic judgments on these matters, and (2) whether it enables a reformulation of our roughly formulated background values that seems to improve our understanding of this background and to clarify its relationship to practice.

I think we can see at least the roots of a reflective equilibrium account in these remarks of Popper's. Superficial appearances notwithstanding, the justificational approaches of Kuhn and Popper can thus be seen to share many important features.

Of course important differences remain. For, as noted above, Popper's program has a specifically elaborated structure—one that establishes a prior (though not an *a priori*) principle, a set of tasks for it to achieve, and a general account of how to determine the acceptability of this and other scientific norms in the light of prior practice.[9] The Hempel-Kuhn account, on the other hand, seems only to manifest the last, general feature.

Indeed, under this reconstruction, Popper's justificational strategy can be characterized in the same terms that he uses to characterize science— namely, as guided by an attitude of "modified essentialism" (1972, p. 202). Just as on Popper's account, scientists should proceed *as if* they were out after *the* true account of the "deep structure" of nature, so should the metascientist proceed as if there are essential features of science to be revealed in as systematic a way as possible—ideally in terms of one or several basic characteristics such as falsifiability: yet in the full light of a strong skepticism with regard to any attempt to "absolutize" these results. Popper's reasons for thus clearly structuring his metascientific program concern more the maintenance of a clear critical standpoint than the revelation of true essences. It is this feature of Platonic essentialism that he most highly values and that he endeavors to infuse into science and metascience. Popper wishes to critically assess and understand scientific practices and norms rather than simply to accept them. He regards his justification of such norms as value precision and simplicity in science, in terms of maximization of falsifiability, as distinctly superior to the mere acceptance of received scientific opinion (1972, p. 204). In the Socratic critical spirit, he would rather honor scientists because he believes they are right, as opposed to the other way around.

The Relative Merits of Popper and Kuhn

We have seen that both Kuhn's and Popper's positions can be augmented and supplemented by placing them in the pragmatic normative setting of reflective equilibrium. Such a construction is certainly not alien to either position, since both are evolutionary and non-*a priori* in outlook, and since each is firmly committed to a justificatory strategy that derives prescriptions from preexisting conventions.

But it is clear that there is a difference in the structuring of the programs—or, instead, a difference between a clear structure in one case and *no* evident structure in the other. The value of Popper's structuring may be seen, I think, in the light of much the same rational criteria as lend at least initial credence to the falsifiability criterion itself. Of course there are important differences between the ways one "tests" metascientific as opposed to scientific accounts—the former's refutability can less easily be justified in terms of providing needed input from the world of sense experience, since their "tribunal" typically comprises case value judgments rather than observations. But there seem to be solid rational grounds for desiring maximal criticizability and assessability in *any* constructive systematization or theory, and the remarks made by Popper, Feyerabend, and others concerning the eschewal of dogmatist tendencies bring this common rational norm to light. The attainment of truth in science, or of an adequate codification in metascience, cannot be hindered by processes that work to obscure the potential failings of systems constructed with such aims in mind. Indeed, this seems the very norm that underlies our favorable valuation of clarity and transparency of language and argument, and of the utility we attach to making presuppositions clear.

The structuring of Popper's program gives it the clear virtue of enhanced assessability. It attempts to indicate clearly how we are to make inferences regarding lower-order norms and decisions, and it tries to bring out its own most basic valuational presuppositions and their role in the resulting systematization. Popper's is a far more *aggressive* strategy. If successful, it would provide an elegant, systematic justificatory account of science. Even given its shortcomings—or, as many contend, its failures—it provides significant advantages over Kuhn's account: for these partial or total failures are more clearly detectable and the lessons learned from them more evident. That an intuitively plausible attempt to clarify our concept of science fails leaves us with a gain in understanding of that concept in the process: what seemed workable did not work out, and hence something else must be tried. (Of course I am only assuming the worst for Popper's program, since its seeming advantages even in the face of failure seem to speak most strongly for the general strategy it embodies—a final, negative verdict on falsifiability is far from settled.)

I think it is fair to say that Popper sees in Kuhn's approach much the same overconcession to relativism that he and Lakatos decry in Duhem's conventionalist view of science (see, e.g., Popper, 1959, Ch. 4; Lakatos, 1970, pp. 104-5). That is, he fears (with Plato) that the matter of justification will simply be circumvented in the process. And I think he is correct: for Kuhn's program never seems to attempt, much less complete, any concrete *task* of justifying. Even with Hempel's supplementary remarks, Kuhn's position still seems to suffer substantially the same afflictions noted at the outset. We are directed to collect historical data regarding the criteria scientists actually use to guide their work, codifying these criteria in a limited way by attempting to formulate and order them to some degree. Indeed, this sort of limited ordering might seem to count as what Goodman regards as justification, except that Goodman's own work on the problem of confirmation builds on his own and others' serious and complex efforts to deepen our understanding of this aspect of science by seeking a relatively small set of fundamental principles on which a highly systematic codification is based. And it is not clear that codification without this sort of structuring can be regarded as justification at all: the "getting down to fundamentals" seems an important feature of our original notion of justification that is worth preserving in any reconstrual of it—and it is preserved in the albeit relativistic accounts of Goodman and Rawls.

Something seems evidently gained from Kuhn's inquiries, despite the absence of programmatic structure. The reason for this, I believe, is that Kuhn's work fits, albeit implicitly, into a guiding background program: the program, or set of programs, against which he is reacting—including that of Popper. We learn from Kuhn, I think, precisely insofar as his more careful look at the practice of science presents problems for the normative proposals of these earlier programs. As Hempel notes, this is a valuable antidote to philosophical tendencies to explicate the norms of science without attending closely to scientific practice: it is necessary to canvass the practice quite thoroughly if our reflective equilibrium account is to be ideally self-critical. But there seems to be a limit to the potential yield of this sort of negative critique, unless it is part of a continuing positive attempt to revise and replace older programs in light of their failures.

Kuhn, I suspect, has too easily abandoned the hope of attaining a systematic justification of science that might promise to make substantial positive improvements in our understanding and appraisal of science. It may well turn out that a lot of what scientists do resists codification, or rests on "tacit knowledge," and is beyond justification (cf. ftn. 4). However, we want to be sure that this is a genuine result, and not the mere consequence of having ceased to try.

In addition to valuing the assessability that structure brings to justification, I believe Popper also fears the loss of a necessary degree of *independence* of our critical perspective of science from science itself. In this, again, I agree with him. But the degree and nature of critical independence is unclear given the pragmatic rejection of essentialism he and Kuhn both expound. The achievement of genuine critique by such a "circular" account is a fundamental problem that any "bootstrap" approach to justification and rationality must confront. The circle of justification in which Kuhn moves seems too small, but just what is to be gained by Popper's attempts to widen the circle? The gain, I think, comes to this: Popper's approach sets science in a wider context of intuitive rational and ethical valuations: he makes it clear that certain background assumptions, albeit themselves not fully systematized or clear, are being made, and that science is being made answerable to them in metascience. This appeal to the wider background of rationality is also evident in Hempel's criticisms of Kuhn's metascientific account—in his insistence that a causal connection must be shown between norms and rational deliberation if the norms are properly to be regarded as rational guideposts. Also, behind the Goodmanian notion of justification as codification we saw the rational valuation of a more systematic and efficient codified practice over its uncodified ancestors.

This general background of value judgments regarding rationality is important for assessing questions of justification, and I believe the value of thus widening the circle can be seen in the light of the reflective equilibrium account outlined above. These norms to which the metascientist appeals are not regarded as fixed, but instead as a general, evolving normative background which, despite its shifting character, is still our "ultimate arbiter" of justification. And just as the basic aims of science are clarified by attempted codifications, successful or unsuccessful, tested against this background, so our notions of rationality are improved by our attempts to pull ourselves up by a variety of bootstraps. The task of showing the rationality of scientific practice is inextricably interrelated with the task of learning what rationality is; in part, from science. The best one can ask for on this pragmatic view of justification is an increasingly more systematic understanding of the important relationships among elements of this background and the various areas of rational inquiry: science, theories of value and rational choice, and so on. Each individual area is better understood, and better rationally justified, by its increasingly more systematic interrelation with this background and with other attempted formalizations.

In this general context the need for structure seems just as important. Without some effort to systematize and test, there seems no evident way to make much of a gain over what we start with: namely, a rough set of intuitions about rationality, proper scientific procedure, and so on, and a rough idea of how to order practice in accordance with them. Moreover, one wants to perceive as clearly as possible the relationships between our judgments of scientific rationality and other contexts in which parallel judgments about the rationality of other sorts of behavior are made: for some of these judgments broach ethical concerns that seem well outside the immediate context of evaluating science, and yet are such that the input of the lessons learned from science cannot be overlooked. These lessons play an important role in the attainment of a broader understanding of human rationality and in the clarification of norms that govern other important psychological, legal, and political contexts—but given this, the scientific input cannot be too uncritically accepted.

Notes

1. The present context is not affected, I believe, by whether one takes systems, theories, or hypotheses as the entities subject to falsification.

2. Popper and others also challenge Kuhn's account of normal science, arguing that it is more revolutionary than Kuhn claims. I shall not concern myself here with these and other albeit important matters of historical fact, as my purpose is instead to inquire as to how certain facts come to be regarded as salient in the first place. It seems to me that disputes about the adequacy of a historical analysis make sense only after the (perhaps conflicting) criteria of selection have been clarified.

3. Popper's earlier contention that evolutionary biology is unfalsifiable and hence metaphysical has been regarded by critics as a serious weakness for the falsifiability criterion construed even as stating only a necessary condition. However, Popper has recently changed his views, and now contends that evolutionary biology is falsifiable. See his "Natural Selection and the Emergence of Mind," *Dialectica* 32 (1978): 339-355.

4. Hempel (1981, p. 402) limits the applicability of Goodman's account only to cases in which "fairly precise rules can be stated that are acknowledged and observed in scientific practice." This would make justifiable, in this sense, only "fairly narrow and specific inductive problems, such as measuring a quantity, or testing a statistical hypothesis . . ." However, I see no evident reason thus to limit ourselves. All that seems necessary are some at least roughly formulable principles that are offered to systematize an already existing judgmental practice (which may or may not already comprise stated rules on the basis of which presystematic judgments are made). If justification consists in a "fit" between principles and intuitions and in an improved practice, why should such justification presuppose *precise* formulation? However, Hempel's point can be granted, I think, without severely undermining hopes that the Goodmanian view can be applied to the broader contexts of justification considered in this paper. For if one demands, with Hempel, sufficient *explication* prior to justification, such explication proceeds in the very same reflective fashion on Goodman's account. Indeed, as Goodman points out: "The interplay . . . between rules and particular inductive inferences is simply an instance of this characteristic

dual adjustment between definition and usage, whereby the usage informs the definition, which in turn guides extension of the usage." (1973, p.66)

5. Rawls (1971), pp. 17-22; cf. Goodman (1973), pp. 62-67. The background of this view is most fully developed by Morton White in *Toward Reunion in Philosophy* (Cambridge: Harvard University Press, 1956), see especially pp. 254-268. In essence it is the extension to normative contexts of Quine-Duhemian epistemology—an epistemology, by the way, that Popper generally accepts, but whose relativistic implications he tries to combat in the manner indicated in this essay (see the concluding section). An excellent account of Goodman's views on justification may be found in Richard Rudner's article on Goodman in *The Encyclopedia of Philosophy*, vol. 3, ed. Paul Edwards (New York: Macmillan, 1967), pp. 370-371.

6. The intuitive judgments are "presystematic" in a logical rather than a temporal sense—as they continue to function in the assessment of a system after it has been initially developed.

7. Popper's primary objective is to explicate the notions of "testability" and "empirical content" in terms of falsifiability (1959, pp. 92, 113), that is, by determining the magnitude of these features in terms of the classes of "basic statements" that constitute "potential falsifiers" (I shall pass by the weighty technical difficulties involved here).

8. See especially "The Aim of Science" in Popper (1972).

9. Indeed, Popper's falsifiability criterion seems to mediate the background concept of justice with particular evaluative case judgments in much the same way that Rawls's principles of justice mediate our concept of justice with presystematic case judgments regarding the proper conduct of institutions and practices. In both cases there is the same generation of systematic first principles from background concept as well as the testing of the adequacy of the resulting systematizations by presystematic judgments in the light of which the principles are always regarded as subject to revision. And both programs draw similar benefits in virtue of this structuring. See Rawls (1971), especially pp. 11-22, 578-587; see also Rawls's "Justice as Fairness," *Philosophical Review* 65 (2) (April 1958) and "Outline for a Decision Procedure for Ethics," *Philosophical Review* 1 (1951).

10. The imposition of normative structure is not without its dangers. It is possible to be *too* bold in making conjectures before having adequately attended to the complexities of scientific practice. Popper may well be guilty of this, and we must regard with caution his attempt to lay down a supreme rule from the start (1959, p. 54). Among historically oriented philosophers of science there is concern that philosophical programs of this sort are fundamentally wrongheaded because they are initially launched from too fragmented an understanding of science, with the result that philosophical explications of, say, simplicity are in their view too belabored and artificial. Many doubt that any scientific utility can be served by current philosophical endeavors to codify or justify scientific practice.

Thus there appears to be a growing divergence—manifest in the debate between Popper and Kuhn—between two schools of thought in philosophy of science: one confronting the traditional problems of the logical empiricists, and the other abandoning much of the other's results and directing most of its energy to close historical study. By way of rapprochement, I should like to make the following two points.

(1) I do not see that the major advantages of an aggressive and self-critical learning strategy hinge on its boldness or grandness. If there is too much distance between the comprehensive programs of a Popper or Lakatos and actual scientific practice, or if it is the case that too little of scientific practice is understood by anybody to

enable the derivation of a relatively small set of governing criteria pertinent to *all* scientific inquiry, there still seems to be an important place for at least limited conjectures aimed at more modest justification and understanding of certain aspects of scientific practice. (See, e.g., Shapere's "The Character of Scientific Change," in *Scientific Discovery, Logic, and Rationality*, ed. T. Nickles (Dordrecht, Holland: D. Reidel, 1980), and "What Can the Theory of Knowledge Learn from the History of Knowledge?" *The Monist* 60 (1977): 488-507. His notion of progressive understanding of science through a "rational feedback mechanism" seems to conform to the general reflective equilibrium idea.)

(2) I do not believe the value of philosophical codifications of science rests fundamentally on their current or long-run scientific utility. For even where it is of little use to scientists to grasp or value such systematizations, there remains the other side of the reflective equilibrium dynamic: that is, the improved understanding and accounting of the structure and aims of science that those with a philosophical interest in metascience can achieve. (This group could include the philosophically motivated scientist, though he might not thus be thinking qua scientist.) Given that I have chosen to exploit certain parallels in Rawls, this point can be missed: for although an improved reflective understanding of justice is one of Rawls's goals, the improved governance of practice, the clear understanding of his principles of justice by those they govern, are crucially important features in his contractarian account. The scientific codification, on the other hand, might have little use for the "governed"—but these formulations and systematizations (as well as those of rational choice) can serve other legitimate philosophical objectives.

11. I view the backgrounds of rational, ethical, and aesthetic valuation as inextricably interrelated. However, I shall not broach this issue here.

References

Goodman, Nelson. *Fact, Fiction, and Forecast*. 3rd ed. Indianapolis, Ind.: Bobbs-Merrill, 1973.

Hempel, Carl G. "Turns in the Evolution of the Problem of Induction." *Synthese*, 46 (1981): 389-404.

Kuhn, Thomas S. *The Essential Tension: Selected Studies in Scientific Tradition and Change*. Chicago: University of Chicago Press, 1977.

Lakatos, Imre, "Falsification and the Methodology of Scientific Research Programmes," in Lakatos and Musgrave, 1970. Portions of this piece appear as "The Popperian versus the Kuhnian Research Programme," in this collection.

Lakatos, Imre, and Musgrave, Alan, eds. *Criticism and the Growth of Knowledge*. New York: Cambridge University Press, 1970.

Popper, Karl R. *The Logic of Scientific Discovery*. New York: Basic Books, 1959. (Originally published, 1934.)

Popper, Karl R. *Conjectures and Refutations: The Growth of Scientific Knowledge*. London: Routledge & Kegan Paul, 1963.

Popper, Karl R. *Objective Knowledge: An Evolutionary Approach*. Oxford: Clarendon Press, 1972.

Rawls, John. *A Theory of Justice*. Cambridge: Harvard University Press, 1971.

The Conservatism of "Pseudoscience"

Roger Cooter

For some time historians have recognized that the driving underground of hermeticism, alchemy and magic in the seventeenth century was integral to the process of rationalizing and legitimating the then newly-altered social order and its intellectual controls. Only recently, however, has it begun to be realized just how "socially imposed and self-consciously accepted" was the distinction between "correct" and "incorrect" knowledge made by the second generation in the Scientific Revolution, and how effectively that distinction served to obfuscate the social and ideological origins of modern science and its methodology.[1] As we are now beginning to understand, it was chiefly through the drawing of that distinction that the new way of knowing nature was able to establish itself as a universal and value-transcendent touchstone of truth, reason and rationality.

This paper, extending the discussion that has taken place on the role of pseudoscience in the seventeenth century, aims at formulating a general thesis on the role of pseudoscience in modern society. By "role" what is meant here is not the significance of a particular pseudoscience for those involved with it (although that will be partly the concern when we turn to the historical focus of this paper, the classic Victorian pseudoscience of phrenology). "Role" here refers, rather, to the function that the appellation "pseudoscience" serves for those who deploy the term. And the thesis I wish to put forward is simply this: that at least since the beginning of the consolidation and ossification of the capitalist order in the seventeenth century, the label "pseudoscience" (or the appropriate synonym) has played an ideologically conservative and morally prescriptive social role in the interests

A longer form of this paper appears as "Deploying Pseudoscience: Then and Now," in *Science, Pseudoscience, and Society*, ed. Marsha P. Hanen, Margaret J. Osler, and Robert G. Weyant (Waterloo, Ontario: Calgary Institute for the Humanities and Wilfrid Laurier University Press, 1980). Used with permission.

of that order. Further, I want to suggest that whenever and wherever "pseudoscience" is labelled, restraint is placed upon the creation of reality.

A convenient starting point is the definition of pseudoscience offered by Jeffrey Blum in *Pseudoscience and Mental Ability: The Origins and Fallacies of the IQ Controversy*. For the most part, this definition is characteristic of the way the term is deployed by today's liberal-minded. Pseudoscience, he says, is the "[p]rocess of false persuasion by scientific pretense." Its existence is dependent on the joint occurrence of (1) grossly inadequate attempts at verification and (2) the successful dissemination of the unwarranted conclusions drawn from those attempts at verification. "The label 'pseudoscience,' " he continues,

> becomes pertinent when the bias displayed by scientists reaches such extraordinary proportions that their relentless pursuit of verification leads them to commit major errors of reasoning. Simple tautological fallacies, a refusal to consider obvious alternative explanations, and the deliberate or unconscious falsification of data constitute some of the basic features of pseudoscience. [2]

Blum also tells us how pseudosciences work: usually they are associated

> with individual cranks working in isolation to promote their eccentric theories. Most instances of pseudoscience conform to this description and involve nothing more than the development of an idiosyncratic cult. When social circumstances are favorable, however, the perpetrators may acquire sufficient resources and respectability that they appear to the society generally as legitimate scientists rather than cranks. [3]

The obvious historical inadequacies of this use of "pseudoscience" scarcely need dwelling upon. As it is difficult to regard Newton as a crank because of his study of alchemy, so, as Barry Barnes has demonstrated, it is difficult to regard Arthur Jensen's scientific work on heredity as any less a product of a disinterested search after truth than that of Jensen's scientific opponents W. Bodmer and L. Cavalli-Sforza. [4] As for the isolation of the "crank," while it is hardly the case, at least before the second half of the nineteenth century, that anyone in scientific discovery worked in close professional association, [5] on the other hand it can no more be said of the "crank" than of the orthodox scientist that he or she operates in a social or cultural vacuum. Nor do I think that with this example of the use of the label "pseudoscience" we need to deliberate any further on its consistently perjorative nature. It is rather to the unstated ideological implications of the deployment of the label that I would like to turn.

Although recent debate on the issue of ideology and science has been stormy, complex and highly politicized, a number of gains have been made through it. One of these is the overcoming of much of the former difficulty of agreeing on what is meant by the use of the term "ideology." Historians of science of many different political persuasions would now generally agree that by "ideology" we mean the partial view of nature and human nature expressed by a group or class which informs perception and conceptualization. It is not meant that ideology must be false nor that the reality of nature is necessarily distorted through ideology. The usage is Marxist-informed but it is non-perjorative and non-epiphenomenal, denoting simply the social, political, metaphysical, and theological or philosophical superstructure that must accompany every economic system.[6] According to this view, science as an intellectual formation and ideology as worldview can never be separate realities or autonomous "things" merely interacting, but must always be mutually constitutive of each other or interpenetrating to form a seamless web.

For many commentators on pseudoscience, however, ideology retains the implication of distortion and falsification: "ideology" is the external social and political thing that lurks in the shadows and infiltrates objectivity; it is what can render science into pseudoscience and what is held to distinguish science from "scientism."[7] This narrow and trivial conception of ideology is precisely that which serves to deflect attention from the real ideological function of Blum's and others' exposés of pseudoscience, for in deploying this conception of ideology they conceal what is implicit in their exposés: the belief that there actually is out there somewhere a real, objective (nonideological) body of truth called Science. By dogmatically asserting the existence of an imagined opposite—truncated "ideological" pseudoscience—one judges as one names and obscures the fact that modern science itself was and remains the product of men in the activity of historically specific social relations. As Lukács pointed out as long ago as 1923, modern science and its methodology were the chief instruments in reifying, and hence making to appear as objective, the social relations upon which our science was reared.[8]

When armed with this narrow view of ideology the student of pseudoscience is restricted from entering into any kind of relativist perspective on scientific truth, such as that provided by ethnomethodologists of science.[9] Completely foreign to Blum is the latter's Wittgensteinian interest in showing how scientific meaning and conceptualization are dependent upon the social context (and political interests) in which they are embedded. Denied access to the perception of scientific truths as cultural artifacts socially negotiated and organized in specific milieus, commentators such as Blum merely take "correct science" as the subject to explain "false science." Hence they commit themselves not to a sociology of knowledge, but to a sociology of error.

Of all so-called pseudosciences, none perhaps has committed so many to a sociology of error as phrenology, for none has been more frequently relied upon to elaborate the difference between true and false science. Yet no body of scientific knowledge is perhaps better suited to lead one into a sociological understanding of knowledge than is phrenology, for when its history is considered non-teleologically and non-evaluatively, no body of scientific thought so readily permits the labelling "true" and "false" to be revealed as socially negotiated. For both these reasons phrenology lends itself to my cautionary tale on the deployment of the term "pseudoscience." For these reasons too, it is appropriate to broach the subject here by bouncing off from its actual history its received historiography.

Until very recently most historical accounts of phrenology were written within the positivist use/abuse framework. [10] The aim, intentionally or unintentionally, was not the writing of history, but the legitimating of the wisdom of value-free modern science. Consequently these accounts have given scientistic capital to dozens of other attacks on perceived scientifically-related bogies. Karl Dallenbach, for instance, in an article in the *American Journal of Psychology* in 1955 sought to deride the scientific pretensions of psychoanalysis by likening it to phrenology, and the same path has more recently been trod by the Nobel scientist turned literateur P.B. Medawar.[11] In his essay on the IQ racket, the neurobiologist Steven Rose refers to Hans Eysenck as operating in "the manner of a nineteenth-century phrenologist" as if this comparison utterly vanquished Eysenck's credibility. [12] In fact it invalidates nothing, since Eysenck himself in his popular works on the "facts and fictions in psychology" justifies his scientificity by a similar contrast with phrenology. [13]

The positivist motives of both the left and the right in so deploying phrenology are patently obvious, and it has been partly because of this that revisionist historians of society and medicine have felt a strong need to rescue the subject. [14] These historians have performed a good service; they have shown (much in the manner of Darnton's splendid study of mesmerism in France) [15] that in its early-Victorian social context phrenology was very far from being the outlandish nonsense that Whig historiography implied through its teleological evaluation of the past from the present. By observing the interest in phrenology of countless illustrious contemporaries from A. R. Wallace, Robert Chambers and Samuel Smiles, to Comte, Spencer and Marx, they have shown that phrenology's impact and significance at both popular and intellectual levels was far deeper than had been previously assumed. [16]

Yet, paradoxical as it may seem, it cannot be simply inferred from this that phrenology was therefore scientifically legitimate in its context. Since the contextual evidence for phrenology being regarded in Britain as "a sick man's dream" [17] is in fact overwhelming, the Whig historians of phrenology were perfectly justified in calling phrenology pseudoscientific. From the moment of its first appearance in Britain, Gall's theory of brain and science of character were regarded as physiologically, craniologically and philosophically visionary as well as socially, morally and theologically dangerous. Thomas Brown, in a well-known review of the theory in the *Edinburgh Review* in 1803, set a precedent for many far more passionate commentators with his conclusion that it was an unconvincing "species of physiognomy" unjustified by anatomy. Before 1820 phrenology was regarded with disdain in nearly every medical and literary journal while in popular prose and poetry, theatricals and cartoons it was satirically lambasted. And this was far from being just an initial British reaction to a Franco-Germanic theory. A sustained stream of irreproachable evidence rationally deducing phrenology's scientific failing flowed from Britain's most distinguished contemporary scientists and philosophers. Though nationalism was not absent from these critiques, it was not essential to them and was not in fact much relied upon. More typical was the dispassionate critique such as that presented by the physiologist Peter Mark Roget in his article on "Craniology" written in 1818 and published in the sixth edition (1824) of the *Encyclopedia Britannica*. In this, the case against phrenology was divested of emotional anti-materialism and anti-fatalism and was conducted on the basis of supposedly sound, neutral physiological evidence. Gall's basic physiological premises and his cerebral anatomy were even praised, while his theory of cerebral localization and the attendant craniology were torn to shreds.

By 1842, when Roget's article was republished unaltered in the seventh edition of the *Britannica* under the new title of "Phrenology," the scientific and scholarly backing for its opinions had been impressively expanded. In books, in articles in leading scientific and literary journals, dictionaries and encyclopedias, and in lectures delivered in Royal Societies, the doctrine was attacked in the 1820s and '30s by such eminent authorities as Dugald Stewart, Charles Bell, Sir Everard Home, Francis Jeffrey, Sir William Hamilton, James Cowles Pritchard, John Bostock and James Copland among many others. And, as if these attacks weren't enough, at the same time came the well-substantiated opposition to the doctrine by the French physiologists Francois Magendie and J.P.M. Flourens, the Berlin physiologist Karl Rudolphi, and the professor of physiology at Columbia University, Thomas Sewall. Whig historians, then, however misguided in their motives, were thus quite correct to call phrenology "pseudoscientific" in the sense of it being "discredited knowledge" in its own time. Their "only" mistake was to suppose that because of this phrenology must lack serious historical worth.

Just how wrong this supposition is becomes manifest as we swing round to what the revisionists essay, although it should be added that the revisionists neither understand nor appreciate the implications of the Whigs' error. Their primary intention after all is with the writing of accurate history, not with the lessons or the sociological insights to be derived from it. This aside, what they have amply proven is that while the highly-qualified opposition to phrenology was being mounted, the phrenology movement was advancing by leaps and bounds. There is not space here to spell out the many facets of this movement, nor is it so necessary given the many recent publications. Suffice it to say that the institutionalization of phrenology was significant both in the narrow sense of formal organizations at national and local levels and in the sociological sense of establishing a knowledge/behavior base for standardized collective reference and action.[18]

For present purposes it is important to note only that the intellectual leadership of this movement was heavily weighted with medical men and that it was they who were most prominent in popularly disseminating phrenology among the self-styled "thinking classes" between the 1820s and 1940s. One finds, for example, that in the national Phrenological Association (which was formed belatedly in 1838), of the 218 members for whom we have occupations, the largest proportion—33%—were medical men. Local phrenological societies reflect a similar or greater preponderance of this professional group, and according to one phrenological statistician, at least a third of the one-hundred writers on phrenology in Britain that could be identified in 1838 were either physicians or surgeons. [19]

Given that phrenology was a physiological psychology, it is hardly surprising that much of the movement's early momentum should have come from the medical profession. Nor is it remarkable that by the late 1820s most leading medical spokesmen and educators were actively supporting the science along with the majority of medical journals (the *Lancet* chief among them). [20] After all, in undermining the Cartesian proscription on the scientific study of mind by uniting mind with neurology on the one hand and the biology of adaptation on the other, phrenology seemed to transform abstract metaphysical conceptions of mind into organic realities which were fully amenable to the understandings and practical interests of medical men. Moreover, phrenology's veracity seemed derived only through strict adherence to the empirical methodology of Bacon. This intellectualist explanation should not of course be taken as the full reason for the medical professions' support of phrenology; in fact this explanation is not really intellectualist at all, but compellingly social, since phrenology actually purported to maximize the competences of medical men to predict and control behavior. Here, however, it merely wants to be emphasized that in the context the reasons for supporting phrenology could be held to be entirely rational and scientific. Thus it is quite mistaken to transcend the historical context and brand the thought of antiphrenologists as "scientific" and that of the phrenologists as "scientistic." As suggested by the fact that it was Britain's most distinguished scientists and philosophers who conducted and mightily sustained the attack on phrenology as pseudoscience, antiphrenologists no less than phrenologists might be seen as having been anxious to defend interests above and beyond what normally would be called scientific.

What these interests were becomes clear when we socially map the participants.[21] In brief (and comparing here only the scientific-medical opponents and supporters of phrenology), what stands out in bold relief are the differences in age, economic and social security, religious affiliation and politicization. The supporters of phrenology were on the average twenty years younger than their opponents. They were socially marginal and economically insecure: few had upper middle class or aristocratic backgrounds, had had elitist educations or belonged to elitist institutions, as did many of the antiphrenologists; and while some managed to struggle into university positions during the period in which they advocated phrenology, these were few in number and were generally confined to the less prestigious postings. In terms of religion: as opposed to the antiphrenologists' High Church involvement and firm attachment to socially undisturbing Paleyian natural theology, virtually all of the supporters of phrenology were either avowed materialists, proponents of "the grander view" of the Creator (sometimes bordering on token deism), or were vocal dissenters. These positions on religion were at the same time polticial postures for which party labels serve as insufficient guides. For the most part the supporters of phrenology can be seen as Radical partisans, while the antiphrenologists can be seen as having a more conservative attachment to Whig reformism.

The fuzziness of this last distinction reminds us, however, that the social and ideological conflict between phrenologists and antiphrenologists cannot be reduced simply to one of class nor exclusive cosmologies and worldviews. To refer to exclusive cosmologies might imply that the phrenologists were providing the intellectual rationales for social structures and relations totally different from those of the antiphrenologists. But this was not quite the case, even though between the two groups elements of a genuine conflict between aristocratic and bourgeois interests can on occasion be isolated. The antiphrenologists for the most part shared with the phrenologists an optimistic uniformitarian faith in gradual progress, in social homogeneity, stability, and hierarchical order, and shared with the phrenologists a non-literalism in Biblical matters and a view of a lawful rational universe presided over by a benign Deity.[22] Thus, although at first sight the reaction to phrenology by the orthodox scientific-philosophical community (especially in Edinburgh) bears a striking resemblance to the reception given to Velikovsky's ideas by our own scientific community,[23] in actuality the outlook of phrenologists and antiphrenologists was very far from being parallel to that of today's catastrophists and uniformitarians. Though the conflict involved metaphysical and metaphorical sound and fury, it is an over-simplification to say that worldviews were in collision. The history of the conflict is far more subtle, involving in part the rhetorical build-up of caricatures which were then used by marginal men offensively and defensively for the rationalizations of the industrial secular form of bourgeois capitalism. These caricatures distorted and polarized the views of phrenologists and antiphrenologists, making the phrenologists appear as progressives for an open society allied to the mobile fortunes of commerce, and the antiphrenologists appear as the retainers of a static and privileged aristocratic society allied to inherited agrarian wealth. What in most cases were differences between different factions of the same class over *degrees* of social re-ordering and the pace of social change were, through the dialectical spiral of the phrenology debates in the context of the further emergence of urban industrial society, construed as competing class views over *kinds* of social orders. The point is historically important but is not of central concern to us here, for caricature or not, the consequence of entering into debate on phrenology was the same: the impossibility of any participant remaining politically, philosophically, or theologically detached. No matter whether the advocacy or denunciation was polemical or factually dispassionate, that advocacy could never be intellectually autonomous or be *just* scientific and medical. [24]

Once we disengage phrenology from the clutches of the use/abuse model and perceive it in this way as a component in a social and ideological strategy, we can identify the conservatism in the contemporary labelling of it as "pseudo." The labelling can be called "conservative" not because at a political level the phrenologists happened to be more radical, but more fundamentally, because those in or identifying with the scientific and philosophical elite who deployed the label were seeking to discredit what they saw to varying degrees as socially threatening and, by this process, to legitimate and defend certain human and social interests and the neutrality of the knowledge that served them. Would space permit, the thesis could be amplified by revealing how phrenologists themselves attempted to separate a "pure" version of their science from a purported "corrupted" ("pseudoscientific") version once *their* sociopolitical interests had become institutionalized and required protection from below. Even without this elaboration, however, it should be clear that it is the basic defense of interests, rather than the specific political color of those interests that warrants the use of the term "conservative."

To sum up: I've been arguing that the deployment of the label "pseudoscience" always acts to conserve because, almost by definition it seems, the label appertains to perceived forms of sociopolitical deviance or to what at any moment in history may be deemed morally illegitimate.

More specifically, I have argued that the deployment of the label "pseudoscience" acts conservatively because, unwittingly, it legitimates the partial view of reality that is mystified and mediated through science. To identify and attack "pseudoscience" is both to support the reproduction of the capitalist structures and relations that emerged constitutively with modern science and to reproduce the social process of concealment originally conducted in the seventeenth century; isolation and exploitation of "pseudoscience." Through the deployment of "pseudoscience" the partial view of reality asserted by modern science remains thoroughly mystified under the socially-constructed metaphysic of "truth" and "certainty."

Because all post–seventeenth century attacks on "pseudoscience" must similarly conserve and protect the ideology embedded in science, it does not matter whether the deployment comes from the left or from the right or whether the social movements which frequently toss up the stipulation "pseudoscience" have explicitly or implicitly to conserve a great deal or very little. Although it has been necessary here to compress and abridge, the sociopolitical vicissitudes of phrenology's history illustrate how new demands upon the bourgeois order always call forth new deployments of "pseudoscience" which work overtly and covertly in the interest of conserving that order.

Since all knowledge of external nature is made by men and is socially constructed, the identification and criticism of any particular body of knowledge as "pseudoscientific" must count as a defense of some other body of knowledge. In the case of phrenology, as in the contemporary situation with IQ testing, claims and counter-claims of "pseudoscientificity" (i.e. claims and counter-claims of the social infiltration of the other's knowledge) constitute the defense of human and social interests in the name of an "objective world" (however politically or otherwise defined). To label "pseudoscience" in the interest of creating this "objective world" must be seen as deeply conservative, for it is a constraint on man's creative activity—a corset on any alternative view of man making, not only his own history, but also his own reality.

Since bodies of scientific knowledge are thus criticized by identifying the presence of social interests in them, and since that identification is essential to stipulating that certain bodies of knowledge are not knowledge at all, but are "ideology," "pseudoscience," "error," etc.,[25] it follows that the study of "pseudoscience" has value not for its own sake nor for the sake of legitimating as "objective" what is actually political in modern science, but for the sake of revealing more about the science and society that negotiationally *defines* pseudoscience through its interactions with it—both past and present. It is only through this kind of understanding that we can hope to remove the iniquities that are falsely and misleadingly attributed to "pseudoscience." Speaking practically, insofar as the historical aim is description rather than prescription it would be preferable to have the term "pseudoscience" replaced in our vocabularies with something like "unorthodox science" or "non-establishment science," for this would more correctly imply relative differences in scientific social perception rather than categorical methodological opposites or matters of truth and falsehood. If we must use the label "pseudoscience" than let us at least acknowledge that that usage, whether from the left or from the right, is historiographically as ideological and covertly mystifying as its usage by any figure in history. For what does the focus on "bad" science accomplish, but the obfuscation of the ideological power of "good" science? The task before the social historian of "pseudoscience," surely, is not the reproduction of this essentially political process but its avoidance and exposure.

Notes

1. Everett Mendelsohn, "The Social Construction of Scientific Knowledge," in Mendelsohn, et al., eds., *The Social Production of Scientific Knowledge* (Dordrecht, Holland and Boston: D. Reidel, 1977), 3-26, esp. 17-20; see also David Dickson, "Science and Political Hegemony in the 17th Century," *Radical Science Journal*, 8 (1979), 7-37.

2. Jeffrey Blum, *Pseudoscience and Mental Ability: the origins and fallacies of the IQ controversy* (New York: Monthly Review Press, 1978), 12-13.

3. Ibid., 19.

4. Barnes, *Scientific Knowledge and Sociological Theory* (London: Rutledge, 1974), 134.

5. See Susan Cannon, "Professionalization," in her *Science in Culture: the early Victorian period* (New York: Dawson, 1978), 137-165.

To anticipate the critique that follows, note that Blum's defense of "real science," portrayed as collective enterprise, is a recapitulation of Bacon's ideal of scientific organization and is therefore, unbeknownst to Blum, a defense of capitalist practices and forms over alternative practices and forms. On the formal similarity between Bacon's scientific method and the capitalist labor process see Dickson op. cit. (note 1), 11-12.

6. Such a usage, derived from Althusser, may be found in Mary Hesse, "Criteria of Truth in Science and Theology," *Religious Studies*, 11 (1975), 396. My own usage is derived more from Henri Lefebvre, "Ideology and the Sociology of Knowledge," in his *The Sociology of Marx*, trans. Norbert Guterman (Harmondsworth: Penguin, 1966), 59-88 and reprinted in Janet L. Dolgin, D.S. Kemnitzer and D.M. Schneider, eds., *Symbolic Anthropology* (New York: Columbia University Press, 1977), 254-269; Clifford Geertz, "Ideology as a Cultural System," in David E. Apter, ed., *Ideology and Discontent* (Glencoe: Free Press, 1964), 47-76; and the writings of Bob Young, of which see: "Science Is Social Relations," *Radical Science Journal* 5 (1977): 65-129, which contains a useful bibliography of sources on science and ideology, including Young's several other works. See also Jorge Larrain, "Ideology and Science," in his *The Concept of Ideology* (London: Hutchinson, 1979), 172-211.

7. Throughout this paper "scientism" refers to the transformation of positivism (the natural philosophy that combines the traditions of empiricist and rationalist thought) into social philosophy on the basis of which society is explained and interpreted. Scientisms are extrapolations from well-established scientific domains to social domains, but since science itself is constructed under social conditions and subsumes these conditions, scientisms may be said only to accomplish overtly for social legitimation what sciences do covertly. See: Luke Hodgkin, "A Note on 'Scientism,' " *Radical Science Journal*, 5 (1977), 8; and David Dickson, "Technology and Social Reality," *Dialectical Anthropology*, 1 (1975), 34-37. Cf. the orthodox distinction between science and scientism in F.A. van Hayek, "Scientism and the Study of Society," in his *The Counter-Revolution of Science: studies on the abuse of reason*

(Glencoe: Free Press, 1955), 129–142; George Eastman, "Scientism in Science Education," *The Science Teacher*, 36 (1969), 19–22; and Robert B. Fisher, "Science and/or Scientism," *Science, Man and Society* (Philadelphia: W.B. Saunders, 1971), 43–44.

8. Georg Lukacs, *History and Class Consciousness* (1923; new ed. London: Merline, 1971), 7–11, 89–98 et passim. The relevant passages are cited in Gareth Stedman Jones, "The Marxism of the Early Lukács," *New Left Review*, 70 (1971), reprinted in Gareth Stedman Jones, ed., *Western Marxism: a critical reader* (London: New Left Books, 1977), 13–14. See also Robert M. Young, "The Historiographic and Ideological Contexts of the Nineteenth-Century Debate on Man's Place in Nature," in Young and Mikulas Teich, eds., *Changing Perspectives in the History of Science* (London: Heinemann, 1973), 398–399, 402, 405, 414, 430–434. (Note: reification should properly be understood as implying the process whereby the history of a thing is separated from its human origin and quality, i.e., as "misplaced concreteness" or the assuming of the objective entity existence of that which is actually "metaphysical.")

Lest there be any question, I am not saying that science is nothing but ideology, only that science is ideology as well as whatever else it is. If it is universally true that $E = mc^2$, it will remain so whatever the dominant ideology in any society, though it will have different meanings and uses.

9. See: David Bloor, "Wittgenstein and Mannheim on the Sociology of Mathematics," *Studies in the History and Philosophy of Science*, 4 (1973), 173–191; H.M. Collins and Graham Cox, "Recovering Relativity: did prophecy fail?" *Social Studies of Science* 6 (1976), 423–444; H.M. Collins, "The TEA Set: tacit knowledge and scientific networks," *Science Studies*, 4 (1974), 165–186; idem, "The Seven Sexes: a study in the sociology of phenomenon, or the replication of experiments in physics," *Sociology*, 9 (1975), 205–224. Particularly relevant applications of the "relativist" approach to the study of "pseudoscience" (though unfortunately unavailable to me during this paper's preparation) are H.M. Collins and T.J. Pinch, "The Construction of the Paranormal," and R.G. Dolby, "Reflections on Deviant Science," both in Roy Wallis, ed., *On the Margins of Science: the social construction of rejected knowledge*, Sociological Review Monographs (University of Keele: 1979).

I have been mostly informed in this paper by the historical applications of ethnomethodology in the studies by Barry Barnes and Steven Shapin. They have relied heavily on the insights of anthropology (especially that of Mary Douglas) to underline the fact that the question of truth and falsehood in science is neither here nor there; "truth," they argue, is relative to the social context of its use, and to question it historically is to avoid the more important problem of how scientific truth is constructed and how it functions in society. "Truth," as Mary Douglas reminds us in the words of Sir James Fraser, "is only the hypothesis which is found to work best." *Purity and Danger: an analysis of concepts of pollution and taboo* (London: Routledge, 1966), 24, quoted in Shapin, "Homo Phrenologicus: anthropological perspectives on an historical problem," in Barnes and Shapin, eds., *Natural Order: historical studies of scientific culture* (London/Beverly Hills: Sage, 1979), 44. For general criticism of the relativist position (on the grounds that not all truths are context-dependent because some a la Pareto's "residues" are universal and fundamental) see Steven Lukes, "On the Social Determination of Truth," and "Relativism: cognitive and moral," both in his *Essays in Social Theory* (London: Macmillan, 1977); idem, "Some Problems about Rationality," in Bryan R. Wilson, ed., *Rationality* (Oxford: Blackwell, 1970), 3–44, esp. at 39ff.

My own position is that what is argued as scientific "truth" or "certainty" is context-dependent (is invented, not discovered), but there is nevertheless a real world of non-ideological facts (external nature) outside one's consciousness (as Feuerbach argued against Hegelian Idealism). I agree with Marx, that "The human significance of nature only exists for social man" and therefore only through social change is the human significance of nature changed. (For relevant sources on this philosophical problem see Melvin Rader, *Marx's Interpretation of History* [New York: Oxford University Press, 1979], 108–109 and Alfred Sohn-Rethel, "Science as Alienated Consciousness," *Radical Science Journal*, 2/3 [1975], 80n. See also Raymond Firth, "The Sceptical Anthropologist?" in Maurice Bloch, ed., *Marxist Analysis and Social Anthropology* [London: Malaby Press, 1975], 29–60 [esp. at 31]; and Karl Figlio, "Chlorosis and Chronic Disease in Nineteenth-Century Britain: the social constitution of somatic illness in a capitalist society," *Social History*, 3 [1978], 168.)

10. For a review of the recent literature on phrenology and a fairly complete bibliography of past writings on the subject see my "Phrenology: the provocation of progress," *History of Science*, 14 (1976), 211–234.

11. Karl M. Dallenbach, "Phrenology versus Psychoanalysis," *American Journal of Psychology*, 68 (1955), 511–525; P.B. Medawar, "Further Comments on Psychoanalysis," in his *The Hope of Progress* (London: Methuen, 1972), 57–68.

12. Steven Rose, "Scientific Racism and Ideology: the IQ racket from Galton to Jensen," in Hilary and Steven Rose, eds., *The Political Economy of Science: ideology of/in the natural sciences* (London: Macmillan, 1976), 112–141, esp. 117–124.

13. Hans Jurgen Eysenck, *Fact and Fiction in Psychology* (Harmondsworth: Penguin, 1965) 130–131 (where the essay by Dallenbach cited above comes in for praise); *Sense and Nonsense in Psychology* (Harmondsworth: Penguin, 1958), 61 (which in the same breath links in mesmerism); and *Uses and Abuses of Psychology* (Harmondsworth: Penguin, 1954), 28–29.

14. See, in particular, David de Giustino, *Conquest of Mind: phrenology and Victorian social thought* (London: Croom Helm, 1975); John D. Davies, *Phrenology, Fad and Science: a 19th century American crusade* (New Haven: Yale University Press, 1955); Alastair Cameron Grant, "George Combe and His Circle: with particular reference to his relations with the United States of America" (Ph.D., Edinburgh, 1960); T.M. Parssinen, "Popular Science and Society: the phrenology movement in early-Victorian Britain," *Journal of Social History*, 7 (1974), 1–20; Angus McLaren, "Phrenology: medium and message," *Journal of Modern History*, 46 (1974), 86–97; Owsei Temkin, "Gall and the Phrenological Movement," *Bulletin of the History of Medicine*, 21 (1947), 2–15; and, from a different and deeper social perspective, the papers by Steven Shapin cited below (note 21).

15. Robert Darnton, *Mesmerism and the End of the Enlightenment in France* (New York: Schocken, 1970).

16. None of this is any longer in doubt and is strongly reinforced by the evidence given in my "The Cultural Meaning of Popular Science: phrenology and the organization of consent in nineteenth century Britain" (Ph.D., Cambridge, 1978). Because this study is forthcoming by Cambridge University Press, and because the point of this paper is not to marshal evidence but to argue on the basis of that evidence, I have in what follows avoided where possible detailed citations. The evidence upon which the rest of this paper is based is taken from the first two chapters of my study.

17. The dismissive phase of Bentham's: *The Works of Jeremy Bentham*, John Bowring, ed. (1843), vol. 7, 433–434.

18. The literature for the sociological sense of "institutionalization" is cited in Steven Shapin, "Social Uses of Science, 1660–1800," in G.S. Rousseau and Roy Porter, eds., *The Ferment of Knowledge: Studies in the Historiography of Eighteenth-Century Science* (Cambridge: Cambridge University Press, 1980), 93–139.

19. [Hewett Cotrell Watson], *Phrenological Journal*, 11 (1838), 263.

20. See my "Phrenology and British Alienists, c. 1825–1845," *Medical History*, 20 (1976), 1–21, 135–151.

21. Such a map, based on 28 major medical supporters of phrenology in Britain and 14 major scientific opponents, is discussed in Cooter, "Cultural Meaning of Popular Science," op. cit. (note 16), Ch. 1, pt. iv, with biographical details in Appendix A.

That the phrenology debates were about social and ideological interests was first made clear in Steven Shapin, "Phrenological Knowledge and the Social Structure of Early Nineteenth-Century Edinburgh," *Annals of Science*, 32 (1975), 219–243. Since then Shapin has moved beyond the prosopographical and rhetorical terrain of the phrenology debates and into their scientific substance to further illustrate the social and ideological strategies involved; see Shapin, "The Politics of Observation: cerebral anatomy and social interests in Edinburgh phrenology disputes," in Wallis, op. cit. (note 9), pp. 139–178. In "Homo Phrenologicus," op. cit. (note 9), with a heavy reliance on anthropological writings, Shapin has sought to work out "a social epistemology appropriate for the history of science" by ascertaining through the historical case of phrenology in the context of its use in Edinburgh the nature of the links between natural knowledge and social context; while in his " 'Merchants in Philosophy': the politics of an Edinburgh plan for the diffusion of science, 1832–1836," in J.B. Morrell and Ian Inkster, eds., *Metropolis and Province: British science 1780–1850* (London: Hutchinson, forthcoming), he endeavors to illuminate the political and logistic problems involved in the active diffusion of scientific ideas by a careful local study of the involvements of Edinburgh phrenologists in the Edinburgh Philosophical Association.

22. It is difficult to confine Roget (1779–1869), for example, to an aristocratic orbit. Though his uncle was Sir Samuel Romilly, and though he became a Fellow of the Royal Society in 1815 and was the secretary and editor of the Society's proceedings between 1827 and 1848, he was by association and marriage more closely allied to bourgeois mercantile sources of wealth and utilitarian philosophy. He is more representative of the members of the Manchester Literary and Philosophical Society or the Royal Institution, or the Society for the Diffusion of Useful Knowledge, of which he was, respectively, a member, a Governor, and a committee member. He was intimate with Bentham and other Philosophical Radicals; was Physician to the Manchester Infirmary in 1805 and later to the Northern Dispensary in London; and helped found the Manchester Medical School and the University at London. See D.L. Emblem, *Peter Mark Roget: the word and the man* (New York: Crowell, 1970).

23. See Robert Mcaulay, "Velikovsky and the Infrastructure of Science: the metaphysics of a close enounter," *Theory and Society*, 6 (1978), 313–342. Useful parallels can also be drawn from the literature on the early nineteenth-century debate in geology between catastrophists and uniformitarians; from that on the early twentieth-century disputes between the "biometricians" and "Mendelians" and between the Darwinians and Lamarckians; and from the literature on the current debate between hereditarians and environmentalists over IQ. See, respectively, Walter F. Cannon, "The Uniformitarian-Catastrophist Debate," *Isis*, 51 (1960), 38–55; Donald MacKenzie and Barry Barnes, "Scientific Judgement: the Biometry-Mendalism controversy," in Barnes and Shapin, op. cit. (note 9), 191–210; Arthur Koestler, *The Case*

of the Midwife Toad (London: Hutchinson, 1971); Jonathan Harwood, "The Race-Intelligence Controversy: a sociological approach," *Social Studies of Science*, 6 (1976), 369–394 and 7 (1977), 1–20; and idem, "Heredity, Environment, and the Legitimation of Social Police," in Barnes and Shapin, op. cit. (note 9), 231–251.

24. Cf. G.N. Cantor, "The Edinburgh Phrenology Debate: 1803–1828," *Annals of Science*, 32 (1975), 195–218 and idem, "A Critique of Shapin's Social Interpretation of the Edinburgh Phrenology Debate," ibid., 32 (1975), 245–256.

25. Shapin, "The Politics of Observation," op. cit. (note 21), p. 140.

Popper, Kuhn, Lakatos:
A Crisis of Modern Intellect

A controversy agitating philosophers of science has made that some-times narrow specialty a brilliant illumination of the crossroads to which modern intellect has come. The controversy has provided counter-culture anarchism on the one hand, and blind obedience to existing elites or experts on the other, with their most sophisticated rationales. It has also inspired what seems the most persuasive presentation possible of the rationalism that has characterized the West since Newton published his *Principia*. And in the end it seems likely to revive insights that were ancient before Copernicus was born.

The controversy began in 1962 with the publication of Thomas S. Kuhn's *The Structure of Scientific Revolutions*. That book made the autonomy or, as moderns typically construe that quality, the rationality of science a mat-ter of debate. It raised the question, "Do scientific laws and theories chiefly articulate the peculiar interests and extra-scientific beliefs of scientists or are they descriptions of reality based firmly on empirical fact?"

Non-Euclidean geometries, atomic theory, and Maxwell's field theory had disturbed the conventional modern view of science before 1900. Still, from Newton's time to the publication of Einstein's relativity paper of 1905 the question stated above would in most quarters have received a confident answer. That answer would have been: Unlike philosophy and theology, which deduce their conclusions from premises sanctioned only by author-ity or faith, science *induces* its laws and theories from empirical facts, and in consequence those laws and theories accurately describe reality. Einstein's paper, soon aided by the quanta hypotheses emanating from Max Planck's radiation studies, had on that long established confidence the effect of a large rock thrown into a quiet pool.

Excerpted with permission from *The Intercollegiate Review*, Spring, 1974, 99-110.

With Einstein's sharp modification or overthrow of what had seemed the most scrupulously induced and abundantly verified laws of all time, those of Newtonian physics, David Hume's simple but devastating criticism of laws induced from observed facts came once more into prominence. What reason, Hume had asked nearly two centuries earlier, had anyone to suppose that future observations would resemble past observations? Hence, what reason to place any trust in universal, predictive laws induced from observed facts?

Over generations, the accumulating success of Newtonian physics had seemed to Anglo-Saxon empiricists to make Hume's questions academic. Empiricists had completely failed to justify induction logically, but a triumphant science seemingly inducing its universal laws had appeared to guarantee induction's soundness. On the Continent, Kant had answered Hume with a far more knowledgeable conception of scientific method and laws than the empiricists' induction notion. But Kant's reply was so deeply implicated in Newtonian ideas of time and space, which Kant considered phenomenally final, that it fell with them. Within a decade, reflective people generally were wondering what scientific laws and theories really signified.

Concomitantly, an idea put forward somewhat earlier in France gained not a few adherents. According to the physicists Poincaré and Duhem, when science went beyond statements of empirical fact to construct laws and theories, these were simply "conventions," accepted for reasons in the main esthetic. Through many heads in the 1910s and 1920s there passed the unsettling thought that physical theory might really have no more claim to truth than metaphysics or theology, to which modern rationalists long had granted little claim at all.

No one of course denied modern science's contribution to such achievements as dynamos and X-rays, but just as Newtonian theory, now upset, had served to predict multitudes of planetary and other phenomena, so theories equally fallible might have served to produce dynamos and X-rays. In taking sun and star sights with their sextants, mariners were still assuming a geocentric universe and safely making port. In brief, it became apparent that a theory's truth, *i.e.*, its validity as a description of reality, could not be inferred from what was or seemed to be its practical success. Pretty quickly, reflective scientists and laymen retreated from the centuries-old belief that scientific laws and theories constituted an accumulating store of finally proven, settled truth.

The embarrassment thus occasioned showed itself not only in the popularity of French conventionalism but also in the acceptance by a majority of physicists, with Einstein and Planck notable exceptions, of the view of science called instrumentalism. This view totally surrendered science's claim to truth. It declared scientific laws and theories mere instruments for technological manipulation of nature, tools that no more described nature than a pipewrench describes a pipe.

Instrumentalism thus exalted as the goal and finest product of science the technology and hardware that ordinarily interested men like Einstein and Planck only to the extent that technology and hardware might assist in the pursuit of truth. Had instrumentalism endured as the accepted view of science, it might well have diverted from science just the kind of men who usually have been its greatest practitioners. For through the arcane mazes of mathematics and laboratory apparatus such men have, since Copernicus' time, pursued truth in a spirit often comparable to that of saints at prayer. Happily for science, instrumentalism and conventionalism soon faced an attractive rival. This received its most notable expression in Karl (now Sir Karl) Popper's *Logik der Forschung* of 1934, translated in 1959 as *The Logic of Scientific Discovery.*

Popper wanted to assert the authority of empirical observation against conventionalism, which implied that scientific theories were accepted mainly by agreement. He also wanted to relieve science of the embarrassment inflicted by the induction notion, an embarrassment intensified by Logical Positivism, which considered induction the hallmark of science.

Stated as it was generally understood, Popper's argument ran:

Induction could not possibly be justified, because, as elementary logic tells us, the singular statements we base on observation can never justify universal statements. No number of true reports of white swans can ever justify "All swans are white." Hence science's universal theories were not inductions at all; they were bold hypotheses which, unlike metaphysical hypotheses, entailed empirically testable consequences.[1] Tests could never finally prove theories true, since tests, too, were observations yielding only singular statements, but tests could often *falsify* theories, as one report of a black swan falsifies "All swans are white." Science therefore could and did progress toward truth by rejecting hypotheses that were falsified and by holding to those which the severest tests corroborated, meanwhile recognizing even these to be corrigible. Science did not accumulate a store of finally proven, settled truth, but by bold conjecture and self-criticism it could and did improve its theories' *approximation* to the truth.

Popper's verdict on induction (and Logical Positivism) could not rationally be challenged. At the same time, his conception of a science not absolutely true but both autonomous and rational, a science standing on logic and observation only, seemed satisfactorily to distinguish science from metaphysics and to justify belief in scientific progress. With Popper's *Logic*, modern rationalism had to some degree retreated—but to an apparently unassailable position. Falsificationism, as Popper's view came to be called, gave grounds for sober confidence in science as a purely logico-empirical and ever closer pursuit, though not final capture, of truth. Which brings us to 1962 and the heated controversy launched by Kuhn's *Structure of Scientific Revolutions.*

A professor of the history of science . . ., Kuhn declared that falsificationism described, not the strategy actually employed by scientists, but one suggested by science textbooks. These, he wrote, pictured early science as a continuous effort toward modern science. Actually, he went on, the long history of mature sciences like astronomy or physics presented us with long periods of "normal science" wherein basic theories or, as he also called them, paradigms, like Ptolemy's geocentrism or Newton's theory of gravitation were contentedly accepted. While "protosciences" like psychology or sociology persisted in debating basic theory, agreement on basic theory typified mature sciences and enabled them to press forward into esoteric depths and details, aiming to bring additional areas of experience within the basic theories' compass. And here Kuhn advanced a major point:

In the resulting "normal" effort, which Kuhn called "puzzle-solving," scientists actually tested only their own ingenuity, not their basic theory.[2] For examples, Kuhn cited Ptolemaic astronomers' ingenious eccentrics and epicycles, which for centuries had reconciled Ptolemaic geocentrism to observed planetary movements. Conversely, he pointed out that Newtonian astronomers had long recognized the perihelion of Mercury as an anomaly within Newtonian theory, but had taken it as a reflection on their puzzle-solving ingenuity rather than as a falsification of the theory. Instead of promptly rejecting basic theories that were contradicted by observation, as falsificationism envisioned, scientists normally displayed an ingenious loyalty to such theories.

Only, Kuhn continued, in periods of "extraordinary science" leading to scientific revolutions did the basic theories of mature sciences incur the criticism that Popper had taken as the distinctive mark of science generally. Yet even here, Kuhn declared, falsification did not occur. This because the new paradigms that then emerged differed so radically from the old ones as to be incommensurable with them.

Those astronomers, Kuhn wrote, who had called Copernicus insane for saying that the earth moved "were not either just wrong or quite wrong," because by "earth" they meant a stationary body.[3] The earth *they* used in puzzle-solving *could* not move. In Kuhn's view, Copernicus' and other novel theories not only changed the meanings of words but in effect changed the world that presented itself for theoretical interpretation. Consequently, old theories were not in any real sense falsified; a historian could not find "a point at which resistance [to a new theory] becomes either illogical or unscientific."[4] Instead, when scientists in a given science went over to a revolutionary basic theory they did so via a process not fully explicable in terms of observation and logic. The process was in part arbitrary and akin to religious conversion.

Kuhn's well documented book portrayed scientists as singleminded puzzle addicts, normally contented with whatever basic theory set up as solvable puzzles, and no more falsifying such theories than they proved them. Implicitly denying the autonomy of science, Kuhn announced that the explanation for scientists' choice of basic theories lay in the realm of sociology and individual psychology. Adding fuel to his fire, he suggested that the idea that successive theories carried science nearer to the truth might have to be given up.

Once more then, science, the beacon light of modern intellect, threatened to emerge as the stuff of dreams—or as a set of directions for puzzle-solving. Hydrogen bombs and voyages to the moon remained and could not be doubted. But, once more, technological results cannot prove the truth of theories, and little thought of utilitarian ends had sustained the labor of the great architects of modern science. Rightly or wrongly, Copernicus, Kepler, Galileo, and their later peers had thought themselves in pursuit of a true description of reality. Most of them had plainly said so. Moreover, laymen had come to accord scientific theory much of the awed respect once given the pronouncements of prophets, oracles, and priests.

It should be evident, and Kuhn himself has come close to remarking it, that his argument projected into the field of science the historicism that after World War I began to relativize modern views of politics, art, and morals.[5] Suspecting moral judgments like those too easily made by both sides in that war, historicist historians resolved to reject "all standards outside the object," *i.e.*, outside historical facts themselves, and to confine attention to facts alone. This positivistic rejection of philosophy and its standards inevitably made all men, ideas, and movements incomparable with each other, in Kuhn's phrase, incommensurable. Within historicism, not only moral but intellectual judgments became impossible. Everything was viewed on its own terms as no more than a product of its place and time.[6] Kuhn viewed scientific theories in the same way and seemed to urge an equally unlimited relativism in science.

Kuhn's theses raised a host of issues in the philosophy of science and occasioned more than one formal symposium. Foremost among these was one held in London in 1965 and chaired by Popper. There, such luminaries in the field as Stephen Toulmin, John Watkins, Imre Lakatos, Paul Feyerabend, and others, as well as Popper and Kuhn themselves, presented papers.[7]

Feyerabend, . . . whose command of much modern physics has not prevented him from embracing notions typical of the counterculture, explicitly welcomed the unlimited relativism which Kuhn had seemed to friend and foe alike to endorse. Mankind's progress, wrote Feyerabend, easily assuming such progress's reality, results from allowing everyone to follow his own inclinations, and now in science as elsewhere that happy course is justified; nothing compels assent, hence everything deserves respect, even "the most outlandish products of the human brain." Formerly considered an austere and demanding mistress, science now stood revealed as "an attractive and yielding courtesan who tries to anticipate every wish of her lover."[8]

In brief but adamant opposition to that, Popper's paper condemned Kuhnian relativism as part of a general relativism imperilling modern civilization. Lakatos observed that Kuhn, by spreading the impression that even in science no way existed to judge a theory except by the number and vocal energy of its supporters, had endorsed the credo of student revolutionaries, that truth lies in power. Popper and Lakatos clearly saw themselves as rational men confronted by a socially dangerous sophistry.

Now science's long immunity to historicist relativism has rested on science's presumed possession of a solid basis in empirical fact. However abstract an hypothesis might be, scientists have seemed able eventually to derive from it some definite predictions about the behavior of observable objects. And reports of that behavior, accepted as hard facts corroborating or contradicting the hypothesis, have been taken as a happily non-philosophical and therefore reliable standard for scientific judgments. In the modern view, philosophy itself, and politics and morality, could claim no such standard; empirical science could.

Defending science's claim to such a standard, Watkins frontally attacked Kuhn's assertion of the incommensurability of rival paradigms. How, Watkins asked, could admittedly incompatible paradigms, say Ptolemy's and Copernicus', also be *incommensurable*? He went on to assert that in fact crucial experiments based on stellar parallax and star shifts had governed scientists' final choice between competing astronomical theories.[9] Popper and Toulmin adduced similar evidence against Kuhn's incommensurability thesis. But Kuhn did not lack a reply.

In a paper written after reflection on his critics' papers, Kuhn observed that too vague a use on his part of "paradigm" had caused confusion.[10] He had used "paradigm" to denote both the entire outlook shared by a scientific community and one element in that outlook, namely basic theory. What he wished to declare incommensurable was an element that *preceded* theory. For an example he cited what had been instrumental in Galileo's re-formulation of mechanics, viewing a ball rolling down one inclined plane and up another as similar to the swing of a pendulum. (Aristoteleans had viewed pendulums as instances of obstructed falling.) Such perceptions of similarity, along with perceptions of dissimilarity, constituted ways of viewing the world that were required by laymen as well as scientists for perceiving a world at all, instead of a chaos. They conditioned observation and hence conditioned the very data used in evaluating as well as constructing theory. They figured prominently in what, abandoning "paradigm" as spoiled for his purpose, he had come to call "disciplinary matrices."

175

Kuhn's point was the familiar one that different cultural traditions and education made different men perceive the same world differently, and, more fundamental, that theories interpreted culturally conditioned *perceptions*, not the raw or objective facts of the exterior world itself.[11] This meant that science's empirical basis shifted fluently with varying disciplinary matrices or worldviews, that the firm empirical basis long attributed to science was a fiction. This brings us to the heart of Kuhn's position and, in a philosophic view, close to the heart of science itself. . . .

Kuhn meanwhile has denied part of the relativism he has been charged with. He has written that in his view one scientific theory is not as good as another "for doing what scientists normally do," *i.e.*, for setting up puzzles and solving them. But in complete accord with our analysis above, he has added:

> Nevertheless, there is another step, or kind of step, which many philosophers of science wish to take and which I refuse. They wish, that is, to compare theories as representations of nature, as statements about "what is really out there." Granting that neither theory of a historical pair is [absolutely] true, they nevertheless seek a sense in which the later is a better approximation of the truth. I believe nothing of that sort can be found.[12]

Also:

> The notion of a match between the ontology of a theory and its "real" counterpart in nature now seems to me illusive in principle.[13]

Let us be clear about where such statements leave this undoubtedly learned educator and, presumably, his students. They leave them saying, not merely that Einstein's cosmology is no truer than Newton's, and Newton's no truer than Ptolemy's, but that Ptolemy's in turn is no truer than the theory that the earth is flat. With this, Kuhn might seem on the point of endorsing Feyerabend's anarchic notion that even "the most outlandish products of the human brain"—including not only flat earthism but, say, belief in witches and the Protocols of the Elders of Zion—deserve respect.[14]

But Kuhn takes the somewhat different tack that has been steered by the historicism which, as earlier indicated, he projects into the field of science. Suspecting the normative judgments about historical events that philosophic or theological standards made possible, and taking historical events as the only realities, historicism ended by treating existing or "the current" events and trends themselves as standards. In politics, art, and morality historicists came to seek and defer to, not what they declared was good or true, but what was currently established. They ended by making what Burckhart once called the worst normative judgment of all, "the approval of the *fait accompli*."[15]

Kuhn appears to make just such a judgment when he urges that, however much existing science serves the merely puzzle-solving passion of scientists, however distant from real knowledge it persists in being, we not only should accept it as knowledge but should derive our very conception of reason from it. Instead of requiring science to measure up to reason, he would have us redefine reason by whatever a scientific elite, admittedly moved largely by "the value system, the ideology," current in science, currently accepts.[16] He writes:

> To suppose, instead we possess criteria of rationality which are independent of our [the current?] understanding of the scientific process is to open the door to cloud-cuckoo land.[17]

That seems clearly to dismiss as "cloud-cuckoo" the land that most of us would call the land of reason. Is Kuhn not peremptorily dismissing the land wherein value systems and ideologies are critically examined, not accepted merely because they are current? And is not that land also the land wherein a tyrant is known to be a tyrant, even if currently successful, and wherein error is seen to be error, however firmly established?

Taken as seriously as its roots in the traditional modern worldview deserve, Kuhn's sophistic seems sure increasingly to expose the inadequacy of modern rationalism and the unreality of the modern notion of a purely logico-empirical or autonomous, philosophy-free science. Intellectual (and social) anarchism on the one hand, and blind historicist subordination to existing authorities on the other, may well be that exposure's first consequences. But *their* consequences will no doubt, as in ages past, turn men once more to the reason that sets men apart from other animals.

Notes

1. Einstein similarly called physical concepts "free creations of the human mind . . . not, however it may seem, uniquely determined by the external world." Einstein and Infeld, *The Evolution of Physics*, New York, 1938, p. 31.

2. Kuhn, *The Structure of Scientific Revolutions*, 2nd (enlarged) edition, Chicago: University of Chicago, 1970, pp. 77-80.

3. *Ibid.*, p. 149.

4. *Ibid.*, pp. 158-59.

5. *Ibid.*, p. 208.

6. As the German historicist Friedrich Meinecke put it, historicism aimed at "a tolerant and loving sensitivity to all humanity." *Cf.* Meinecke's 1928 essay in *Varieties of History*, Fritz Stern ed., New York: Meridan, 1956. With Hitler's rise to power Meinecke perceived the need for judgments and abandoned historicism.

7. These papers, plus a review of them by Kuhn, have been published as *Criticism and The Growth of Knowledge*, Imre Lakatos and Alan Musgrave, eds., Cambridge, England, 1970.

8. *Criticism and The Growth of Knowledge*, op. cit., p. 229.

9. *Ibid.*, p. 36.

10. *Ibid.*, pp. 234, 271. Also *Scientific Revolutions* (1969 Postscript), pp. 181-82. Dudley Shapere's sharp criticism of *Scientific Revolutions* in *Philosophical Review*, July, 1964, turned on the same confusion.

11. *Criticism and the Growth of Knowledge*, pp. 266-77, and *The Structure of Scientific Revolutions* (1969 Postscript), pp. 187-98.

12. *Criticism and The Growth of Knowledge*, p. 265.

13. *The Structure of Scientific Revolutions* (1969 Postscript), p. 206.

14. The Middle Ages are often reproached for belief in witches, but it was at the break-up of medieval civilization, in the 16th century, following Pope Innocent VIII's 1484 bull against witches, that witch hunting occurred.

15. Such historicism was a main target of Popper's *The Poverty of Historicism*, Boston, 1957, especially sections 22 and 23. Stanley Rosen, *Nihilism*, New Haven: Yale University, 1969, cites Heidegger's embrace of Naziism as a prime example.

16. *Criticism and the Growth of Knowledge*, p. 238.

17. *Ibid.*, p. 264.

Suggested Readings

As noted in the general introduction, this is not intended as anything like a complete bibliography. The works mentioned are generally standard sources, for the most part easily available, which may be of help at the next stage of the reader's research.

Sources on demarcation and related issues:
Duhem, Pierre. *The Aim and Structure of Physical Theory*. Translated by P. Wiener. Princeton: Princeton Univ. Press, 1963.
Feyerabend, Paul K. *Against Method: Outline of an Anarchistic Theory of Knowledge*. London: NLB, 1975.
_____. *Science in a Free Society*. London: NLB, 1978.
Gutting, Gary, ed. *Paradigms and Revolutions: Appraisals and Applications of Thomas Kuhn's Philosophy of Science*. Notre Dame: Univ. of Notre Dame Press, 1980.
Hacking, Ian, ed. *Scientific Revolutions*. New York: Oxford Univ. Press, 1981.
Kuhn, Thomas S. *The Structure of Scientific Revolutions*. 2nd ed. Chicago: Univ. of Chicago Press, 1970.
_____. *The Essential Tension: Selected Studies in Scientific Tradition and Change*. Chicago: Univ. of Chicago Press, 1977.
Lakatos, Imre, and Musgrave, Alan, eds. *Criticism and the Growth of Knowledge*. New York: Cambridge Univ. Press, 1970. An especially good anthology, including contributions by Popper, Kuhn, Lakatos, and Feyerabend.
Laudan, Larry. *Science and Values*. Berkeley: Univ. of California Press, 1984. An excerpt from Laudan's book appears as "Dissecting the Holistic Picture of Scientific Change," in Janet A. Kourany, ed. *Scientific Knowledge: Basic Issues in the Philosophy of Science*. Belmont, Calif.: Wadsworth, 1987.
Popper, Karl R. *Conjectures and Refutations: The Growth of Scientific Knowledge*. London: Routledge & Kegan Paul, 1962.
_____. *The Logic of Scientific Discovery*. New York: Basic Books, 1959.
_____. *Objective Knowledge: An Evolutionary Approach*. Oxford: Clarendon Press, 1972.
Suppe, Frederick, ed. *The Structure of Scientific Theories*. Urbana: Univ. of Illinois Press, 1977.

Articles of note:

Churchland, Paul M. "Karl Popper's Philosophy of Science." *Canadian Journal of Philosophy* 5 (1975): 145-156.

Doppelt, Gerald. "Kuhn's Epistemological Relativism: An Interpretation and Defense." *Inquiry* 21 (1978): 33-86.

Laudan, Larry. "Progress or Rationality? The Prospects for Normative Naturalism." *American Philosophical Quarterly* 24 (1987): 19-31.

Hempel, Carl G. "Turns in the Evolution of the Problem of Induction." *Synthese* 46 (1981): 389-404.

Naess, Arne. "Why Not Science for Anarchists Too: A Reply to Feyerabend." *Inquiry* 18 (1975): 183-194.

On "reflective equilibrium" see:

Daniels, Norman. "Wide Reflective Equilibrium and Theory Acceptance in Ethics." *Journal of Philosophy* 76 (1979): 256-282.

Goodman, Nelson. *Fact, Fiction, and Forecast.* 3rd ed. Indianapolis, Ind.: Bobbs-Merrill, 1973.

Rawls, John. *A Theory of Justice.* Cambridge: Harvard Univ. Press, 1971. Especially Ch. 1, section 9, pp. 46-53, and Ch. 9, section 87, pp. 577-587.

On relativism and the sociological approach:

Barnes, Barry. *Scientific Knowledge and Sociological Theory.* London: Rutledge, 1974.

Mendelsohn, Everett et. al. *The Social Production of Scientific Knowledge.* Dordrecht, Holland: D. Reidel, 1977.

Meiland, Jack, and Krausz, Michael, eds. *Relativism: Cognitive and Moral.* Notre Dame: Univ. of Notre Dame Press, 1982.

Tiryakian, Edward A., ed. *On the Margin of the Visible: Sociology, the Esoteric, and the Occult.* New York: Wiley, 1974.

Wilson, Bryan R. *Rationality.* Evanston, Ill.: Harper & Row, 1970.

Young, Robert M., and Teich, Mikulas, eds. *Changing Perspectives in the History of Science.* London: Heinemann, 1973.

SECTION **III**

PARAPSYCHOLOGY

Introduction

Parapsychology seems to stand out as in some way unique among those topics that are commonly dismissed as "pseudoscientific"; the case of parapsychology seems importantly different. Many who scoff at Atlantis, Bigfoot, or the Bermuda Triangle nevertheless take parapsychological research regarding clairvoyance, precognition, telepathy, and psychokinesis quite seriously, though the reverse is rarely the case. The great bulk of the literature on other topics labeled "pseudoscience" appears in the form of glossy paperbacks designed for quick and perhaps uncritical reading. But parapsychology is represented by numerous tomes of research materials as apparently dry and dispassionate as any in astronomy or sociology. Other topics have their passionate and sometimes eloquent defenders. But parapsychology can boast scores of dedicated researchers who have devoted years and decades to diligent study and investigation.

Of course none of these characteristics will settle the status of parapsychology as science or pseudoscience. But they do serve to distinguish parapsychology at least superficially from other pursuits dismissed as pseudoscience, and thus may indicate that parapsychology presents a particularly troublesome case for attempts at demarcation.

Some, of course, will claim that these differences are to be expected; that parapsychology differs from other pursuits that have been labeled pseudoscience simply because it has *wrongly* been labeled pseudoscience. Although perhaps a fledgling science, or a science in which a great deal of both empirical and theoretical work remains to be done, its defenders will claim, parapsychology is nevertheless a *science*.

In the opening piece of this section, RUTH REINSEL presents the case for parapsychology. She offers a historical review of important results obtained concerning clairvoyance, precognition, telepathy, and psychokinesis, and discusses skeptical attacks and methodological modifications that have been made in response to criticism. In the final section of her paper, Reinsel outlines contemporary trends in parapsychological research and makes some theoretical suggestions of her own.

One of the simplest charges leveled against any putative case of precognition, clairvoyance, or the like is that what is really at issue is merely a case of coincidence. But what does it mean to call something a coincidence? And what would justify or defeat a charge of coincidence, in general or in the case of parapsychology in particular? These are the issues GALEN K. PLETCHER attempts to address in the second selection. To attribute something to coincidence, Pletcher argues, is not to explain it or even to attempt to explain it; it is instead to maintain that no special explanation is required. His charge against parapsychology is that it has not yet shown that its "data" stand in need of special explanation, and here Pletcher faults attempts at explanatory theory as well as case studies and laboratory work in parapsychology. Until parapsychology can offer more consistent laboratory results, he maintains, or until it can present a more satisfactory explanatory framework, it will not have adequately responded to the simple charge of coincidence. A careful evaluation of Pletcher's conclusion may call for a review of the laboratory results surveyed in Reinsel's opening piece.

In the next two sections, ANTONY FLEW and JANE DURAN raise a number of important philosophical difficulties regarding parapsychology and parapsychological phenomena. ANTONY FLEW discusses in some detail three respects in which parapsychology differs "from all the established high-status sciences." Its domain, first of all, is only negatively defined: "psi-gamma," for example, "is at this time defined as precisely not the product of any means we can think of." Second, there is no repeatable demonstration that parapsychology has its own peculiar and genuine data to investigate: "The black record of fraud would not carry nearly so much weight against what might seem to be strong cases of psi; if only we possessed some repeatable demonstration of the reality of such phenomena." Third, Flew claims, "there is no even half-way plausible theory with which to account for the materials [parapsychology] is supposed to have to explain." Flew's is a wide-ranging discussion, as intriguing in its subsidiary treatment of disembodied minds and Hume's position on miracles as in its main line of argument. At no point does Flew simply label parapsychology a pseudoscience. But he does propose that if affiliation with the American Association for the Advancement of Science is to be thought of "as a recognition of actual achievement rather than of good intentions, then the Parapsychological Association is not yet qualified for admission, and ought now to be politely disaffiliated."

JANE DURAN pursues further a number of philosophical reasons that have been given for holding that precognition, clairvoyance, and telepathy are not possible at all. Using the work of C. D. Broad, she considers a set of "basic limiting principles"—part of our everyday intellectual frame-work—that seem at least at first sight to be inconsistent with the claims of parapsychology. On careful consideration, however, she rejects a number of the charges; some of the philosophical arguments that have been leveled against parapsychological phenomena prove inadequate. But Duran holds that one particular charge, evident in Broad's first "limiting principle" regarding causality, is a telling one. "As long, at least, as our ordinary notions of causality remain intact, there seem to be strong philosophical reasons for holding that telepathy, clairvoyance, and precognition are not possible." In a final section Duran argues that even if these were accepted as possible, they could not legitimately be construed as means of knowledge.

At this point in the readings it will be clear that one standard philosophical objection to putative parapsychological phenomena—perhaps the strongest philosophical objection—is that such phenomena would violate basic principles regarding causality. Precognition, for example, would seem to demand "backward causation"; an effect (such as the vision of an airplane crash) would have to precede its cause (presumably the crash itself). But the notion of "backward causation," both Flew and Duran insist, is simply incoherent. Duran claims that part of what we *mean* by "*A* causes *B*" is that *A* precedes *B* temporally, and Flew maintains that "it must be irredeemably self-contradictory to suggest that the (later) fulfillments might cause the (earlier) anticipations."

It is this objection in particular that BOB BRIER and MAITHILI SCHMIDT-RAGHAVAN attempt to answer in the piece included here. With an eye both to a parapsychological experiment of their own and to proposals in physics regarding particles that travel backward in time, and drawing on the work of John Stuart Mill, Brier and Schmidt-Raghavan argue that a scientific view of causality need impose no temporal requirements of the sort urged by Flew and Duran. The claim that "backward causation" is logically impossible, Brier and Schmidt-Raghavan maintain, is based on confusions between asymmetry and atemporality and between changing the past and affecting the past. But they do speak of a "revision of the concept of cause," and in that regard the reader is advised to consider again Jane Duran's comments in the preceding piece concerning changes in the meaning of "cause."

The last recourse of the critic, it has been said, is to charge fraud. But unfortunately—within parapsychology as within other fields—this charge has at least on occasion been justified. Some cases of fraud have been mentioned in Reinsel's opening review and in Flew's work. In the final piece of this section, J. B. RHINE outlines and discusses the falsifications of Walter J. Levy, Jr., discovered in 1974. But Rhine also suggests categories and presents examples of "fraud-proof evidence of psi," and maintains that "these fraud-proof types of evidence, if now given precedence for a showdown on the establishment of psi, will fill the need for the scientific mind that considers it." Rhine closes with some general proposals regarding the future of parapsychological research.

The readings of this section pose a number of issues in the philosophy of science that, though perhaps related to the problem of demarcation, also deserve careful consideration in their own right. What is demanded of an explanation, and is parapsychology able to supply it? What are natural laws, or scientific laws, and what is their role in explanation? What is the status of theoretical entities such as "psi"? What role do basic intuitive principles properly play in science? What is meant by saying that one event causes another, and does that allow a logical possibility of "backward causality"? Issues regarding causality and correlation, first broached in our initial examination of astrology, arise here as well.

It is also clearly of value to address again the issue of demarcation with parapsychology in particular in mind. *Is* parapsychology a pseudoscience? We began by saying that parapsychology seems to stand out as in some way unique among those topics that are commonly dismissed as "pseudoscientific." But precisely how does it differ?

Parapsychology: An Empirical Science

RUTH REINSEL

Although the historical roots of scientific parapsychology are tangled with nineteenth-century spiritualism, the history of parapsychology is to a large extent the story of its struggle to disassociate itself from the occult. The occult and parapsychology can be distinguished both in the choice of methods and in the choice of subject matter. The occult relies on traditional bodies of arcane lore interpreted with a healthy dose of personal intuition. Because of this combination of tradition and intuition, what are sometimes called the occult "sciences" should more properly be called the occult "arts." Parapsychology, on the other hand, relies heavily on objective evidence, quantitative evaluation, and the experimental method. The occult includes topics of popular appeal but of questionable validity, such as astrology, numerology, and magic. Parapsychology, on the other hand, restricts itself to the study of ways of gaining knowledge or affecting the world around us that do not involve the five normal senses. Since no known process or energy can account for the accumulating mass of parapsychological data on ESP, precognition, and psychokinesis, [1] these abilities are termed "paranormal." This, it is to be hoped, will be only a temporary designation, since it is possible that some explanatory principle remains to be discovered that will satisfactorily reconcile these phenomena with conventional science.

If dated from the founding of the Society for Psychical Research in London in 1882, parapsychology will soon celebrate its first century of development. Philosophers have been intimately connected with the development of parapsychology from its inception—witness the membership rolls of the English and American Societies for Psychical Research, which include William James, Henry Sidgwick, C. J. Ducasse, C. D. Broad, and H. H. Price—and it seems appropriate now for parapsychologists and philosophers together to evaluate the progress that has been made in this area. In

what follows I shall present a brief historical review, a survey of methodological difficulties and skeptical objections overcome, and a discussion of
current techniques in parapsychological research. No honest appraisal of a
century of research in parapsychology can, I think, conclude that parapsychology is anything less than it purports to be: a solidly empirical investigation of certain anomalies in our psychological makeup that we call, for
want of a better term, "psychical abilities."

The Beginnings of Parapsychological Research

The early period of psychical research, dated from 1882 with the founding
of the Society for Psychical Research, was closely associated with the
nineteenth-century spiritualist movement. Psychical researchers were concerned with the rampant fraud and trickery that were practiced by many, if not
all, mediums. But if survival after death could be proven beyond doubt, this
would have profound implications for philosophy, religion, and science, as
well as the concerns of daily life. Therefore the reputable philosophers and
scientists of the Society for Psychical Research felt a need to investigate
mediumistic communications in the hope of separating the veridical phenomena, if such existed, from the fraudulent.

The investigations of the Society for Psychical Research were conducted
with rigid controls and under conditions that seemingly excluded all
possibility of fraud. Some of their investigations did produce evidence
strongly in favor of survival after death. But the problems were only beginning. Even among psychical researchers the survival hypothesis—that
returning spirits of the dead were at work—was challenged by other explanations: clairvoyance, telepathy, and coincidence. And even when the conditions of clairvoyance and telepathy seemed to have been excluded, the
possibility of coincidence remained. Just how striking, unusual, or improbable must a coincidence be before one accepts the even more improbable
explanation that a paranormal form of communication is involved?

The events of our everyday life are determined by so many interacting
variables that it is difficult to specify exactly how often two independent
events will be associated by chance alone. For example, how often might it
occur that a person A will dream of a person B's death on Tuesday night and
learn Wednesday morning that B did indeed die during the night? To
objectively evaluate the hypothesis that this dream presented a paranormal
event one would have to know: (1) how often A dreams of B's death, when B
does *not* die, (2) whether B's death was highly probable in its own right, and
(3) whether A had access to any information through normal sensory channels—such as overhearing a conversation, later forgotten, or awareness of
body language or faint chemical stimuli (pheromones) that might indicate
that B was in poor health—which might have acted as a subliminal stimulus,
giving rise to the dream. To provide evidence of a paranormal event, it would

also be essential for A to have communicated the dream to a third party (preferably in writing, noting date and time) before being apprised of B's death.

The need to bring paranormal events into the laboratory where they could be observed under controlled conditions and systematically explored led to the establishment in 1927 of the Parapsychology Laboratory at Duke University by William MacDougall and J. B. Rhine. Concerned with the need to evaluate statistically the probability of occurrence of psychic events in the laboratory, the first task of these men and their colleagues was to find a population of discrete events for which the probability of chance occurrence could be fixed objectively, and then compared to the observed value when purportedly paranormal influences were operating.

The Zener cards used in many early investigations at the Duke Parapsychology Laboratory are an example of such a population of discrete events with fixed probability. These cards consist of five each of five symbols (circle, square, cross, star, and waves). When each card is placed face down on the table, and a subject is asked to guess what the unseen card might be, the subject has a one-out-of-five chance of being right. Thus scores for pure guessing in such a clairvoyance task would hover around the twenty percent level. An occasional score falling more than two standard deviations beyond mean chance expectation would also be expected on the basis of chance alone, but these occurrences would tend to be infrequent (falling as they do at the tail ends of the normal curve) and the circumstances surrounding their occurrence would show no consistent pattern. However, if scores significantly different from chance occurred repeatedly over many trials, and furthermore seemed to be lawfully related to situational or personality variables, then one might hypothesize that some non-chance factor was influencing the results. It is just such a pattern of non-random interactions in the body of ESP data collected to date that has convinced many observers that the factor called "psi" is manifesting itself.

By 1934 Rhine and his associates had used the ESP cards in over 90,000 trials with gifted subjects. They were reluctant to publish the results of these studies until they had established to their satisfaction that the astronomically high levels of extra-chance scoring obtained with their subjects were not just a fluke, but were indeed repeatable over many sessions and could be verified by independent observers. Having refined his experimental techniques and subjected the data to rigorous statistical evaluation, it was not until 1934 that Rhine finally published his monograph *Extrasensory Perception*.

This volume described, among other work, the now famous Pearce–Pratt series of experiments. Pratt, then a graduate student at Duke, served as the experimenter, conducting trials at fixed intervals by means of synchronized watches while Pearce, as subject, was in some different building from 100 to 250 yards away. At intervals of one minute, Pratt would take a card from the top of the deck and place it face down on the desk in front of him. Pearce, in the library building across campus, would record his guess. Then the procedure would be repeated. Pratt did not look at the faces of the cards until all 25 had been used as targets. He then recorded the sequence of the 25 target cards, placed his record in a sealed envelope, and delivered it immediately to Dr. Rhine. Pearce, in the meantime, was on his way back from the other building with the sealed record of his guesses. Over the whole series of 1,850 clairvoyance trials run under varying conditions of distance and experimental controls, Pearce obtained 558 "hits" where only 370 would be predicted by chance. He averaged 8.3 correct per run of 25 cards, where chance alone would give an average of 5 correct. [2]

Another experiment described in Rhine's 1934 publication was the Ownbey–Turner series, which was conducted to determine whether physical distance would interfere with telepathic rapport. Over a period of eight days, Miss Ownbey, at the Duke Parapsychology Laboratory, attempted to "send" 25 cards at five-minute intervals to Miss Turner, who was 250 miles away. Twenty-five calls were made daily, with both agent and percipient keeping separate records. Miss Turner's records were delivered to Dr. Rhine, who independently checked them against Miss Ownbey's records of the actual target cards. The results for the first three days were extremely successful with 19, 16, and 16 hits where chance would predict a mean score of 5 correct calls out of 25. Over the next few days the scoring rate declined somewhat, but the overall scoring rate averaged 10.1 per 25 for the 200 trials. Scoring was as good, or better, at a distance of 250 miles as when agent and percipient were in close proximity to each other.

The early attempts to determine the conditions that affect the manifestation of ESP included an investigation of drugs that affect central nervous system arousal level. An experiment was conducted in which George Zirkle, the subject, was administered five grams each of sodium amytal and caffeine in identical capsules. Although the agent/experimenter (another graduate student) was not blind to which drug was being given at which time, Zirkle was unaware of the order in which the drugs would be administered, and neither experimenter nor subject had any knowledge of the kind of effects the drugs might be expected to have on ESP scoring. With the agent and percipient in separate rooms approximately 8 to 12 feet apart, a baseline series of 50 trials drug-free was made on the same day as the drug experiment. In this baseline condition, ESP scoring averaged 13.5 hits out of 25 cards. Then five grams of sodium amytal (a depressant) in a blank capsule was administered to Zirkle. One hour later he was very sleepy and his ESP scoring rate declined to 7.8 per 25. Another test was made two hours later. At this point Zirkle had to be physically kept awake, was having difficulty walking straight, and was experiencing impairment in vision and memory functions; his ESP scoring declined further to 6.2 per 25. At this point Zirkle was administered five grams of caffeine. When tested one hour later he was much more alert and his motor coordination had returned; his scoring also returned to 9.5 hits out of 25 trials. Thus ESP scoring rates seem to reflect the effects of depressant and stimulant drugs on the central nervous system. Drug tests with other experimental subjects followed a similar pattern.

Psychological patterns began to emerge from the data. Scoring would be above chance when the subject's interest or motivation was high, but would decline as the subject became bored or fatigued. Physical illness would also cause scores to fall to chance level or below. ESP scoring seemed to be highly influenced by subjects' attitudes and expectations. In early tests of the effects of distance on telepathy, subjects tended to believe that ESP could not work at distances over a few feet. Scores thus tended to show a decline with distance. However, when the task was presented to the subjects as a challenge, ESP scores were often better using screens or separate rooms at some distance than they were in the same room. Thus physical distance seemed to be less important than "psychological" distance.

Once the explorations of clairvoyance and telepathy[3] were well under way, Rhine (1938) reported the results of the first experiments on precognition using a technique known as the "psychic shuffle." Subjects were given a response sheet with 25 blanks and asked to write in their guesses of the order in which the 25 cards in the ESP deck would occur after they were shuffled. The catch was that this shuffling had not yet taken place, so the subjects were in the position of having to predict the order of future events. Preliminary results using this technique gave odds of less than one in a million against a chance explanation of the results. However, problems of interpretation soon arose. Perhaps precognition, with its implied violation of temporal causal relations, was not the most parsimonious interpretation of these results. Possibly while the experimenter shuffled the cards, he might have used clairvoyance to determine the precise instant at which the cards approximated the order of the subject's previous responses, and then conveniently stopped shuffling. Alternatively, perhaps while shuffling the cards, the experimenter used psychokinesis (PK) to cause them to take on the sequence of the subjects' calls. Ensuring that experimenters were blind to subjects' responses until after card order had been determined, and introduction of a mechanical device to shuffle the cards, eliminated the clairvoyance hypothesis, but still left the mechanical device open to PK action.

In an attempt to counteract the possibility of PK influence on the card shuffler, an automatic device was incorporated that would randomly cut the cards after the shuffling process was completed. With this modification, precognition scores declined to the chance level. However, closer inspection of the distribution of the scores revealed a curious non-random pattern: a concentration of hits at the beginning of the run, with a gradual decline (hence the name "decline effect") and a brief flurry of successes at the very end of the run. This pattern, which has been found over and over again in psi scores from many different experiments and researchers, is one of the most consistent patterns to emerge from psi experiments. The occurrence of the "decline effect" in the precognition data was interpreted as showing a high initial level of psi, reduced by fatigue and declining interest and motivation, attributable to what is essentially a monotonous and repetitive task. Thus the presence of a decline effect in the subjects' precognitive guesses implied that some non-random factor was indeed operating, and this was taken as evidence for the existence of precognition. Later work from the Duke Parapsychology Laboratory gave results that were less difficult to interpret (e.g., Anderson, 1959).

For many people the ability to influence what is going on in someone else's mind, or to obtain information clairvoyantly, may not be as hard to accept as the thesis that it is possible to gain information about future events before they occur. However, just as distance in space does not appear to affect clairvoyance or telepathy (Osis, 1965), neither does distance in time appear to affect success on a precognition task. Anderson (1959), at the Duke Parapsychology Laboratory, worked with a subject who mailed in her ESP responses from Paris, France. Half of the subject's ESP responses were scored against targets selected immediately (which was in fact a few days after the distant subject had made her calls). The other half of the subject's ESP calls were scored against targets that were not selected until one year later. Although the targets selected immediately gave chance results, scores from the year-interval test were independently significant, with odds of 500 to 1 against chance. Scores in the two conditions taken together were also significantly above chance.

Further work from the Duke Parapsychology Laboratory concentrated on the possibility that the outcome of a random throw of the dice might be affected by the human mind. Preliminary tests with a young gambler who claimed to be able to influence the fall of the dice inspired Rhine to develop a more formal test of the phenomenon. Using hand-thrown dice, Rhine had 25 subjects make 24 throws each. Results were significantly above chance, with a probability of over a billion to one (Rhine and Rhine, 1943). More controlled conditions were soon implemented, including the use of a mechanical dice-thrower and corrections for imperfections in the dice that might cause one face to come up more often than the rest. Also, if the subject was allowed to determine for himself the target face for each trial, his later success could be interpreted as precognition of the outcome. To control for both these possibilities, the choice was taken out of the subject's hands, and target faces were assigned in a predetermined random order with an equal number of trials for each face. The PK ability of various subjects seemed to be unaffected by these variations in technique, and overall probability values as high as $p =$

.008 were obtained (that is, eight chances in a thousand of getting such a result, if only chance factors are operating) (Rhine, 1943).

Skeptical Objections and Methodological Considerations

The first reports from Rhine's laboratory were greeted skeptically in many quarters. The early criticisms of laboratory psychical research centered on the following possibilities:

(1) Subjects might have had an opportunity to cheat while left alone with the targets.
(2) Some form of normal sensory cuing (such as markings on the targets, or slight variations in the experimenters' facial expression, body language, or tone of voice) might account for the extra-chance scoring levels, with no extrasensory information transfer involved at all.
(3) Non-randomness in the target sequences might coincide with a subject's response bias tendencies, yielding a spuriously above-chance score.
(4) The experimenter(s) might have inaccurately recorded the subject's responses.
(5) Data analysis might have involved inappropriate use of statistical tests.
(6) In order to prove that ESP exists, the experimenter(s) might have deliberately fabricated the evidence.

These methodological criticisms prompted further refinement of experimental techniques. Current parapsychological methodology emphasizes the need to incorporate rigid controls in order to exclude all explanations other than the operation of ESP. The following precautions have become standard methodological procedure in experiments accepted for publication in the reputable journals of the field.

1. Strict procedures are followed to ensure that the subject has no opportunity to cheat. The subject is never left alone with the targets in a clairvoyance experiment, and is not allowed to communicate with the agent in telepathy experiments. Targets are concealed from the subject by a screen or an opaque envelope, or better yet, they are kept in a distant place to which the subject has no access.

2. In order to eliminate sensory cues, targets are handled as little as possible, so that random scratches or markings do not become the basis for the subject's responses. The targets are prepared by an independent assistant who has no contact with the subject. Thus the experimenter who presents the sealed targets to the subject is "blind," that is, has no knowledge of which target is in which envelope, and cannot systematically influence the subject's response in the direction of a higher score.

3. Elaborate procedures are followed to ensure random selection and presentation of targets. Random number tables or other random sources (REG determination, or computer algorithms) are the source for target

sequences, and targets or conditions are presented to the subjects in a counterbalanced sequence.

4. Scoring must be double-checked later by an independent scorer blind to the hypothesis of the experiment, and ignorant of the experimental group to which the subject belongs. Alternatively, two observers can make simultaneous but independent records of the subject's responses and later compare them for errors in recording.

5. The question of the appropriateness of the statistics used to evaluate ESP data was submitted to the American Institute of Mathematical Statistics for evaluation. In 1937 the president of that organization issued a public statement that verified the adequacy and integrity of the statistical techniques used by Rhine and his colleagues at the Duke Parapsychology Laboratory (Camp, 1937).[4]

6. The charge of experimenter fraud has often been made against parapsychological data as a kind of last resort, after all other criticisms have been answered or invalidated. Regrettably, however, the charge of experimenter fraud has not been without validity in some few cases, notably those of W. J. L. Levy and S.G. Soal (see the paper by Rhine in this volume [Rhine, 1974], and Markwick, 1978). Although fraudulent manipulation of data may have occurred on only one occasion in these investigators' research careers, that is sufficient to bring into question all of their published research. Parapsychologists are generally among the first to repudiate those among their ranks who are found guilty of such unethical practices, first because this represents unacceptable procedure in science, but perhaps also since these occurrences cast doubt on the integrity of all other researchers in what is already a highly controversial field. But does the fact that two experimenters have been found guilty of fudging data mean that *all* the published evidence for the existence of ESP is fraudulent? To contend that, over the decades, scores of highly respected researchers with otherwise impeccable reputations have conspired in a gigantic hoax to falsify evidence and deceive the public and their fellow scientists is even less parsimonious than the alternative hypothesis: that ESP exists.

To summarize the crucial elements of methodology: no matter what the form of psi being investigated, double-blind methods are essential. That is, the assistant who prepares the targets has no contact with the subjects. The experimenter who interacts with the subjects does not know what the targets are on any given trial. Scoring is double-checked by an assistant who is blind to the hypothesis of the experiment, had no contact with the subjects, and does not know to which experimental group or condition the subjects belonged.[5] Finally, it is important in parapsychological research, as in other fields of scientific endeavor, to await independent replication of a finding before drawing any definite conclusions.

In the "hard" physical sciences, the criterion of the repeatable experiment is interpreted to mean that anyone can follow the same experimental procedures and get the same results every time. ESP experiments, on the other hand, tend to be frustratingly inconsistent. A novel procedure will often give highly significant results when it is first introduced, only to seem to "fail" later on, when results decline to chance, or when the effect reverses direction completely. If psi is indeed a natural and lawful phenomenon, it should be consistent and predictable once one can specify the laws and conditions that govern its occurrence. To many scientists, this lack of repeatability in psi experiments is the most powerful argument against the existence of psi.

But this argument may be somewhat too extreme (Rhine, 1959). The repeatability of psi experiments is actually higher than is generally assumed. Certain findings have been demonstrated experimentally over and over again. Take, for example, the effect of belief in ESP on scoring levels (the "sheep–goat effect," reviewed by Palmer, 1971); or the positive effects on ESP scoring of high interest and motivation, and the negative effects of withdrawal, apathy, and negativism (Nicol and Humphrey, 1955; Rao, 1962; Scheerer, 1948; Shields, 1962). On the other hand, it is incorrect to assume that the repeatability criterion is met by all fields of scientific endeavor *except* parapsychology. Epstein (1980) points out that the social or "soft" sciences in general (including many fields within psychology) experience difficulties in replicating experiments. Epstein suggests that this problem may be inherent in the nature of the sampling assumptions within the experimental method itself. If repeatability in psychology in general is less than perfect, the problem in ESP experiments is even more pronounced. Psi scores have been shown to be affected by the subject's personality, attitudes, mood and motivation, and the nature of the subject's social interaction with the experimenter and the testing situation. Experiments in which these variables are not taken into account generally show chance results. The high variance in the data caused by the effects of these uncontrolled variables may in many cases obscure the operation of the underlying psi processes.

Current Techniques in Parapsychology

The combined weight of experiments such as those described in the first section—including the weight of much more recent work with progressively tighter methodological controls—is strongly against a "random-chance" explanation. As it became obvious that merely demonstrating the existence of non-random scoring tendencies in the data would not yield any immediate understanding of the psi process (or processes), interest turned to a search for the conditions under which psi was manifested and for the kinds of personality attributes that correlated with psi success (see, e.g., Carpenter, 1977; Honorton, 1977; Palmer, 1977). As researchers became dissatisfied with repetitious card tests that failed to hold subjects' interest, new instrumentation and more sophisticated experimental methods were introduced.

One of the recent developments in parapsychological research has been the attention given to the importance of an alteration in the normal state of consciousness in order to make the psi signal "heard" above the "noise" of the other sensory stimuli against which it must compete. Honorton (1977) has provided a complete review of all major research dealing with alterations in state of consciousness in the direction of "internal attention states."

Relaxation is one of these states that has been quite extensively investigated as a psi-conducive technique. For example, L. W. Braud and Braud (1977) conducted a clairvoyance experiment where 100 subjects were to attempt to describe art prints in sealed envelopes. Subjects listened to a tape of suggestions selected to promote relaxation, calmness, feelings of coolness in the forehead, and heaviness and warmth in the extremities. The tape was designed to test the theory that a reduction in arousal of the autonomic nervous system would facilitate psi scoring. Apparently the tape was successful, because 36 of the 100 subjects made direct hits (first choice on ranking the possible targets) when asked later to identify, from among several art prints, the one that had been their concealed target ($p = .0055$). In total, 63 of these 100 subjects placed the actual target in the top half of their rankings of the art prints ($p = .0047$).

Earlier research by the same team of Braud and Braud (1974) reports two studies using the same tape of relaxation suggestions in a telepathy (GESP) paradigm. In the first study, 16 subjects listened to the same tape-recorded instructions. Subjects recorded their impressions of the ESP target and then ranked their degree of relaxation on a ten-point scale. Highly relaxed subjects scored significantly better on the free-response GESP task than did less relaxed subjects ($p < .04$; that is, less than 4 chances in 100 of this result occurring by chance alone). In the second study, involving 20 subjects, 10 of the subjects listened to the relaxation tape, while the remaining 10 heard a tension-inducing series of instructions. The relaxation group showed significantly more psi success on the ESP task than did the tension group ($p < .025$). The degree of relaxation was monitored by EMG recording devices (though no feedback was given) and the degree of both objective (EMG) and subjective (self-rated) relaxation correlated significantly with psi success (correlations of .49 and .53 respectively, $p < .05$ in both cases).

In line with the hypothesis that reducing external sensory distractions and redirecting attention internally will facilitate psi, a large number of studies involving the ganzfeld technique have been reviewed by Honorton (1978). The ganzfeld is a mild form of sensory deprivation where the subject sits in a comfortable reclining chair, with eyes covered by screens that exclude patterned visual stimulation (usually ping-pong balls cut in half and taped over the eyes), with the visual field flooded with red light. The subject simultaneously listens to a tape of white noise. Thus patterned stimulus input is totally excluded and the subject can turn his attention inward and report his ongoing mentation into a tape recorder. An agent sits in another room, without contact with the subject, and views a target picture while trying to "send" information about the picture to the subject (percipient). At the conclusion of the session, the percipient is shown several pictures, one of which was the actual target, and he is asked to rate the pictures for the degree of correspondence to his ongoing imagery and associations.

The ganzfeld has proven to be one of the most successful techniques for eliciting psi and has stood up under the test of numerous replications. Seven of the eight studies conducted in Honorton's lab were independently significant, giving a combined significance level of 7.9×10^{-8}. Honorton demonstrates in his review that of 23 studies from 11 different laboratories (not including his own), 14 studies demonstrated overall significant psi scoring rates. This constitutes a replication rate of 54 percent where only 5-percent-successful replications would be expected if the results were just flukes caused by chance variations of the normal curve.

A major methodological advance over the early work with ESP cards and dice was made when Helmut Schmidt, an electronics engineer by profession, developed a series of automated machines known as random event generators (REGs). These devices are based either on radioactive decay processes (the most completely random processes yet discovered) or on random electronic-noise sources. The Schmidt machines operate by generating random binary electrical pulses that serve as targets, in the manner of an electronic coin flipper: a positive pulse is "heads" and a negative voltage is "tails." The Schmidt REGs have several advantages over earlier experimental techniques: (1) they are completely random, so that their target sequences do not suffer from the problem of "pseudorandomness" inherent in printed random number tables or in computer algorithms; (2) they allow for automatic target and response recording, thus excluding the possibility of experimenter error in recording the results; and (3) they provide immediate feedback to the subject on whether his guess was right or wrong, thus allowing for experimental study of the application of operant conditioning learning techniques to ESP (see Tart, 1976).

One of these random event generators was used by Schmidt (1969a) in a clairvoyance test to compare the responses of six subjects to a prearranged random target sequence. Over 15,000 trials, the results of these tests were highly significant, with odds of over a million to one against chance. The Schmidt REGs have also been very useful in studying precognition. The subject makes his response a fraction of a second before the random target is electronically generated; thus many trials can be accumulated over a short period. Schmidt (1969b) used his random event generator to test precognition with three gifted subjects. Over a total of 63,066 responses, the results from these three subjects were highly significant, with odds of 2,000 million to 1 against chance.

In another experiment also using one of the Schmidt REGs, large numbers of subjects were tested in groups. Five hundred trials were run in both precognition and PK test modes, and results were significantly above chance in both cases (Schmidt and Pantas, 1972). In Part 2 of this same experiment, a single subject meditated for approximately twenty minutes before making a series of either precognition or PK calls. With 500 trials in each mode, extreme deviations from chance expectation were found in both conditions, and no difference in scoring rate was found between the precognition and PK conditions. This overlap between the PK and precognition abilities has profound theoretical implications: it may be that we do not merely gain "knowledge" of our future, but actively create it.

Along with experiments on dice and electronic random event generators, much recent PK work has used natural or biological systems as targets. Interest has focused on whether the reports of "psychic healing" might not be founded in a PK ability to influence living systems. Thus Smith (1972) had a noted psychic healer (Mr. Estebany) hold a sealed flask of enzyme solution (trypsin), and found significant increases in the rate of enzyme catalysis when the flask was held by the healer, as opposed to the rate in a control (un-treated) condition. To control for the effects of temperature on trypsin activity, a sensitive thermistor was placed between Mr. Estebany's hands as he held the trypsin sample; the water bath in which the control sample was placed was kept at the same temperature. Replications of this experiment have been moderately successful (see Edge, 1980); the limited success is attributed to the difficulty the healer has in establishing a caring personal relationship with a flask of enzyme solution. In Smith's experiment, a control condition where identical flasks of enzyme solution were exposed to a high-intensity (1,300 gauss) magnetic field for three hours showed definite results in the same direction as, but stronger than, the effects produced by the healer. It is therefore possible that psychic healing produces its effect by inducing a magnetic field; however, using a magnetometer, Smith found no indication of any unusual magnetic field around Estebany's hands while he was "heal-ing" the enzyme solution.

Grad (1965, 1976) has worked with the same healer who claimed to be able to heal through the laying on of hands. In this work, Grad had available to him the facilities of a biomedical research laboratory with large numbers of experimental animals (mice). For the psychic healer to administer the "treatment," mice were placed into a specially constructed iron cage with a solid bottom and a wire mesh top. The healer held one hand below the cage and the other one covering the top of the cage. At no time did the healer actually touch the mice themselves in any of the experiments. In control conditions, mice were exposed to warm temperatures by placing a heating pad under their cages for an equivalent amount of time.

In one experiment, mice were fed an iodine-deficient diet that produced large goiters in two control ("untreated") groups of mice, whereas mice whose cages were held by the healer showed significantly less goiter development. In another experiment, which has been successfully replicated several times, equal-size patches of skin were removed from the backs of 300 mice. Mice whose cages were held by the healer showed a much faster rate of wound healing than did the control animals ($p < .01$). Mice whose cages were held by skeptical medical students healed the most slowly of all three groups.

This PK explanation of psychic-healing phenomena has also been studied with plants. A state of "need" in the plants was induced by treating the seeds with a one percent sodium chloride solution. Salt treatment damages the seed and tends to inhibit sprouting and subsequent plant growth. However, seeds treated with saline solution that had been previously held by the healer for fifteen minutes showed significantly more subsequent growth than control seeds treated with solution not held by the healer. A chemical analysis of the hand-held saline solution revealed no difference between the hand-held and the control saline solutions in sodium concentrations or pH value.

In another experiment along these lines, Watkins and Watkins (1971) tested the ability of nine reputedly psychic subjects to hasten the resuscitation of anesthetized mice. Five of these subjects claimed to have healing abilities, and eight of the nine scored well on laboratory PK tests. On the average, the experimental mice (those concentrated on by the psychic subjects) awoke in 87 percent of the time taken by the control mice. Eight of the nine psychic subjects were able to induce a significantly shorter duration of anesthesia in their mice, whereas none of the three control subjects could produce this effect ($p < 10^{-5}$).

Another experiment, conducted by Schmeidler (1973), attempted to determine whether the effects of PK could be voluntarily controlled and localized in space. This time, instead of using living systems as targets, a talented psychic attempted to influence temperature-sensitive devices (thermistors) which were sealed into vacuum bottles and connected to an automatic recording device. The subject was asked to raise and lower the temperature of specific thermistors on demand, following a predetermined, counterbalanced sequence. The subject was successful at producing temperature changes in the desired direction over repeated trials. Sixteen trials made up an experimental session, and in seven out of ten sessions the temperature deviations of the sixteen trials were statistically significant. In five of these seven sessions, the probability was less than one in a thousand of chance fluctuations being responsible for the results. Curiously, a drop in the temperature recorded by one thermistor was often accompanied by a rise in air temperature in a different part of the room, as if the heat energy had somehow been transferred from one part of the room to another. The results suggest that PK may operate in accordance with the laws of thermodynamics.

Conclusion

This brief review has only scratched the surface of the issues in collecting and interpreting parapsychological data. There is a great deal of evidence, which it has not been possible to review here in detail, that the psi process bears a lawful relation to personality and attitude, interest and motivation. Psi seems to be related to a host of other psychological variables. Spontaneous cases show that ESP may lie dormant for many years, a small voice that is not heeded above the demands of coping with everyday life, until a moment of overwhelming personal crisis. It is not possible to duplicate in the laboratory the crisis situation and the intense emotional involvement evident in spontaneous cases. Thus ESP experiments offer at best a shadow of the true nature of psi. And yet the scientific method has had great success in unraveling other mysteries of nature. Can it not be fruitfully applied to mysteries of human nature as well?

Of course, any experimental evidence is only as good as the conditions under which it was obtained. This is the reason for the meticulous attention to details of methodology in parapsychological research, given the importance of what we are trying to demonstrate.

If one accepts the evidence as valid, then how is it to be interpreted? At least four different aspects of psi functioning have been identified (clairvoyance, telepathy, precognition, and psychokinesis). But curiously they often seem to occur in the same individuals, and to be influenced by the same conditions. Once the existence of these discrete forms of psi had been demonstrated, it was no longer possible to interpret unequivocally the results of any ESP experiment in terms of only one psychic ability. The evidence in support of precognition using the "psychic-shuffle" technique could be explained, with somewhat less violation of causality in temporal relations, as clairvoyance by the card shuffler, or alternatively as PK effect on the cards or on the mechanical card-shuffling device. Psychokinesis results could be interpreted as involving precognition in the correct call of the die face before it actually came to rest. Telepathy could be seen either as unassisted clairvoyance on the part of the percipient, or conversely as a PK process exerted by the agent on the percipient's brain cells, causing him to "perceive" the correct answer.

These overlapping theoretical explanations of psi phenomena give rise to a feeling that psi phenomena might not actually be discrete classes of events, involving different processes and mechanisms. Instead, psi ability might be a unitary phenomenon. Our thinking about the different aspects of psi may be more a consequence of the different methods we use to elicit them than an indication of any fundamental differences in the nature of the psi process itself.

Those who have tried to theorize about the nature of the psi process have taken many different approaches. At first the search centered on physical conditions and variables. But tests with Faraday cages (Vasiliev, 1963) have ruled out most of the frequencies of the electromagnetic spectrum as "carriers." Psi does not obey the inverse square law; thus one cannot speak of it as a kind of "mental radio." Psi appears to be independent of space and time; therefore no presently known form of physical energy can account for the observed data. Speculation has turned to the esoteric fringe of contemporary physics (Oteri, 1975). Paradoxically, some theorists in quantum mechanics do not seem to feel that ESP violates any physical laws at all!

Building on a quasi-physical approach, Murphy (1945) and Roll (1966a) have speculated on the existence of interpersonal "psi fields." More within a mainstream psychological tradition, Roll (1966b) has proposed that psi information may be mediated by the personal memory structure. Stanford has developed a theory relating to unconscious motivation (1974a,b). W. G. Braud (1975) described a "psi-conducive syndrome" characterized by relaxation and imagery, and Honorton (1977) has emphasized the importance of "internal attention states." These theories are based on a noise-reduction model, which allows the "weak" psi signal more access to processing capacity (Irwin, 1978a,b).

Yet the final word has not yet been uttered in these debates. As the human race begins to explore outer space, this aspect of our inner space still remains a mystery. We search for other forms of intelligent life elsewhere in the galaxy, but we have not yet fulfilled Socrates' command to "Know thyself."

Notes

1. A brief note on terminology seems in order. ESP consists of telepathy (mind-to-mind communication) and clairvoyance (obtaining information that is not known to another human mind). Precognition refers to gaining knowledge of the future, and psychokinesis (PK) refers to physical effects of the "mind-over-matter" variety. Collectively these various abilities are known as "psi."

2. These experiments, described in further detail by Rhine and Pratt (1954), have been criticized by Hansel (1966) on the grounds that under the experimental conditions described, it was hypothetically possible for cheating to have occurred, and that the experiments were not sufficiently stringent in their supervision of the experimental subject Pearce. Hansel's charges were refuted by Rhine and Pratt (1961) and again by Pratt (1969). The reader who wishes to make up his or her own mind is referred to this very interesting exchange.

3. In spite of the evidence from his own laboratory which proved the existence of ESP, Rhine (1945, 1974a) has questioned whether telepathy actually exists. He notes that most experiments on telepathy (mind-to-mind communcation) are carried out under conditions where the percipient's clairvoyance (ESP not involving information

gained from another human mind) could actually be responsible for the results, with the agent's contribution being superfluous or irrelevant. Rhine lists several criteria that must be met to demonstrate conclusively that pure telepathy exists. Rhine's main requirement is that targets cannot be written down, spoken aloud, or exist in pictorial form, or in any other form amenable to clairvoyant perception. The only record of the target that can exist is a purely mental impression. McMahan (1946) and Birge (1948) have made ingenious attempts to meet these requirements, but the evidence for pure telepathy remains inconclusive. Along with Rhine, most parapsychologists today regard this issue as an "untestable hypothesis" and are content to allow for the possibility of clairvoyant perception in telepathy experiments. In recognition of the difficulty in distinguishing between the two, telepathy is commonly referred to as GESP (General Extra-Sensory Perception).

4. This vindication has not prevented the same criticism from being raised on subsequent occasions (see, for example, Hansel, 1966). When it comes to evaluating the occurrence of chance and coincidence against other, less parsimonious hypotheses (such as the occurrence of ESP), the need for parapsychologists to have a thorough grounding in the mathematical assumptions and techniques of statistics cannot be overemphasized.

5. In addition to these standard methodological considerations, special precautions are required depending on what form of ESP one wishes to investigate. Space does not permit more detailed coverage of the special methodological considerations for telepathy, precognition, and psychokinesis, but the interested reader will find a more in-depth discussion of these issues in Schmeidler (1977).

References

Anderson, Margaret. "A Precognition Experiment Comparing Time Intervals of a Few Days and One Year." *Journal of Parapsychology* 23 (1959): 81–89.

Birge, William R. "A New Method and an Experiment in Pure Telepathy." *Journal of Parapsychology* 12 (1948): 273–288.

Braud, Lendell W., and Braud, William G. "Further Studies of Relaxation as a Psi-Conducive State." *Journal of the American Society for Psychical Research* 68 (1974): 229–245.

Braud, Lendell W., and Braud, William G. "Clairvoyance Tests Following Exposure to a Psi-Conducive Tape Recording." *Journal of Research in Psi Phenomena* 2(1) (1977): 10–21.

Braud, William G. "Psi-Conducive States." *Journal of Communication* 25 (1975): 142–152.

Braud, William G. "Psi-Conducive Conditions: Explorations and Interpretations." In *Psi and States of Awareness*, edited by B. Shapin and L. Coly. New York: Parapsychology Foundation, 1978, pp. 1–41.

Camp, Burton H. Statement in notes section. *Journal of Parapsychology* 1 (1937): 305.

Carpenter, James C. "Intrasubject and Subject-Agent Effects in ESP Experiments." In *Handbook of Parapsychology*, edited by B. B. Wolman. New York: Van Nostrand Reinhold, 1977.

Edge, Hoyt. "The Effect of the Laying on of Hands on an Enzyme: An Attempted Replication." In *Research in Parapsychology 1979*, edited by W. G. Roll. Metuchen, N.J.: Scarecrow Press, 1980, pp. 137–139.

Epstein, Seymour. "The Stability of Behavior. II. Implications for Psychological Research." *American Psychologist* 35 (1980): 790–806.

Grad, Bernard. "Some Biological Effects of the 'Laying-on-of-Hands': A Review of Experiments with Animals and Plants." *Journal of the American Society for Psychical Research* 59 (1965): 95–127.

Grad, Bernard. "The Biological Effects of the 'Laying-on-of-Hands' on Animals and Plants: Some Implications for Biology." In *Parapsychology: Its Relation to Physics, Biology, Psychology and Psychiatry*, edited by Gertrude R. Schmeidler. Metuchen, N.J.: Scarecrow Press, 1976, pp. 76–89.

Hansel, C. E. M. *ESP: A Scientific Evaluation.* New York: Scribner's, 1966.

Honorton, Charles. "Psi and Internal Attention States." In *Handbook of Parapsychology*, edited by B. B. Wolman. New York: Van Nostrand Reinhold, 1977, pp. 435–472.

Honorton, Charles. "Psi and Internal Attention States: Information Retrieval in the Ganzfeld." In *Psi and States of Awareness*, edited by B. Shapin and L. Coly. New York: Parapsychology Foundation, 1978, pp. 79–100.

Irwin, Harvey J. "ESP and the Human Information Processing System." *Journal of the American Society for Psychical Research* 72 (1978a): 111–126.

Irwin, Harvey J. "Psi, Attention and Processing Capacity." *Journal of the American Society for Psychical Research* 72 (1978b): 301–314.

McMahan, Elizabeth A. "An Experiment in Pure Telepathy." *Journal of Parapsychology* 10 (1946): 224–242.

Markwick, Betty. "The Soal-Goldney Experiments with Basil Shackleton: New Evidence of Data Manipulation." *Proceedings of the Society for Psychical Research* 56 (1978): 249–277.

Murphy, Gardner. "Field Theory and Survival." *Journal of the American Society for Psychical Research* 39 (1945): 181–209.

Nicol, J. Fraser and Humphrey, Betty M. "The Repeatability Problem in ESP-Personality Research." *Journal of the American Society for Psychical Research* 49 (1955): 125–156.

Osis, Karl. "ESP over Distance: A Survey of Experiments Published in English." *Journal of the American Society for Psychical Research* 59 (1965): 22–42.

Oteri, Laura., ed. *Quantum Physics and Parapsychology.* New York: Parapsychology Foundation, 1975.

Palmer, John. "Scoring in ESP Tests as a Function of Belief in ESP. I. The Sheep-Goat Effect." *Journal of the American Society for Psychical Research* 65 (1971): 373–408.

Palmer, John. "Attitudes and Personality Traits in Experimental ESP Research." In *Handbook of Parapsychology*, edited by B. B. Wolman. New York: Van Nostrand Reinhold, 1977, pp. 175–201.

Pratt, J. G. "Addendum." In *Extrasensory Perception*, edited by Gertrude R. Schmeidler. New York: Atherton, 1969, pp. 54–57.

Rao, K. Ramakrishna. "The Preferential Effect in ESP." *Journal of Parapsychology* 26 (1962): 252–259.

Rhine, J. B. *Extrasensory Perception.* 1934. Reprint. Boston: Branden Press, 1964.

Rhine, J. B. "Experiments Bearing on the Precognition Hypothesis. I." *Journal of Parapsychology* 2 (1938): 38–54.

Rhine, J. B. "Dice Thrown by Cup and Machine in PK tests." *Journal of Parapsychology* 7 (1943): 207–217.

Rhine, J. B. "Telepathy and Clairvoyance Reconsidered." *Journal of Parapsychology* 9 (1945): 176–193.

Rhine, J. B. "How Does One Decide about ESP?" *American Psychologist* 14 (1959): 606–608.

Rhine, J. B. "Telepathy and Other Untestable Hypotheses." *Journal of Parapsychology* 38 (1974*a*): 137–153.

Rhine, J. B. "A New Case of Experimenter Unreliability." *Journal of Parapsychology* 38 (1974*b*): 218–225.

Rhine, J. B., and Pratt, J. G. "A Reply to the Hansel Critique of the Pearce-Pratt Series." *Journal of Parapsychology* 25 (1961): 92–98. Also in *Extrasensory Perception*, edited by G. R. Schmeidler. New York: Atherton, 1969, pp. 47–54.

Rhine, J. B. and Pratt, J. G. "A Review of the Pearce-Pratt Distance Series of ESP Tests." *Journal of Parapsychology* 12 (1954): 165–177.

Rhine, J. B., and Rhine, Louise E. "The Psychokinetic Effect. I. The First Experiment." *Journal of Parapsychology* 7 (1943): 20–43.

Roll, William G. "The Psi Field." *Proceedings of the Parapsychological Association* (1957–1964) 1 (1966*a*): 32–65.

Roll, William G. "ESP and Memory." *International Journal of Neuropsychiatry* 2 (1966*b*): 505–521.

Scherer, Wallace B . "Spontaneity as a Factor in ESP." *Journal of Parapsychology* 12 (1948): 126–147.

Schmeidler, Gertrude R. "PK Effects upon Continuously Recorded Temperature." *Journal of the American Society for Psychical Research* 67 (1973): 325–340.

Schmeidler, Gertrude R. "Methods for Controlled Research on ESP and PK." In *Handbook of Parapsychology*, edited by B. B. Wolman. New York: Van Nostrand Reinhold, 1977, pp. 131–159.

Schmidt, Helmut. "Clairvoyance Test with a Machine." *Journal of Parapsychology* 33 (1969*a*): 300–307.

Schmidt, Helmut. "Precognition of a Quantum Process." *Journal of Parapsychology* 33 (1969*b*): 99–109.

Schmidt, Helmut and Pantas, Lee. "Psi Tests with Internally Different Machines." *Journal of Parapsychology* 36 (1972): 222–232.

Shields, Eloise. "Comparison of Children's Guessing Ability (ESP) with Personality Characteristics." *Journal of Parapsychology* 26 (1962): 200–210.

Smith, M. J. "The Influence on Enzyme Growth by the 'Laying-on-of-Hands.' " In *The Dimensions of Healing: A Symposium*. Los Altos, Calif.: Academy of Parapsychology and Medicine, 1972, pp. 110–120. See also *Human Dimensions* 1 (1972): 15–19.

Stanford, Rex G. "An Experimentally Testable Model for Spontaneous Psi Events. I. Extrasensory Events." *Journal of the American Society for Psychical Research* 68 (1974*a*): 34–57.

Stanford, Rex G. "An Experimentally Testable Model for Spontaneous Psi Events. II. Psychokinetic Events." *Journal of the American Society for Psychical Research* 68 (1974*b*): 321–356.

Tart, Charles T. *Learning to Use Extrasensory Perception*. Chicago: University of Chicago Press, 1976.

Vasiliev, L. L. *Experiments in Mental Suggestion*. Church Crookham, Hampshire, England: Institute for the Study of Mental Images, 1963.

Watkins, Graham K., and Watkins, Anita M. "Possible PK Influence on the Resuscitation of Anesthetized Mice." *Journal of Parapsychology* 35 (1971): 257–272.

Coincidence and Explanation

GALEN K. PLETCHER

Skeptics sometimes dismiss putative paranormal occurrences as "coincidences." A seemingly clairvoyant dream, for example, will be partly explained by the usual mechanisms of dream formation, and the residue will be dismissed by a general waving of hands and the attribution of coincidence. Persons sympathetic to belief in paranormal occurrences fume against this kind of treatment, sometimes seeing in it a kind of unreasoning intransigence not unlike that which plagued Galileo. I wish to examine some features of coincidence in this paper. I believe that the attribution of some phenomenon to coincidence is a more serious claim than is usually appreciated, even though if not carefully defended it can indeed be an irresponsible dismissal. My general position, only part of which is developed in this paper, is that in the absence of improvement in consistency in the results of laboratory investigation of ESP phenomena, nothing short of a new (or drastically revised) explanatory framework will provide a *rational* basis for deciding that a seeming coincidence has any more significance than the skeptic sees in it (namely, none).

To call something a "coincidence" explains nothing. It does exactly the opposite: it asserts that the fact that two events are closely related—in time, and in other ways—does not need to be explained. It says more than that the relation between them cannot (at present) be explained, for if we felt that the failure of an explanation were only a matter of our (always changing) grasp of the external world, we would not call it a "coincidence." Of course we are sometimes mistaken in taking some phenomenon to be a coincidence. But then, on learning more about the situation, we no longer call it "coincidental."

My thinking about these matters has been greatly helped by conversations with several of my colleagues at Edwardsville, although they are not to blame for faults that may remain. I am particularly grateful to Professor Ronald J. Glossop for sustained and painstaking commentary, both oral and written.

It follows that to call something a coincidence is to express (even if only implicitly or perhaps even unwittingly) the opinion that it is misguided to search for an explanation (in the proper sense) of the coinciding of the phenomena at issue. It expresses the view that the explanations for each of the occurences will not involve reference to the other; and furthermore, it implies that there is not some level of explanation that will ultimately account for both of the phenomena. This is a very great deal to say. In fact, it might occur to us to wonder whether we ever are in fact justified in calling some set of events a coincidence.

We will find, I think, that most claims that two events are coincidental have reference to some suggested possibility about the relation of the events. If we say that a putatively clairvoyant dream is only coincidentally related to the event or state of affairs seemingly related to it, we do not mean that there is *no* connection between them, for the very good reason that this claim would usually be false. A putatively clairvoyant dream about some relative of mine gets at least part of its material from my knowledge of and acquaintance with that relative. If I express the opinion that it is a coincidence that the circumstances I dreamed about do in fact obtain, I do not mean to say that there is no connection between the fact that my cousin is a Caucasian and the fact that the person in my dream was also a Caucasian. But I do mean to deny, when I say that the whole thing is a coincidence, that there was a relationship between that state of affairs and my dreaming of it, such that an explanation for the occurrence of one would involve reference to the other.

It is at this point that disagreement arises between the skeptic and the person who would allege that the dream is in fact clairvoyant. The latter maintains that there *is* some relationship between the state of affairs and the dream. Moreover, the claim usually made is that the relationship is (somehow) causal, and that the dream could therefore be a legitimate basis for a knowledge claim concerning the state of affairs dreamed about. That a knowledge claim is usually not made would be attributable, according to those arguing for clairvoyance, to the fact that the dream is not ordinarily *taken* as a basis for a knowledge claim, and hence no belief is entertained on its basis. When the existence of a correspondence *is* discovered, it is on the basis of some better-established way of gaining knowledge, and we can then claim to know the state of affairs on that latter basis alone.

The "believer" in this little story will attribute the concurrence of dream and reality to clairvoyance. To what extent does that explain the phenomenon of resemblance between state of affairs and dream? It seems to offer no explanation at all, but instead to suggest a classification of the phenomenon in one of many possible ways. To call a phenomenon "clairvoyance" assimilates it to a group of such phenomena in which (1) the *usual* modes of perception do not seem to be in effect, but (2) we do not wish (for one or more reasons) to consider the resemblance a coincidence. In this way, "clairvoyance" and other names of ESP phenomena are (as para-psychologists J. B. Rhine and J. G. Pratt have recognized) "mere descriptive terms that were applied to the phenomena as they came to be identified in the early stages of the developing science" of parapsychology (J. B. Rhine and Pratt, 1957, p. 9). Such a naming operation brings us no closer to under-standing the congruence of dream and state of affairs than we are when we call them a "coincidence." But there is one crucial difference: to attribute the phenomenon to clairvoyance is to claim that the fact that the two events are closely related—in time, and in other ways—*does* stand in need of explan-ation. It is to claim, moreover, that to treat them as if they were two discrete and unrelated occurrences would be to miss some of the relationships that are to be found in the world.

But we do not know, at this point, what such relationships are. It may be that we can agree that to call a phenomenon "parapsychical" (or some synonymous term) is to say of it that it "has been shown by experimental investigation to be unexplainable wholly in terms of physical principles" (J. B. Rhine and Pratt, 1957, p. 6). But we do not know of it, as Rhine and Pratt go on to claim, that it is "a mode of perception that is independent of the senses" (p. 7). At least, we do not know this so long as our means of identifying such phenomena is that they are not accountable by means of the physical principles known to us. We can see this by means of an example.

A student sees the papers and sundry notes on my bulletin board, and suddenly becomes very interested in their "order," as he calls it. I say, in response to a question, that they are in no particular order, but that I have tacked up various bulletins, memos, calendars, and cartoons as seem to me noteworthy. He persists, and eventually it emerges that the colors of those papers on the bulletin board that are other than white are arranged alphabetically left to right. They are arranged, let us say, in the order of blue, gold, green, pink, and yellow. He now presses me to explain how they got that way. I will explain to him that this blue sheet is the official listing of Univer-sity holidays and, because of its size, fits just above this Pogo cartoon, while this green sheet was put here because . . . and so on. I will utilize explanations entirely unrelated to the colors of the sheets of paper. He, however, will press for explanations having *reference* to their colorations. For me, their color arrangement—that one feature of their configuration on the bulletin board—is coincidental. Once I have explained how I came to place each individual sheet where I did (relying on when they were received, their importance, their sizes, and the like), I will be satisfied. He, however, will urge that such an explanation of the placement of the sheets cannot fully explain their color configuration. He will maintain that it couldn't just have *happened* that they are so arranged, whereas I will maintain that it not only *could*, but it *did*, "just happen."

The difference in the two positions just outlined is that the student regards a certain feature of the arrangement of the papers on my bulletin board as standing in need of an explanation additional to those I have given, whereas I regard this same feature of the arrangement of papers as *derivative* from the explanations I have given, and not requiring additional explanation beyond those. When we try to evaluate the claims of spontaneous clairvoyance, and of other paranormal phenomena, we are in a very similar situation. If X has a dream in which a state of affairs is featured in some detail, and then later encounters a remarkably similar state of affairs in waking life, X may very will insist that some explanation *for the resemblance* is called for, in addition to those explanations that we would *all* agree can be found for the dream, and for the state of affairs. Y, a "skeptic," may well regard the resemblance as derivative, secondary, not something that calls for a special explanation unique to it. It is for this reason that they are at loggerheads over the proper assessment of the situation. Each can say, echoing Rhine and Pratt, that a certain feature of the situation is not explainable in terms of the usual principles of explanation, just as my student and I agree about that. But X and the student feel that something is "left over" which needs to be explained, whereas Y and I do not agree that it is a *deficiency* that there is no explanation for what we regard as derivative features of the phenomena. It is for this reason, primarily, that Rhine and Pratt are wrong in going on to assert that this inexplicability must be due to something independent of those principles of explanation. It seems instead to be due to what has been selected from the situation as standing in need of (separate) explanation. Hence, it does not follow from the fact that putative cases of clairvoyant dreams cannot be wholly explained in terms of physical principles that they must involve a mode of perception that is independent of the senses.

Consider one other example from Rhine and Pratt. They continually speak of psi "capacities," even though they admit that "psi" is simply a "general term to designate the whole range of parapsychical phenomena" (p. 9). It then seems natural to them to say this:

> The occurrence of interaction between psychical and physical systems [in mind and body] implies to the logical mind a basic unity suggesting that the phenomena of parapsychology and physics are both of [*sic*] the same all-embracing universe. If so then a larger scope of reality is still to be disclosed than has been as yet revealed. Physics, then, is not unrelated to psi and its operation. We can say, rather, only that psi is not describable in terms of physical processes. (p. 11)

True, but vacuously so. Phenomena to be designated as "psi" have been picked out precisely on this basis. We do not have something independently identified which is then found to act other than physical laws would lead us to expect. The phenomena are isolated *because* they (or some feature of them) act otherwise than physical laws would lead us to expect. We still have two very important possibilities open to us other than the one adopted by Rhine and Pratt, namely (1) that physical laws can be revised so as to account for these phenomena, and (2) that some features of the world have been arbitrarily, and unjustifiably, selected as *warranting* explanation.

I shall say nothing in this paper about the problem in the first possibility, but I do wish to emphasize the importance of the second possibility. We have seen that the disagreements concerning the appropriate evaluation of putatively paranormal phenomena stem partly from disagreements about what phenomena, and what aspects of phenomena, require a "non-derivative" explanation—that is, what phenomena require to be explained in their own right, and not just as consequences of other phenomena. I have emphasized this in opposition to the assumption that everyone agrees that something peculiar has happened, calling for special explanation, when someone has a putatively clairvoyant dream, or when a better-than-chance run is established in some laboratory card-calling tests. The problem with this assumption is that there are all kinds of "coincidences" that could be picked out as standing in need of explanation. There are presently as many students enrolled in one of my classes as there are days in the current month. Seventy is the mandatory retirement age in my institution, and 1970 was the year I began teaching there. And so on and so on.

Why do some cases seem so much more interesting than others? One possible response (implied in the quote from Rhine and Pratt in the middle of page 171) must, I think, be rejected. It may be said that we are interested in spontaneous phenomena that have been shown by laboratory tests—where we can specify exactly what chance, and extra-chance, results would look like—to occur with some regularity in the world. We know from card-calling experiments, it may be said, that clairvoyance is possible in certain laboratory situations. Hence all other clairvoyant cases are rendered more believable, or at least more noteworthy. This reasoning is backwards. The spontaneous cases came first, and the laboratory work was initiated in order to have a more orderly method of assessing the likelihood of such phenomena. Nor would it do, I think, to say that, however we may have come to be interested in clairvoyant phenomena, laboratory work has now shown that interest to be justified. I shall return to this point.

Why, then, are we struck by cases of seeming clairvoyance? My feeling is that we do not proceed on the basis of reasons in counting such cases as important. Instead, they strike us as important and then we, taking that importance for granted, attempt to explain the thing that struck us. I should like to devote more attention to the phenomenon of our *being struck* itself. It would seem that being correlated in time or in space cannot (*logically* cannot) be the sole criterion for our picking out of the welter of events in the world those whose relationships deserve to be investigated. If so, it must be largely a psychological matter that something strikes us as "more than coincidental."

But this seems to me to involve issues in psychology and the nature of operant conditioning that are beyond my competence. Moreover, it may be that such speculations would never go very far toward resolving the issue. For these reasons, I shall devote the remainder of this paper to an examination of some problems facing someone who wishes to provide reasons for believing that some set of events that others regard as only coincidentally associated are really related in some more interesting way.

Such an argument will be hard to make out by reference to the general nature of coincidence. For one thing, skeptics say things like this: "The number of events in which you participate for a month, or even a week, is so huge that the probability of noticing a startling correlation is quite high, especially if you keep a sharp outlook" (Gardner, 1972, p. 69). Again:

> the world in which we live is an extremely complex sample space, in which it is doubtful whether there are any "laws of chance" which apply to many of the single events occurring in it. Coincidences are certainly to be expected, and the sheer number may be felt to build up a case for a force or agent which is metaphysical, supernatural, or at least not part of the current corpus of science. But the mere accumulation of instances has less to do with probability than with the striking force of coincidence. ... Science has not ignored some underlying order; [instead,] it has not yet devised ways of protecting us against spurious evidences of order. (Skinner, 1977, p. 11)

Some people regard it as unsurprising that the world contains surprising coincidences. Moreover, it is troublesome that (as I have earlier pointed out) the coincidences that can be noted are indefinitely numerous. It would seem that so long as we remain at the level of generalities, we will not be able to decide whether coincidences are evidence of some as yet hidden order, or "spurious evidences of order."

Parapsychologists themselves look to what they believe to have been established by laboratory research as a general solution to the separation of genuine coincidences from more significant happenings. Here is an example:

> Until the discovery of ESP, the easiest way to dispose of any puzzling experience now recognizable as involving it was to say it was only a coincidence. ... Against this formula no one attempting to sort out presumptive instances of ESP from other kinds of experiences also coming from unconscious sources had a decisive argument until the actual existence of the ESP ability had been proven in the laboratory. ... Now, knowing ESP does occur, one can claim for it with a much higher degree of assurance than before those experiences that fulfill the definition. (L. E. Rhine, 1967, p. 45)

There seem to me to be some problems with this kind of confidence. I will express them in terms of our continuing discussion of coincidence, but it will be noted that these are traditional, even venerable objections against the reliance on laboratory results. Whether, as I hope, it will be helpful to cast them in this relatively novel light remains to be seen.

Someone who does not believe that there is any kind of underlying order of the sort parapsychologists seek will look at these experiments as evidence of some other misleading factor, just as I look at the fact that a woman *seems* to be sawed in two and then to be whole again as evidence that some other hidden feature(s) of the situation must exist. Nor is the skeptic obliged to turn to fraud as an explanation. There are the well-known possibilities of unconscious recording errors, faulty experimental design, and many other culprits. What the existence of these alternatives illustrates is that the assessment of such phenomena must be based on the assumption of some explanatory framework or other. Parapsychologists seem to see what they are doing in the laboratory as a *preliminary* to the provision of some kind of explanatory framework. But the way they interpret the results shows that they are assuming such a framework all along, and one that varies in a far too ad hoc way to be satisfactory. Rhine and Pratt, in discussing the effects of distance on ESP results, claim that there is no appreciable effect of distance in the Pearce–Pratt series. But they can say this only by explaining away what variation *does* occur on the basis of "psychological factors." This does not seem a warranted procedure, because we have no idea whether psychological factors should be expected to influence psi the way they do more usual perceptions. If psi violates so many other regularities, why hold this one constant? We simply do not have enough of a theory to enable us to discriminate distance variations from psychological factors. (See J. B. Rhine and Pratt, 1957, pp. 67–68; also Murphy, 1961, pp. 154–155.)

It would appear that those characteristics are assigned to psi that are necessary to explain deviations from expectations. When Basil Shackleton's results began to deviate in the experiments with Soal, it was found that his answers corresponded to future targets. He began to "hit" the target that was observed *after* the one that was being observed when he actually made the guess. Soal began to suspect precognition. When the speed of calling was doubled, Shackleton began to "hit" the target *two* places ahead of the current one. This seemed to Soal and Bateman to make good sense. But is this really a case, as Gardner Murphy claims, where the "evidence groups itself according to some inner logic of its own"? The procedure, he tells us, "is to let nature group data for us as she likes" (1961, p. 154). But surely a *choice* is involved when we see the Shackleton results as successful precognition and not (say) unsuccessful (simultaneous) telepathy, or even unsuccessful postcognition?

This kind of theoretical looseness can also be found in the evaluation of spontaneous phenomena. Louisa Rhine describes a case communicated to her laboratory by a woman in California, in which her very strong wish that her husband (stationed in post–World War II Japan) would call her at 11:00 P.M. on a certain night was fulfilled by a call at 11:05 P.M. from the husband, who had just arrived in San Francisco. He, however, knew nothing of her need to speak with him about their son's health, since her letter on that topic, with its request that he call her when he did, had reached Japan only after he had left. He had, as it turned out, simply received his orders to return to the States at an unexpectedly early time, and was calling his wife on arrival. Louisa Rhine says:

> Precognition? ... Possibly it could explain the setting of the date when her husband would telephone. But the rest—her husband's orders to return and the actual timing of his call—must be relegated to coincidence. Even if more might have been involved, it is better in studying ESP to decide conservatively. (1967, p. 44)

But why not suggest that psychokinesis was involved? We are told that the woman wished very much to speak with her husband. If relatively weak wishes about little plasticene cubes can deflect their courses of flight, why might not much stronger wishes alter the course of the dispensing of military orders? At the very least, we can easily imagine other parapsychologists taking this route, rather than agreeing to relegate such a striking coincidence to the scrap heap. The conservatism Rhine recommends is at best inconsistently applied.

An important part of our understanding of the world lies in knowing what *not* to try to explain. We neglect to explain far more than we explain, and this is a result of, as well as a crucial determinant of, the serviceability of the explanations we *do* give. To call something a coincidence is not necessarily to be mean and lunkheaded; it may point to the fact that the explanatory frameworks that are being accepted do not show this event to be something that warrants a separate, non-derivative explanation. To urge that someone treat some such event as not a coincidence after all is to encourage him or her to dislodge at least part of the explanatory framework by which sense is made of the world, and important features of it separated from unimportant features. It is not unreasonable to treat such urgings with the caution and conservatism that our very serviceable explanatory frameworks have partially justified. Demands for repeatability of experimental results have more to them than a desire to fit parapsychology into some pre-existing and questionable scientific mold. We need something to pry us out of our usually justifiable confidence in our ordinary understanding of the world. We *may* not feel any more regret about not being able to explain a putative laboratory demonstration of clairvoyance than we do about failure to account for a lucky run at cards. *Some* unexplained events and states of affairs are inevitable. Demand for repeatability is demand for one of the criteria that we ask a phenomenon to meet before it is considered other than a coincidence.

Suppose my phone rings just as I turn on my office lights in the morning. There is no one on the line. Struck by this, and by the simultaneity of the ringing with the lights' coming on, I turn the lights off and turn them on again. The phone rings again. I do this several times, and the concomitance is always repeated. At some point, I would no longer seriously entertain the suggestion that the conjunction of switching and ringing is coincidental. But now suppose that when I tried the switching the second time the phone had not rung. I continued to make tests, and to be struck by the first time's concomitance, but I could not get the phone to ring again. This would not *prove* that the switching did not cause the ringing in the first case, but it would make the contention less plausible. Moreover, it would not be rational procedure for me to fabricate some complicated explanation for why, even though the switching *did* cause the ringing the first time, it did not on any of the later trials. To insist that in certain special cases where we have no consistent repeatability, and also have no theoretical understanding of a possible connection, we must nevertheless be able to *discover* some theoretical underpinning shows a procedure remarkably divergent from our usual approach to unknown areas. To use ad hominem arguments against people who fail to agree to the strange procedure compounds the divergence. We do not expect the ringing of phones to be connected to the switching on of lights, and we demand something more than one occurrence of seeming connection before we surrender that expectation. We also do not expect seemingly haphazard guesses to conform to a random sequence of symbols with any accuracy, and it seems rational to insist on as much demonstration in this case as we would demand in the other. It is not the fact that events cannot be explained by science that will convince skeptics of the existence of hitherto unnoticed "underlying order." Something has to convince us that that failure of explanation is worthy of serious notice.

References

Gardner, Martin. "Arthur Koestler: Neoplatonism Rides Again" (review of *The Roots of Coincidence*). *World*, 1 August 1972, pp. 87–89.

Murphy, Gardner. *Challenge of Psychical Research*. New York: Harper & Row, 1961.

Rhine, J. B., and Pratt, J. G. *Parapsychology: Frontier Science of the Mind*. Springfield, Ill.: Charles C. Thomas, 1957.

Rhine, Louisa E. *ESP in Life and Lab*. New York: Macmillan, 1967.

Skinner, B. F. "The Force of Coincidence." *The Humanist* 37(3) (May/June, 1977): 10–11.

Parapsychology:
Science or Pseudoscience?

ANTONY FLEW

1

One thing has to be said with emphasis at the start. It is that the case of parapsychology is quite different from most of the others falling within the scope of the Committee for the Scientific Investigation of the Claims of the Paranormal.[1] It is quite different, that is to say, from the factitious, but richly profitable mysteries of the Bermuda Triangle and of the Chariots of the Gods, from astrological prediction, from the extraterrestrial identification of Unidentified Flying Objects, or from most of the other affairs dealt with so faithfully in that committee's useful and entertaining journal *The Sceptical Enquirer*.[2] The crucial difference from these other cases mentioned is that there we either know from the beginning that it is all bunkum, or else we can come to know this very soon after serious and honest investigation has begun.

Thus the moment someone concerned to discover what's what, rather than to produce a best-selling real life mystery, began to probe the Bermuda Triangle story it became apparent that there is no sufficient reason to believe that more ships and aircraft vanish without trace in that area than anywhere else with comparable traffic densities and comparable natural hazards. Again, there just is no good reason to believe that there have been any close encounters of the third kind, nor indeed of the first nor second either. The truth here is that the content of visions, dreams, and misperceptions is always in part a function of the wishes, beliefs, and expectations of the subject. So Chinese, under the old Emperors, used to dream dreams of dragons and Confucian officials; but not of Red Guards, chanting doubleplus good Chairman Mao-think. So too Bernadette Soubirois in her nineteenth century French village had a vision of the Blessed Virgin, as represented in pictures and images in her local church; but not of Shiva the Destroyer, as represented in Indian temple sculptures. So, again and likewise, when contemporary North American readers of science fiction misperceive celestrial phenomena, what they believe that they have seen is neither gods nor a dragon but a spaceship. Such false identifications are, in one of the finest phrases of Karl Marx, "the illusion of the epoch."

Parapsychology, however, is a horse of quite another color. One of the properly uncelebrated silver jubilees of 1978 was that of the publication of my own first book, entitled in English English, with all the brash arrogance of youth, *A New Approach to Psychical Research*.[3] Yet it is just worth saying here that, after reviewing the literature as it then was, I concluded there that, although there was no repeatable experiment to demonstrate the reality of any of the putative psi-phenomena, and although the entire field was buried under ever-mounting piles of rubbish produced by charlatans and suckers; nevertheless one could not with a good academic conscience dismiss the case as closed. Too much seemingly sound work pointing to the genuineness of at least some of these phenomena had been done. Too many honest, toughminded, methodologically sophisticated and often formidably distinguished persons had been involved in this work. Not even the youngest and most wholehearted of Humians could recommend that we commit it all to the flames as "containing nothing but sophistry and illusion."[4] The research had to go on.

With, it must be confessed, precious little participation by the author of "that juvenile work," the research has indeed gone on.[5] In all probability its sum in the years between is as great or greater than the total for all the years before. Yet it is hard to point to any respect in which the general situation is better now than it was then. Certainly there is still no repeatable experiment to demonstrate the reality of any putative psi-phenomenon. Now as then the experts are inclined to construe the night on night regularity of the performance of any stage or screen psychic as proof that the performance is nothing but conjuring. Even worse or—according to taste—even better S. G. Soal's work on Gloria Stewart and Basil Shackleton has been progressively discredited. This won Soal a D.Sc. from the University of London, and was hailed by so tough a nut as C. D. Broad as involving, among other things, "The Experimental Establishment of Telepathic Precognition."[6] Nevertheless, not to put too fine a point on it, Soal, who was in his later years to present the crudely fraudulent Jones brothers as *The Mind Readers*,[7] seems to have been faking the scores.[8]

Having so far, in the present Section 1, labored in the main to distance parapsychology from some wholly disreputable exercises in deception and self-deception, I intend in the remaining three sections to consider three respects in which it appears to differ from all the established high-status sciences. First, its field has to be defined negatively. Second, there is no repeatable demonstration that it does in truth have its own peculiar and genuine data to investigate. And, third, there is no even half-way plausible theory with which to account for the materials it is supposed to have to explain.[9]

2

In his Gifford lectures, Sir Charles Sherrington remarked that the names given to the vitamins were at first "non-committal in order that scientific ignorance should not be cloaked. Under fuller knowledge they are already being christened properly and chemically. Vitamin C is ascorbic acid . . ." [10] It is now usual for parapsychologists to begin by following this excellent example; although here, regretttably, there is no sign of progress towards legitimate re-christenings. "Parapsychology" is thus defined as "the study of the psi-phenomena"; "psi" being the name of the initial letter of the Greek word from which our "psychic" is derived. Psi-phenomena are divided into two fundamental categories: psi-gamma; and psi-kappa. "Gamma" and "Kappa" are again names for Greek letters; the initial letters, respectively, for the Greek words for knowledge and movement. The word "psi-gamma" covers both of what are elsewhere more tendentiously described as "clairvoyance" (clear seeing) and "telepathy" (distant feeling). The word "psi-kappa" substitutes for the equally tendentious "psychokinesis" (movement by the mind).

(i) We speak, or would speak, of psi-gamma when some subject comes up with information; and when that subject's acquisition of this information cannot be put down either to chance, or to peception, or to inference from materials ultimately obtained through sensory channels. These phenomena, or alleged phenomena, are then subdivided in two ways. One distinction is between clairvoyant and telepathic conditions. The idea is to distinguish two kinds of psi-gamma information: that already available to some person other than the subject, and presumably being somehow acquired from that other person; and that not available to any other person but immanent in the non-personal world, and presumably being somehow acquired directly from that non-personal world. The tradition, strongly challenged yet dominant still, takes a Platonic-Cartesian view of the nature of man for granted. So it describes the former as mind to mind, the latter as matter to mind.

The other distinction refers to temporal order. If the information produced in or by the subject is only going to become normally available in the future, then it is usual to speak of paranormal precognition or of precognitive psi-gamma. With appropriate alternations the same formula will give the meanings of "paranormal retrocognition" and "retrocognitive psi-gamma." When there is no such qualifying adjective we may take it that the psi-gamma is neither precognitive nor retrocognitive but simultaneous.

Once these several definitions are given and understood, it must become immediately obvious that it is inept—not to say perverse—to characterize psi-gamma as a new form of either perception or knowledge.

(a) If the word "extra" in the expression "Extra-Sensory Perception (ESP)" is construed as meaning outside of—like the "extra" of "extra-marital sex"—then that expression becomes self-contradictory. It becomes equivalent to "extra-perceptual perception"; and hence, as Thomas Hobbes would have had us add, parallel to "incorporeal substance." If, on the other hand, "extra-sensory" is interpreted as referring to an hypothetical additional sense, then that hypothesis is at once falsified by two decisive deficiencies. First, there is no bodily organ or area masking or local anaesthetization of which suppresses psi-gamma. Second, there is no accompanying sixth mode of sensory experience as different from visual, tactual, gustatory, auditory, and olfactory as each of these is different from all the others. For good measure we may conclude the paragraph by mentioning a further deficiency. It seems that the subjects who come up with the information are unable at the time to recognize the deliverances of this supposed new sense, and to distinguish them from plain ordinary guesses or hunches or imaginings.

(b) Since psi-gamma information is defined as precisely not being acquired through the senses it really is, as has just been urged, perverse to insist upon thinking of psi-gamma in terms of a perceptual model. It is almost equally perverse to think of such information as constituting a kind of knowledge. For the definition stipulates that the subject must not be in a position to know, either on the basis of perception, or on the basis of inference from antecedently available material. If, but only if, subjects were at the time of coming up with the information able to pick out some of the items as coming from a fresh, special and reliably veridical source; then indeed we might quite properly begin to speak of belief in the truth of these items as knowledge, knowledge duly grounded in that source or faculty. But that ability is no part of the accepted definition of "psi-gamma." Nor would it be sensible to require it by adding a further clause. For it appears that such ability is rather seldom claimed, and never in fact found, among those responsible for what are, on the established weaker definition, ostensible cases of psi-gamma.[11]

Since the suggestion that we have here a fresh form of knowledge is, for the reasons given, wrong, I regret that no one took up my proposal to make the temporal distinctions by applying to the Greek noun "psi-gamma" the appropriate member of a trio of more familiar Latin letters: M (for minus, replacing retrocognitive); S (for simultaneous); or P (for plus, replacing precognitive).[12]

(ii) For present purposes the most important feature of the definitions presented in Subsection (i) is that they stipulate what psi-gamma is not, rather than what it is, or would be. In the opening words of one especially thoughtful address: "The field . . . must be unique in one respect at least: no other discipline, so far as I know, has its subject matter demarcated by exclusively negative criteria. A phenomenon is, by definition, paranormal if and only if it contravenes some fundamental and well-founded assumption of science." [13]

(a) There is, I imagine, no disputing but that this must make it harder to establish that there really is psi-gamma: the difficulty of proving negatives is notorious and trite. But some other consequences are less obvious and more disputatious. Take first the points of Subsubsection 1 (i) (b), above, and especially the last two; that subjects are not able to pick out items as coming from a fresh, special, and reliably [11] veridical source; and that it neither is nor ought to be part of the meaning of "psi-gamma," that they should be so able.

From all this it surely follows, as indeed is the case, that psi-gamma can only be identified by subsequent checkups; and hence that it is not an independent source of knowledge. Thus, in the experimental work, the only way of telling whether or not we have any psi-gamma effect is by scoring up the subjects' guesses against the targets, and then calculating whether the proportion of hits to misses is too great to be dismissed as no better than what could have been expected "by the law of averages."

The case is substantially the same with what Broad would have us describe as, sporadic rather than spontaneous, psi-phenomena. There is again no way of identifying information coming telepathically or clairvoyantly save by comparing the hunches, dreams, visions, thoughts or what have you of the subject with whatever it is to which they may or may not correspond; and then estimating as best we can whether or not the degree of correspondence is greater than might reasonably be put down to chance, perception, or conscious or unconscious inference from materials ordinarily available to the subject.

The conclusions that psi-gamma as at present defined can only be identified by subsequent sensory checkups, and that it is therefore not an independent source of knowledge, carry an interesting corollary. This corollary seems to have been noticed only once or twice, and never discussed. [14] It is that, even supposing that we were able to construct a coherent concept of an incorporeal soul surviving the dissolution of its body, we could not consistently suggest that such souls might first learn of one another's existence, and then proceed to communicate, through psi-gamma.

For suppose first that there were such incorporeal Cartesian subjects of experience. And suppose further that there is from time to time a close correspondence between the mental contents of two of these beings, although such a fact could not, surely, be known by any normal means to anyone in either our world or the next. Now, how could either of these two souls have, indeed how could there be, any good reason for hypothesizing the existence of the other; or of any others? How could such beings have, indeed how could there be, any good reason for picking out some of their own mental contents as—so to speak—messages received; for taking these but not those to be, not expressions of a spontaneous and undirected exercise of the imagination, but externally provoked communication input? Suppose these two challenging questions could be answered, still the third would be "the killing blow." For how could such beings identify any particular items as true or false, or even give sense to this distinction?

The upshot seems to be that the concept of psi-gamma is essentially parasitical upon everyday, this-worldly notions, that, where there could not be perception, there could not be "extra-sensory perception" either. It is assumed too often and too easily that psi-capacities not only can be, but have to be, the attributes of something immaterial and incorporeal; mainly for no better reason than that they would be non-physical in the quite different sense of being out of touch with the scope of today's physical theories. Yet the truth is that the very concepts of psi are just as much involved with the human body as are those of other human capacities and activities. In the gnomic words of Wittgenstein: "The human body is the best picture of the human soul." [15]

(b) A second important, but too rarely remarked, consequence of the fact that the definition of "psi-gamma" is negative, is that the concept itself, and not just the best available evidence that it does in fact have some application, is essentially statistical. Consider first a standard experiment in which a subject guesses through a well-shuffled pack of Zener cards—five suits of five identical cards—while an agent, suitably concealed from the subject, exposes to himself, and briefly contemplates, each card in turn. We are, that is, supposing telepathic as opposed to clairvoyance conditions. And suppose that, after this procedure has been many times repeated, it emerges that the subject has scored significantly better, or worse, than the expected chance average rate of one in five. Then on the face of it we have a case of psi-gamma.

But now notice that we have absolutely no way of picking out from the series any single hit, or any collection of particular hits, and identifying this, or these, as due to the subject's psi-capacity, rather than to chance. Or, rather, that is misleading. It is not that we as a matter of fact at this time cannot thus divide the singly paranormal from the singly normal. The crux is that no meaning has been given to this distinction: psi-gamma just is understood as a factor which manifests itself, if at all, only in the occurrence of significant deviations from mean chance expectation over a series of guesses—or over a series of whatever else it may be.

This is also one of the reasons why it is misleading to speak of a subject who puts up a score significantly better, or worse, than mean chance expectation as doing this *by* or *by means of* telepathy, clairvoyance, or other paranormal power. For while it remains possible, at least as far as the present consideration goes, for theorists to hypothesize some so far unrecognized kind of radiation through which information is conveyed to subjects; still "psi-gamma" is at this time defined as precisely not the product of any means we can think of.[16] If the subject used the methods of the conjuror, or cheated by stealing a peek at the target cards, or had some hand in the determination of their values, then the results are on these grounds disqualified as not genuine psi-gamma.

Some have thought to dismiss the contention of this subsection on the ground that it does not apply to sporadic psi-gamma. But it does. Consider, for instance, the person who—"on the night when that great ship went down"—had a dream which both they and the parapsychologists are inclined to rate as telepathic. Their case will rest, not upon any particular correspondence between the dream images and the reality, but upon the total amount of that correspondence. A perfect fit of the whole would be the sum of fits at every particular point. Once again, there is no way of determining, and no sense in asking, which of these particular fits should be scored to chance and which to psi-gamma.

(iii) In this paper I am concentrating on psi-gamma, without asking systematically how much of what I say applies to psi-kappa. But I cannot leave the point about the essentially statistical character of the former without—not for the first time—drawing attention to a most remarkable fact. The fact is that all the evidence for the latter, and almost all the work on it, is similarly statistical. Yet the concept of psi-kappa is not. In the Glossary printed in every issue of the *Journal of Parapsychology* "psi-kappa (PK)" is defined as "the direct influence exerted on a physical system by a subject without any known intermediate physical energy or instrumentation." More popularly, it is the putative power to move something, or at least to impress a force upon it, by just willing, and without touching it or employing any electrical or mechanical device to bring this result about.

Now, it should be immediately obvious that there is no analogue here for those mere chance correspondences which investigators of psi-gamma through their statistical calculations labour to discount. There seems to be no a priori reason why psi-kappa should have to be detected and studied as the production, by a subject just "willing," of a significant surplus of sixes among the falls of dice mechanically rolled ten or more at a time; rather than as the production, by the same subject just "willing," of particular single movements in some highly sensitive and scrupulously shielded physical instrument. On the contrary: it would seem a priori far more likely that subjects would be able to direct their—shall we say?—willpower at a single stationary target than at (presumably at most one or two of the) several dice moving rapidly yet raggedly in midair. For would not such direction require a find-fix-and-strike mechanism comparable with what is needed in an anti-ballistic missile (ABM) defense system?

Of course, nature neither has to be nor is slave to our notions of the a priori probable or improbable. So it may be that in fact it is easier or only possible to deploy the force—May the force be with you, investigators!—against either a confusion of dice spinning in midair or a jostling mob of paramecia. But the observation of the present subsection must still raise questions about experimenters who seem never in the first decade or so, and rarely later, either to have effected tests of the most obvious kind, or to have provided any rationale for their long-sustained refusal to do so.[17] Rhine, I believe, spoke truer than he either knew or would have cared to know when, in 1947, he insisted: "The most revealing fact about PK is its close tie up with ESP. . . ."[18] For the uncanny resemblances between, on the one hand, the methods and findings of the experimental investigation of psi-gamma and, on the other hand, those of psi-kappa, do in truth constitute strong, though much less than decisive, reason for concluding that what we have in both cases is evidence, not so much of some previously unrecognized personal power, but rather of a lot of fraud, self-deception or incompetence—and maybe of some real statistical oddities not significant of causal connections. There certainly is "Something Very Unsatisfactory" about what is supposed to be evidence of putative personal power, yet in which there seem to be no close concomitant variations between the effects alleged and any psychological variables in the supposed effectors.[19]

3

One of the most important similarities between the two main subareas of the field of parapsychological experimentation is that, typically, the work of one investigator cannot be repeated by another, not even when the second is able to use the same subjects as the first.[20] This fact is one of several which give purchase for the representation here of Hume's once notorious arguments about the difficulty, amounting usually to the impossibility, of establishing upon historical evidence that miracles have occurred.[21] These arguments were thus a few years ago redeployed in the present context by G. R. Price.[22] This Price, it has to be said, must not be confused with two others better known in this field: the late disreputable Harry Price, who surely faked some of the Borley Rectory phenomena which he was pretending to investigate; and the most excellent sometime Oxford professor, well known for an almost Kantian integrity.

Hume, it will be remembered, contended "that no testimony for any kind of miracle has ever amounted to a probability, much less to a proof; and that even supposing it amounted to a proof, it would be opposed by another proof; derived from the very nature of the fact, which it would endeavor to establish." Confronted by such a conflict of evidence, and—the interpreter must interject—remembering Hume's unfortunate ambition to develop a psychological mechanics, "we have nothing to do but substract the one from the other, and embrace an opinion, either on one side or the other, with that assurance which arises from the remainder." However, for reasons which are not made altogether clear, "this substraction, with regard to all popular religions, amounts to an entire annihiliation; and therefore we may establish it as a maxim, that no human testimony can have such force as to prove a miracle, and make it a just foundation for any such system of religion." [23]

(i) A miracle for Hume would be much more than a fact "which ... partakes of the extraordinary and the marvellous." For, by the force of the term, "A miracle is a violation of the laws of nature . . . ." (A footnote adds a supplementary clause: "A miracle may be accurately defined, *a transgression of a law of nature by a particular volition of the Deity, or by the interposition of some invisible agent.*") [24] Waiving on this occasion the scholarly question whether Hume himself was in any position to provide an account of laws of nature strong enough to permit this contrast between a miracle and a fact which merely "partakes of the extraordinary and the marvellous," we need first to show that reports of psi-phenomena would lie within the range of Hume's argument.

(a) There is no doubt but that they would, or do. Certainly it is not easy to think of any particular named law of nature—such as Boyle's Law or Snell's Law or what have you—which would be, or is, as Hume would have it, "violated" by the occurrence of psi-gamma or psi-kappa. What that threatens is more fundamental. For the psi-phenomena are in effect defined in terms of the violation of certain "basic limiting principles;" principles which constitute a framework for all our thinking about and investigation of human affairs, and principles which are continually being verified by our discoveries. If, for instance, official secret information gets out from a government office, then the security people try to think of every possible channel of leakage; and what never appears on the check lists of such practical persons is psi-gamma. When similarly there has been an explosion in a power station or other industrial plant, then the investigators move in. At no stage will they entertain any suggestion that no one and nothing touched anything, that the explosion was triggered by some conscious or unconscious exercise of psi-kappa. Nor shall we expect them to turn up any reason for thinking that their, and our, framework assumptions were here mistaken.

It is some of these usually unformulated "basic limiting principles" which both psi-gamma and psi-kappa would, or do, violate, and which C. D. Broad formulated in his much reprinted *Philosophy* article on "The Relevance of Psychical Research to Philosophy." [25] Broad's formulations here are pervasively Cartesian. They thus provide for "the interposition of some invisible agent," if not for "a particular violation of the Deity." What, for instance, psi-kappa would violate is the principle that "It is impossible for an event in a person's mind to produce directly any change in the material world except certain changes in his own brain. . . . [I]t is these brain changes which are the immediate consequences of his volitions; and the willed movements of his fingers follow, if they do so, only as rather remote causal descendants." [26]

A Rylean, of course, would attribute any psi feats to the flesh and blood person rather than to his putative incorporeal mind or soul. But Broad, taking absolutely for granted a fundamentally Cartesian view of the nature of man, is instead so misguided as to conclude that it is the supposed establishment of the reality of the psi-phenomena, rather than this unnoticed and unargued preconception, which "has undermined that epiphenomenalist view of the human mind and all its activities, which all other known facts seem so strongly to support . . ." [27]

(b) In their first response to G. R. Price's paper "Science and the Super-natural" Paul Meehl and Michael Scriven wrote: "Price is in exactly the position of a man who might have insisted that Michelson and Morley were liars because the evidence for the physical theory of that time was stronger than that for the veracity of these experimenters." [28] It is important to appreciate why this is not so. Two of the reasons I shall consider here and now; the third is the subject of Section 4, below.

First, the Michelson-Morley experiment was not one in a long series including many impressively disillusioning instances of fraud and self-deception. Second, there was in that case no reason at the time—nor has any reason emerged since—for suspecting that the experiment would not be repeatable, and repeated; as well as confirmed indirectly by other experiments similarly repeatable, and repeated. It is these two weaknesses together which lay parapsychology wide open to the Humian challenge, each weakness reinforcing the other. The black record of fraud would not carry nearly so much weight against what might seem to be strong new cases of psi; if only we possessed some repeatable demonstration of the reality of such phenomena. We should not be in such desperate need of that repeatable demonstration; if only there had not been so much fraud and self-deception.

This is perhaps the moment to perform the nowadays mandatory genuflexion towards Thomas Kuhn's *The Structure of Scientific Revolutions*. Normal science—here to be construed as contrasting with pseudo-science rather than science in revolution—involves "research firmly based upon one or more past scientific achievements . . . that some particular community acknowledges for a time as supplying the foundation for its further practice." [29] Such an acknowledged achievement, if the acknowledgement and diploma title "science" are to be deserved, must surely embrace some measure of demonstrable repeatability. For—becoming now a brazen and reactionary non-Kuhnian—remember that the aim of science is, after discovering what sorts of things happen, to explain why: the qualification "sorts of" has to go in to cover the point that, unlike history, science is concerned with the type rather than the token. The formula for the repetitive production of a type is at the same time an initial, no doubt inadequate, explanation of the occurrence of any and every particular token of that general type, while in these two aspects together the achievement of that formula constitutes a pledge of more and better yet to come.

So, until and unless the parapsychologists are able to set up a repeatable demonstration, they will at best be making preparations for the future development of a future science—with no guarantees that these aspirations ever will in fact be realized. One moral to draw from this point, and indeed from the whole paper in which it is made, is that, if affiliation to the AAAS is thought of as a recognition of actual achievement rather than of good intentions, then the Parapsychological Association is not yet qualified for admission, and ought now to be politely disaffiliated.

(c) It is sometimes suggested, either that repeatability does not matter, or else that there already is as much of it in parapsychology as there was in say the study of magnetism before electricians learnt how to construct artificial magnets, or as there is now in abnormal psychology.[30] But these analogies break down. Certainly alleged star performers in psi-gamma or psi-kappa are, like natural lodestones or calculating boys, rare. But, when the latter are found, different investigators regularly repeat the same results. The same, unfortunately, is not true with psi.

Nor will it do to dismiss the demand for repeatability as arbitrary or unreasonable. For, if only it could be satisfied, then parapsychology would escape the Humian challenge. But, as it is, any piece of work claiming to show that psi-phenomena have occurred is in effect a miracle story. So, in order to form the best estimate we can of what actually happened, we have to resort to the methods of critical history. This means that we have to interpret and assess the available evidence in light of all we know, or think we know, about what is probable or improbable, possible or impossible. But now, as we saw earlier in the present subsection, psi-phenomena are implicitly defined in terms of the violation of some of our most fundamental and best evidenced notions of contingent impossibility. So, even before any Humian allowance is made for the special corruptions afflicting this particular field, it would seem that our historical verdict will have to be, at best, an appropriately Scottish, and damping: "Not proven."

(ii) Hume started with a general argument about the difficulty of establishing upon historical evidence the occurrence of a miracle. He then proceeded to contend that this difficulty is compounded when the miracle stories in question have "regard to . . . popular religions." So much so that he felt entitled to conclude "that no human testimony can have such force as to prove a miracle, and make it a just foundation for any such system of religion."

Whatever force Hume's worldly contentions here may have must bear equally against the miracle stories of parapsychology. For the Founding Fathers unanimously believed that to establish the reality of what we now call the psi-phenomena would be to refute philosophical materialism; thus opening the way to an empirically grounded doctrine of personal survival, even personal immortality. Frederic Myers, for instance, in his 1900 Presidential Address to the original Society for Psychical Research (London) said in as many words, that their goal was to provide "the preamble of all religions," and to become able to proclaim: "thus we demonstrate that a spiritual world exists, a world of independent and abiding realities, not a mere 'epiphenomenon' or transitory effect of the material world."[31]

Again, Henry Sidgwick in his own second Presidential Address, speaking of the motives of the whole founding group, explained how "it apeared to us that there was an important body of evidence—tending *prima facie* to establish the independence of soul or spirit . . . evidence tending to throw light on the question of the action of mind either apart from the body or otherwise than through known bodily organs."[32] In his third Presidential Address he added: "There is not one of us who would not feel ten times more interest in proving the action of intelligence other than those of living men, than in proving communication of human minds in an abnormal way."[33]

In our own day J. B. Rhine's best-selling accounts of the research at Duke University present it all as proving some sort of Cartesian view of the nature of man, and refuting philosophical materialism. "The thread of continuity," he writes, "is the bold attempt to trace as much as we can see of the outer bounds of the human mind in the universe."[34] Descriptions of familiar flesh and blood creatures guessing cards, or "willing" dice to fall their way, are spiced with references to minds; their powers, frontiers, and manifestations; their unknown, delicate and subtle capacities; and the experimental findings are all construed as striking hammer blows for "spiritual values" in the global battle against "materialism." Always Rhine deplores "the traditional disinclination to bring science to the aid of our value system."[35]

It is, by the way, a noteworthy indication of the enormous power and fascination of the Cartesian picture that, as we have seen, even so acute and so unspiritually-minded a professor as Broad took it as obvious that to establish the reality of paranormal human powers is both to establish the reality of incorporeal thinking substances as the bearers of those powers, and to undermine the plausibility of the epiphenomenalist account of the relation between consciousness and the conscious organism. Yet what reason did Broad have for attributing these putative powers to such unidentified and unidentifiable metaphysical entities,[36] rather than to those familiar flesh and blood creatures who to the philosophically uncontaminated eye are the ostensible performers?

4

In Section 1, I distinguished, as the third peculiarity prejudicing the scientific pretensions of parapsychology, the fact that "there is no even half-way plausible theory with which to account for the materials it is supposed to have to explain." This deficiency bears on the question of scientific status in two ways. For a theory which related the putative psi-phenomena to something else less contentious would tend both to probabilify their actual occurrence and to explain why they do thus indeed occur. Here we have the third reason why to refuse to accept the reality of such phenomena is not on all fours with dismissing the result of the Michelson-Morley experiment. For, even if no one then was ready immediately with an alternative theory, still in that case there was no good reason to fear that such a theory could not be produced. But, in the case of parapsychology now, our investigators have had a hundred years for theoretical cogitation, while there is also reason to believe that at least some of the phenomena alleged are so defined as to be necessarily impervious to causal explanation.

(i) The situation is confused by the fact that most investigators have been, and are, attached to a conceptual scheme whose actual explanatory power they tend vastly to exaggerate. For, as we have seen, most of them, taking the Cartesian concept of soul to be quite unproblematic, are ready to construe any proof of the reality of psi-phenomena as at the same time proof of the existence and activities of Cartesian souls. So, as they come to believe in this reality, they forthwith attribute all such performances to those putative agents. When they leap to this congenial conclusion, not only do they over-look the by now surely notorious difficulties of offering any serviceable description to enable these incorporeal somewhats to be identified, in-dividuated and reidentified through time, they also fail to provide their proposed hypothetical entities with any characteristics warranting the ex-pectation that these could, and naturally would, achieve what for mere creatures of flesh and blood must be simply impossible. If, as C. W. K. Mundle put it in his 1972 Presidential Address to the Society for Psychical Research (London), "materialism is to be rejected in favor of dualism on the ground that materialism cannot explain all kinds of ESP, it needs to be shown that, and how, all kinds of ESP can be explained in terms of immaterial minds." [37]

Where detail is vouchsafed sufficient to yield a piece of discussable and even testable theory, the result is almost if not quite always a fragment, a fragment which could at best serve only to explain one kind of psi-phe-nomenon—simultaneous psi-gamma under telepathic conditions. This applies, for instance, both to Whately Carington's proposals about the association of ideas [38] and to Ninian Marshall's physicalistic postulation of an assimilative force by which all physical things tend to make others more like themselves. [39] But the evidence for straight simultaneous psi-gamma under telepathic conditions now appears to be neither substantially stronger than, nor of a significantly different kind from, the evidence for "precognitive" psi-gamma, simultaneous psi-gamma under clairvoyance conditions, or psi-kappa. So it looks as if our choice is: either to think up a comprehensive theory covering all kinds of psi-phenomena, and presumably a lot else besides; or else to go back in the end to the position which we could not in good academic conscience adopt at the beginning—that of committing the whole pseudo-subject to the flames, in high Humian style, "as containing nothing but sophistry and illusion."

(ii) Already in Subsection 2 (iii) I hinted at the great obstacle in the way of an explanation of psi-kappa: this is the problem of describing some believable "find-fix-and-strike" mechanism for directing the force "at (presumably at most one or two of the) several dice moving rapidly yet raggedly in mid-air"; and when those dice or other larger objects may not be within the sensory range of the "willer." We have now to notice in passing a similar massive obstacle standing in the way of any attempt to explain psi-gamma under clairvoyance conditions. Suppose that someone does spectacularly better than mean chance expectation in guessing the values of the cards in a well-shuffled pack, guessing these "down through," with no one touching that pack until the complete guess-run is later scored. What conceivable mechanism could that subject have employed—unconsciously, of course—to acquire the information needed to achieve such scores? (An appreciation of the force of this question has held many psychical researchers back from accepting clairvoyance even when they have no remaining doubts about telepathy.)

(iii) These are both formidable difficulties for the speculator. But the obstacle barring the way to any explanations which accept the genuiness of P psi-gamma (precognition) is an altogether different kind, and totally decisive. For if the various conditions usually specified as essential to a genuine case of P psi-gamma are in truth all satisfied simultaneously, then what is going on just is not susceptible to explanation in terms of causes or of causally interpretable natural laws. It is significant that when Broad offered a theory "to explain precognition" he added this warning: ". . . notice that, on this theory of 'precognition,' no event is ever 'precognized' in the strict and literal sense."[40]

The crux is that inexplicability is built into what is here rather oddly called "the strict and literal sense" of (paranormal) precognition. For consider, to start with we must have highly significant correlations between what someone says, or does, or experiences and what is later said, or done, or happens. But though necessary this is by no means sufficient. If the subject played any part in bringing about those later ongoings, then that would be enough to disqualify that subject's anticipations as a case of P psi-gamma. Suppose next that the correlation between these anticipations and their later fulfillments could be explained in terms of some common causal ancestry: maybe the guesser is an identical twin of the person choosing the targets; and they share genetically determined patterns of guessing and choosing dispositions. Here too the correlations would be disqualified: it is indeed mainly in order to avoid disqualification by reference to a common causal ancestry that experimenters insist that the targets must be randomly selected.

But now, what possibility of causal explanation is left? If anticipations and fulfillments are causally connected, then either the anticipations must cause the fulfillments, or the anticipations and the fulfillments must both be partly or wholly caused by something else, or the fulfillments must cause the anticipations. The first two disjuncts are, as we have just seen, ruled out by the force of the (expression) term (paranormal) "precognition." The third is radically incoherent. Because causes necessarily and always bring about their effects, it must be irredeemably self-contradictory to suggest that the (later) fulfillments might cause the (earlier) anticipations. By the time the fulfillments are occurring the anticipations already have occurred. It would, therefore, be futile to labour either to bring about or to undo what is already unalterably past and done.[41]

I indicated in the first paragraph of this final Section 4 how a well-supported theory may probabilify the occurrence of whatever it predicts. So I trust that it will not look like a lapse into anti-empirical dogmatism to conclude with a maxim from Sir Arthur Eddington, a leading British physicist of the period between the wars: "It is also a good rule not to put overmuch confidence in the observational results until they are confirmed by theory."

Notes

1. This was set up a year or two ago on the initiative of Professor Paul Kurtz of SUNY at Buffalo, then editor of *The Humanist*, in hopes of doing something to stem the rising tide of popular credulity.

2. Formerly *The Zetetic*, now edited by Kendrick Frazier from 3025 Palo Alto Drive NE, Albuquerque, New Mexico 87111.

3. London: C. A. Watts, 1953. This book was long ago remaindered, and its publishers were later absorbed into the fresh-founded firm Pemberton Books. But two chapters, revised, are still current in philosophical anthologies: see, for one, Note 35.

4. Hume, *An Inquiry Concerning Human Understanding*, XII (iii), *ad. fin.*

5. See the Advertisement, referring to the *Treatise*, added by Hume to what in the event became the first posthumous edition of that *Inquiry*.

6. Compare this article under this title in *Philosophy* (London) Vol. XIX (1966), pp. 261–75. Broad was only perhaps the most distinguished of the many who accepted this work at its face value. That throng—or should I say that rout?—included the author of *A New Approach to Psychical Research*.

7. London: Faber and Faber, 1959.

8. See, most recently, Betty Markwick, "The Soal-Goldney Experiments with Basil Shackleton: New Evidence of Data Manipulation" in *Proceedings of the Society for Psychical Research* (London), Vol. LVI (1978), pp. 250–78. Compare D. J. West, "Checks on ESP Experimenters" in the *Journal* of the same society, Vol. XLXIX (1978), pp. 897–9, also further references given in these articles and in Edward Girden, "Parapsychology," Chapter 14 of E. C. Carterette and M. P. Friedman (Eds.)

Handbook of Perception, Vol. X (New York: Academic Press, 1970), pp. 385–412, especially pp. 396 ff. All this sheds a very bright light upon Soal's "calm, but perfectly devastating reply" to B. F. Skinner's contention that mechanical scoring devices should have been employed, to reduce the possibilities of cheating. (See G. R. Price: "Where Is the Definitive Experiment?", reprinted in J. Ludwig (Ed.), *Philosophy and Parapsychology* (Buffalo, N.Y.: Prometheus, 1978), pp. 197–8.

9. Had I been writing this paper now I should have needed to make some reference to the recent ganzfeld work, which does at least appear to offer more promise of full repeatability than anything else so far achieved.

10. *Man on His Nature* (Cambridge, CUP, 1946), p. 96.

11. See, for instance, C. J. Ducasse in J. Ludwig (Ed.), *loc. cit.*, p. 131.

12. Almost everyone embarking on a discussion of 'the philosophical implications of (paranormal) precognition' has in fact got off on the wrong foot by first thinking of P psi-gamma on the model of either cognition or perception or both. See, for instance, my article "Precognition" in Paul Edwards (Ed.) *The Encyclopaedia of Philosophy* (New York: MacMillan and Free Press, 1967) Vol. I. pp. 139–50; or more fully, my "Broad and Supernormal Precognition" in P. A. Schilpp (Ed.) *The Philosophy of C. D. Broad* (New York: Tudor, 1959), pp. 411–35.

13. John Beloff, reprinted in J. Ludwig (Ed.) *loc. at.*, p. 356: the first printing was in the Society's journal, Vol. XLII (1963), pp. 101–16.

14. See my "Is there a Case for Disembodied Survival?" in the *Journal of the American Society for Psychical Research*, Vol. LXVI (1972), pp. 129–44; perhaps more easily found in J. M. O. Wheatley and H. L. Edge (Eds.) *Philosophical Dimensions of Parapsychology* (Springfield, Illinois: C. C. Thomas, 1976) or—in a revised and retitled version—in my own *The Presumption of Atheism* (New York: Barnes and Noble, 1976).

15. *Philosophical Investigations*, tr. G. E. M. Anscombe (Oxford: Blackwell, 1953), p. 278.

16. So far as I know the first person to make much of this point that psi-gamma is essentially meansless was Richard Robinson. See his contribution to the Symposium "Is Psychical Research Relevant to Philsophy?", originally published in the *Proceedings of the ARistotelian Society,* Supp. Vol. XXIV (1950), but reprinted in J. Ludwig (Ed.) *loc. cit.* Thd relevant passage is at pp. 80-3 in that reprinting.

17. Compare my "Something Very Unsatisfactory," in the *International Journal of Parapsychology*, Vol. VI (1964), pp. 101–5.

18. See E. Girden "A Review of Psychokinesis (PK)" in the *Psychological Bulletin* Vol. LIX (1962), at pp. 353 ff.

19. Compare Note 17 above. The only apparent exception to the generalization in the text is the sheep/goat work initiated by Dr. Gertrude Schmeidler: the "sheep," who believe in the reality of psi-phenomena, usually put up better scores than the "goats," who do not. But, even if the findings here were unequivocal, they would not provide the sort of dramatic concomitant variation reported in—and, apparently, reported only in—the since discredited work of Soal: there when the conditions were suddenly and secretly changed from telepathy to clairvoyance one subject's scores at once dropped to a chance level, recovering equally immediately when the original experimental conditions were restored.

20. Compare, again, Girden *loc. cit.*

21. I develop and defend an interpretation of these arguments in Chapter VIII of my *Hume's Philosophy of Belief* (New York: Humanities Press, 1961). So here I follow without defending that same interpretation.

22. This Humian challenge was first published in *Science* CXXII (1955) No. 3165, pp. 359–67; but it is now more easily found, along with one or two counter-blasts, in Ludwig *loc. cit.*

23. *An Inquiry Concerning Human Understanding*, X (ii); p. 127 in the standard Selby-Bigge edition. I suppose that "the principle here explained"—a principle which so remarkably ensures that "substraction" of the lesser quantity from the greater, neither leaves the greater undiminished, nor diminishes it by the amount "substracted," but instead always yields a zero remainder—is the sum of all the considerations, which Hume has been deploying in Part II of this Section X, for thinking that the evidence for the occurrence of miracles, which, if they occurred, would support "popular religions," is peculiarly rubbishy and corrupt. The upshot should be that it is all this superstitious evidence which is thus annihilated, and hence that there is here nothing to "substract" from the strong contrary evidence that whatever laws are in question do in fact obtain universally. But this is not what Hume actually wrote; which is, it has to be admitted, just muddled.

24. *Ibid.*, pp. 113, 114 and 115*n*: italics original.

25. *Philosophy* for 1949 (Vol. XXIV), pp. 291–309; reprinted in J. Ludwig *loc. cit.*, pp. 43–63.

26. *Ibid.*, p. 46: italics original. This principle is Broad's "(2) Limitations on the Action of Mind on Matter."

27. *Ibid.*, p. 63; and perhaps compare my *A Rational Animal* (Oxford: Clarendon, 1978).

28. J. Ludwig *loc. cit.* pp. 187–8.

29. (Chicago: University of Chicago Press, 1962), p. 10.

30. For the second of these see M. Scriven "The Frontiers of Psychology" in R. G. Colodny (Ed.) *Frontiers of Science and Philosophy* (Pittsburgh: University of Pittsburgh Press, 1962), reprinted in Wheatley and Edge (Eds.), *loc. cit.* pp. 46–75. The relevant paragraphs are in this reprint at pp. 64–5.

31. *Proceedings of the S. P. R.*, Vol. XV, p. 117.

32. *Ibid.*, Vol. V, pp. 272–3.

33. *Ibid.*, Vol. V, p. 401.

34. *The Reach of the Mind* (London: Faber, 1948), p. 50.

35. *Telepathy and the Human Personality* (London: Society for Psychical Research, 1950), p. 36. For critiques of such Cartesian misconstructions compare: first, my "Minds and Mystifications," in *The Listener* Vol. XLVI (1951), pp. 501–2, reprinted in P. A. French (Ed.) *Philosophers in Wonderland* (St. Paul, Minn.: Llewellyn, 1975) pp. 163–7; and, second, "Describing and Explaining," in the book mentioned in Note 3, revised and reprinted in J. Ludwig (Ed.) *loc. cit.*, pp. 207–27.

36. Compare Terence Penelhum "Survival and Disembodied Existence," a potpourri from his book under the same title, cooked up specially for inclusion in Wheatley and Edge *loc. cit.*, pp. 308–29.

37. "Strange Facts in Search of a Theory," in Wheatley and Edge (Eds.) *loc. cit.* pp. 76–97: the sentence quoted is at p. 88.

38. *Matter, Mind and Meaning* (London: Methuen, 1949), pp. 203 ff.

39. ESP and Memory: A Physical Theory," in the *British Journal for the Philosophy of Science* Vol. X (1960), pp. 265–86.

40. *Religion, Philosophy and Psychical Research* (London: Routledge and Kegan Paul, 1953), p. 80.

41. Compare the unusually hard-hitting symposium "Can an Effect Precede Its Cause?", in *Proceedings of the Aristotelian Society*, Supp. Vol. XXVIII (1954), pp. 27–62. By altogether ignoring the second contribution A. J. Ayer contrives to repeat in

his *The Problem of Knowledge* (Harmondsworth: Penguin, 1956), pp. 170–5, the main mistakes of the first.

Nowadays someone is likely to object that modern physics gives hospitability to the notion of backwards causation. Here I can and will do no more about this than quote yet again from Broad, this time pillaging his contribution to J. R. Smythies (Ed.) *Science and ESP* (London: Routlege and Kegan Paul, 1967), p. 195: "May I add that it would not be enough to cite eminent physicists who talk as if they believed this. What is nonsense if interpreted literally, is no less nonsense, if so interpreted, when talked by eminent physicists in their professional capacity. But when a way of talking, which is nonsensical if interpreted literally, is found to be useful by distinguished scientists in their own sphere, it is reasonable for the layman to assume that it is convenient shorthand for something which is intelligible but would be very complicated to state in accurate literal terms."

Philosophical Difficulties with Paranormal Knowledge Claims

Jane Duran

Are precognition, telepathy, and clairvoyance possible? This is, to some extent at least, a philosophical question, and one that has been treated by philosophers. The field of paranormal research in general—whatever may be encompassed by that label—calls for philosophical examination precisely because some paranormal claims seem to clash with our twentieth-century presuppositions about reality.

In this paper I want to review and evaluate some of the philosophical charges that precognition, telepathy, and clairvoyance are not possible. In a latter section I shall concentrate in particular on the claim that any of the phenomena commonly labeled telepathy, clairvoyance, or precognition can be said to amount to knowledge.

Conceptual Difficulties with Psychic Phenomena

In "The Relevance of Psychical Research to Philosophy," the philosopher C. D. Broad [1] lists that he terms *basic limiting principles*, part of the intellectual framework we carry with us in the conduct of our ordinary affairs. Some of these limiting principles appear to conflict with the claims of parapsychology. If they do so conflict, and if we have good reason to stick with such limiting principles, we have good reason to view parapsychological claims with a suspicious eye.

Broad's *basic limiting principles* are these:

(1) *General principles of causation.* Stated crudely, the import of this set of principles is that any event that is said to cause another event (the second event being referred to as an "effect") must be related to the effect through some causal chain, and must precede the effect temporally.

(2) *Limitations on the action of mind on matter.* The only kind of direct change in the material world that can be produced by "mind" is one of changes in brain state. All other putative effects are indirect, not direct.

(3) *Dependence of mind on brain.* A necessary condition of any event in what we refer to as "the mind" is an event in the brain itself. Mental events of differing persons are the direct results of events in differing respective brains.

(4) *Limitations on ways of acquiring knowledge.* If a person makes a knowledge claim that is not spurious, whatever is known could only be known to the epistemic agent as the end product of a causal chain that at some point involved the sense organs of the knower.

It is easy to see that some of the types of events that fall under the rubric of the parapsychological would apparently violate one or more of these principles were they actually to occur. It can be argued that precognition, which may be defined as knowledge of an event before it has occurred, violates the causal principles articulated in (1), and violates the epistemic principles adumbrated in (4). Telepathy, or communication between parties with no physical interaction involved, has been deemed to violate (4). If vision must be causally related to what is seen, clairvoyance—or the ability to visualize events that are not physically present—would seem to violate both (1) and (4).[2]

All of this poses a problem for the possibility of parapsychological phenomena, however, only if principles (1) through (4) are as basic as Broad takes them to be and only if the apparent conflicts cited above are more than merely apparent. Let us take each of these questions in turn, beginning with an examination of Broad's principles themselves.

Broad makes the following claims with regard to his limiting principles.

> These principles do cover very satisfactorily an enormous range of well-established facts of the most varied kinds. We are quite naturally inclined to think that they must be all-embracing; we are correspondingly loath to accept any challenged fact which seems to conflict with them; and, if we are forced to accept it, we strive desperately to house it within the accepted framework. But just in proportion to the philosophic importance of the basic limiting principles is the philosophic importance of any well-established exception to them. The speculative philosopher who is honest and competent will want to widen his synopsis so as to include these facts; and he will want to revise his fundamental concepts and basic limiting principles in such a way as to include the old and new facts in a single coherent system.[3]

In considering the principles as formulated, we must first note that those who are "naturally inclined to think that they must be all-embracing" are in general only the educated minority of western societies. Broad no doubt takes this for granted, but it is worth calling attention to. Many or most of the inhabitants of more technologically primitive societies, and many less educated inhabitants of western nations, are not naturally inclined to think along the lines of these principles, nor would they be inclined to think that such principles were all-embracing. Principle (3), for instance, has received the support it has only within the past few hundred years. Historically, the tendency has been to separate the mental or spiritual from matter, rather than to assert its dependence on it. Then, too, principle (4) conflicts with the long-standing traditions of many, if not all, of the world's major religions. To the extent that a religion has embraced any of the doctrines of its mystical subsects, it has usually held that some knowledge not obtained through the senses is possible.

But even if we grant that to certain educated people the limiting principles seem to have enormous appeal, we must also examine how it is that phenomena such as telepathy, clairvoyance, and precognition violate one or more of these principles, if in fact they do. Let us take each principle in turn.

a. Principle (1) is highly important, and the standard line has it that precognition and clairvoyance would violate this principle. Precognition would apparently demand that an earlier event (a vision or dream of a plane crash, for example) be caused by a later event (the plane crash itself). True clairvoyance would seem to require causal action without any intermediate sequence of events. But our post-Humean model of causality is such that it is simply what we mean by "A causes B" that A precedes B temporally and that B is directly or indirectly the result of A through some sequence of events.

C. J. Ducasse, C. W. K. Mundle, and Antony Flew (in the preceding selection) have each emphasized this point. Ducasse insists on the importance of temporal sequence to causality:

> But I say that the only conclusion that is open to any English speaking, or rather English understanding reader, when he is told that an effect may well be supposed to precede its cause, is either that the asertion is startlingly false, or else that in spite of appearances, it is not really expressed in English, but in some other tongue in which the words "cause" and "effect" also occur, but surely with very different meanings indeed than in English.[4]

Mundle notes that:

> It appears self-contradictory to say that an event which has not yet happened, not yet "come into existence", could be either an "object of non-inferential knowledge" or "a cause-factor influencing what happened earlier." The latter phrase is certainly self-contradictory as the word "cause" is normally used, for it is part of the meaning of "cause" that a cause must precede its effects.[5]

But, it might be objected, is it not possible that usage of words is the crux of the matter here? Could it be argued that the notion of a cause's preceding its effect is strictly conventional? One might find this sort of argument appealing, and might thus propose a revised version of causality. A cause, one might say, is something related to its effect directly or indirectly, preceding it or occurring subsequently, such that the effect is somehow contained in the cause and the usual criteria of one following on the other—"constant conjunction"—apply. On this view, it might not be unnecessarily procrustean to say that my dream of a plane crash was caused by the plane crash (which occurred at a later time) simply because the sequence of events in the dream seemed to be contained in the sequence of events in the plane crash itself. Isomorphism, or congruence, would then be one of our criteria for isolating a cause; if the cause seemed to take place later than the "resultant" event, we might label as the cause whatever later event proved to be most congruent to the earlier occurrence.

It appears to be something like this line that currently appeals to a number of those who believe, however vaguely, that their precognitive dream or visual experience is causally related to the concomitant later event. The problem with all this is that it still ignores other crucial aspects of what it means for a thing to function as a cause. If my dream of a plane crash is, on my view, the result of a given plane crash, and if the dream may be said to have taken place at time t_1 and the crash itself at time t_2, what I am using as my criterion for causality here is the putative isomorphism, and nothing more. Not only is it unclear how a causal chain could work backward in time; in this instance we have nothing resembling constant conjunction because very few dreams turn out to be isomorphic to later events. (It is worth noting that the number is so few that it is commonly held to be a matter of note when there is some congruency.) In other words, temporal precedence, the establishment of a traceable chain, and contiguity over a series of similar repetitions are at the heart of our notion of cause. *Mere* word usage cannot be cited as the crucial factor here. Thus the charge that precognition and clairvoyance violate principle (1) does seem to be well-founded, and not answerable by a proposed change of the meaning of "cause." But what of other supposed inconsistencies?

b. Principles (2) and (3) are linked, and might be held to constitute an argument against the possibility of telepathy.

Some version or variant of principle (3) is today referred to in philosophical circles as an "identity theory"—crudely put, the identity claimed is between minds and brains.[6] Some identity theorists would simply claim that thoughts or beliefs *are* brain states. But the minimum requirement for an identity theory would seem to be that what one *does* as a result of what one is thinking is the result of changes that are first made at the neurological level of brain cells, and that what one *thinks* is somehow the result of activity in these cells or other neurological areas in the first place. But if I were to obtain some piece of information telepathically, I have obviously gotten it in a way not involving the senses (presumably also violating principle (4)), and since the senses themselves are part and parcel of the neurological system, it would appear that principle (3) is violated. Quite simply, telepathy would demand that I know something without the involvement of the brain.

But some philosophers, John W. Godbey Jr. among them, have argued that some sort of identity theory is in fact compatible with this type of parapsychological claim. Godbey's counterargument to the standard objection is disarmingly simple:

> The fact that people can learn about distant facts other than by present sense perception, memory and inference does not show that their minds are not brains. Such data would be consistent with, and in fact would seem to imply, merely that these people can acquire information in ways other than normal. Parapsychological data only demonstrate capacities of the mind which exceed any known capacities of the brain. [7]

The key sentence is the last one. A claim of telepathic experience does not show that the mind is not dependent on the brain; it may merely show that there is more to the principle that minds reduce to brains—or more to brains—than we are now aware of. The general line in Godbey's response is that what is "paranormal" about many of the cases that might appear, on their face, to violate principle (3) is the manner in which the knowledge was acquired. But the fact that the *manner* was "paranormal" does not itself show that the mind is not a brain, nor that the mind is something more than a brain.

Godbey's argument seems well taken. It may be the confusion or ignorance regarding the manner in which the knowledge was acquired that has led some to conclude illegitimately that paranormal manner implies paranormal mind substance. In any case, although objections to some paranormal claims on the grounds that they violate principle (1) appear acceptable, according to our earlier analysis, this is not the case for all objections based on the point that principle (3) or a conjunction of principles (2) and (3) is violated. It is by no means clear that a telepathic claim violates principles (2) and (3), or either one separately.

c. Let us finally consider the fourth of Broad's basic limiting principles. We have acknowledged at an earlier point that objections to both precognitive claims and claims of clairvoyance stem partially from the fear that principle (4) is violated. It now looks as if this principle would be relevant to the claims of telepathy as well. Principle (4) says somewhat long-windedly that knowledge may be had only if it is acquired through the senses. Interestingly enough, philosophers argued for centuries—until a comparatively short time ago—about the relative importance of sensory knowledge in comparison to knowledge obtained through "other" sources; so this principle, too, is one that is readily acceptable only to a certain modern cast of mind. Nevertheless we need to inquire as to whether any sort of claim of the three types already mentioned actually would violate principle (4).

Now it seems to be part of what is meant by a claim that precognition has occurred that the senses are only peripherally involved in the act of precognition, if at all. That is, we might concede from the outset that there is a certain type of knowing before the fact that involves the senses—rapid and unusually acute accumulation of evidence, astoundingly acute induction, and so on—but that is not paranormal in the sense that my knowledge claim about a plane crash is if such a claim is based solely on a dream. The objection regarding the senses is not that a vision makes no use of any sense whatsoever, but simply that it arises in a way deviant from any way that might admit of normal explanation.

We know, for example, that certain intoxicating substances, including alcohol if ingested in sufficiently large quantities, can give rise to visions, yet the ingestion of these substances themselves accounts for the stimulation of the senses and seldom, it seems, is involved with any paranormal claim to knowledge. In the more typical case dubbed "precognitive" the agent is doing nothing that might be thought to be causally related to the vision—it arises spontaneously, as it were. The precognitive dream, although perhaps explicable on one level simply qua dream (unlike a vision) is frequently distinguished, so we are told, by its lifelikeness and clarity.

In these cases, then, whatever the causal chain produced by the senses, it does not seem to have its origin in anything that might explain the content of the vision or dream. Hence the claim that principle (4) is violated. A similar line of argument might be developed for the claims of telepathy or clairvoyance.

Now principle (4), if fully presented, relies on something like the following conception of sensory stimulation. In the case of my vision, I see veridically only when light rays reflected from some object within my line of sight strike my eye, stimulate the retina, and so on. No visual phenomenon not produced in this fashion could correctly be labeled "seeing" in the special case of "seeing" an afterimage or miniature pinkish elephants. Analogous statements could be constructed for the senses of hearing, smell, taste, and touch.

But do claims about precognitive visions really violate principle (4) in this regard? Not, I would claim, if the term "see" which often appears in such claims is taken to indicate only an isomorphism between the content of the vision and the later event "foreseen." It is widely reported in the research on cognition that many putative instances of precognition are so labeled only *after* the external event has taken place; Ducasse reports that "in a large percentage of them, no feeling of conviction distinguished the true ones [visions, auditory hallucinations, etc.] from the false." [8] Two points, then, can be made here. A precognitive (telepathic, clairvoyant) claim may not violate principle (4) because it may be claimed that the vision, auditory experience, or whatever is only some *analog* of a genuine case of seeing, hearing, and so on. But the use of the terms "saw," "heard," and so on in these reports (if in fact they are used) seems to stem not so much from the fact that the agent actually thinks he saw anything in the ordinary sense as from the fact that he had a visual experience that gave rise to "knowledge" in a way analogous to the way in which visual perception gives rise to knowledge in the veridical case.

In any case, it is clear from the above that even if we may back off from claiming that principle (4) is violated in these instances of paranormal claims, we seem to be left with the conviction that principle (1) is violated. In the case of precognition, the reasons for thinking that principle (1) is violated have already been stated. In the cases of telepathy and clairvoyance, it would be the inability to specify the causal chain that would give rise to the sense of violation of principle (1). In these cases, the "cause" may very well precede the "effect." But the only causal chains to which we are accustomed to refer in such cases—causal sequences involving the senses—are absent in the cases referred to as paranormal.

The upshot of our examination of the limiting principles is that it seems almost certain that at least principle (1) is violated in every purported case of paranormal phenomena of the type under scrutiny here. Whether or not other principles may be thought to have been violated, the absence of a specifiable and recognizably causal chain seems to constitute a difficult, if not insurmountable, objection to our giving a coherent account of what it means to make such a claim. As long, at least, as our ordinary notions of causality remain intact, there seem to be strong philosophical reasons for concluding that telepathy, clairvoyance, and precognition are not possible.

Our Notion of Knowledge

Another set of difficulties concerns parapsychological phenomena and our notion of knowledge. Clairvoyance, telepathy, and precognition are commonly thought of as means of *knowledge*—extraordinary means, perhaps, but means of *knowledge* nevertheless. If we were to grant that each of these is possible, we still might question whether they can be regarded as means of knowledge.

The traditional account of knowledge consists of three elements. A person can be said to know that *p* if and only if: (i) *p* is true; (ii) *p* is believed; and, (iii) adequate evidence for *p* is had. This account has come under heavy attack in recent years.[9] Nevertheless, even modified versions of accounts of knowledge are built around the core of this version, justified true belief, or JTB. More recent versions usually attempt to incorporate a fourth condition of some sort.

Claims to knowledge based on precognitive, telepathic, or clairvoyant phenomena might be said to fail even the three-pronged versions of the necessary and sufficient conditions for knowledge. Leaving aside the first condition, conditions (ii) and (iii) cannot be met in the usual straightforward fashion where the claims to knowledge arise from situations involving the paranormal.

We mentioned at an earlier point that even such information as has been alleged to have been received through, for example, an instance of precognition does not seem to present itself to the recipient as immediately believable in the same way that information obtained in a more mundane fashion usually does. As Ducasse points out:

> True guesses or visions would have title to the name of perception and knowledge only if, in their case, the guesser's or dreamer's experience contained some feature—whether sensory or extrasensory—that were more or less a *reliable sign* that the guess or dream is true. But a study by Louisa E. Rhine of more than 3000 spontaneous "precognitive" and "contemporaneous cognitive" experiences, published in the June 1954 issue of the *Journal of Parapsychology*, shows that, in a large percentage of them, no feeling of conviction distinguished the true ones from the false.[10]

Some people may uniformly believe all "information" they receive by means of precognition and the like, some may be uniformly skeptical of it all, but in either case evidential matters seem to play no role in dictating belief for some items of "information" and skepticism for others. One might guess that certain receivers of information imparted in a paranormal manner choose to believe all of it, at least initially, since they are unable to distinguish true from false, or on some other ground. But the criterion of belief has traditionally been thought to be related to the criterion of evidence in some way; it is not merely that *p* is *believed*, but that one has *reasons for believing p*. Indeed, recent epistemology is full of examples of knowledge claims that rely on belief without rational basis and hence fail—for example, claims made on the basis of crystal-ball gazing.[11] So the kind of belief that one might have in the case of precognitive knowledge, for example, fairly clearly fails the test of fulfilling condition (ii) in the traditional three-pronged version of conditions for knowledge.

Condition (iii) is, if anything, even more problematic when viewed in this light. The paranormal experiences do not seem to be distinguishable on the basis of feelings of conviction as they occur. But for such an experience it might well be argued that "feelings of conviction" are all we have to go on. The ordinary standards of evidence clearly do not obtain in precognitive situations—we cannot appeal to the area that is itself under suspicion as being the evidential support for its claim. If one were to find a case in which a certain agent had met with *uniform* success in paranormal experiences, then that might conceivably constitute some sort of evidence (although even this is debatable), but no such situations are forthcoming.

But do conditions (i) through (iii) constitute an adequate account of knowledge? If not, it might be thought that parapsychology's failure to satisfy (i) through (iii) is no particular disqualification. And a justified-true-belief account has been challenged, using examples such as the following proposed by Edmund Gettier:

239

Let us suppose that Smith has strong evidence for the following proposition:
(f) Jones owns a Ford.
Smith's evidence might be that Jones has at all times in the past within Smith's memory owned a car, and always a Ford, and that Jones has just offered Smith a ride while driving a Ford. Let us imagine, now, that Smith has another friend, Brown, of whose whereabouts he is totally ignorant. Smith selects three place-names quite at random, and constructs the following three propositions:
(g) Either Jones owns a Ford, or Brown is in Boston;
(h) Either Jones owns a Ford, or Brown is in Barcelona;
(i) Either Jones owns a Ford, or Brown is in Brest-Litovsk.
Each of these propositions is entailed by (f). Imagine that Smith realizes the entailment of each of these propositions he has constructed by (f), and proceeds to accept (g), (h), and (i) on the basis of (f). Smith has correctly inferred (g), (h), and (i) from a proposition for which he has strong evidence. Smith is therefore completely justified in believing each of these three propositions. Smith, of course, has no idea where Brown is.

But imagine now that two further conditions hold. First, Jones does *not* own a Ford, but is at present driving a rented car. And secondly, by the sheerest coincidence, and entirely unknown to Smith, the place mentioned in proposition (h) happens really to be the place where Brown is. If these two conditions hold then Smith does *not* know that (h) is true, even though (i) (h) is true, (ii) Smith does believe that (h) is true, and (iii) Smith is justified in believing that (h) is true.[12]

If this is an adequate counterexample to "justified true belief" accounts, it appears that they are not entirely adequate. But does this vindicate parapsychological claims to knowledge? I think not. Notice how this sort of example differs from the type of problem posed by putative knowledge obtained through paranormal means. In the case of a Gettier-type example, both conditions (i) and (ii) hold—*p* is true, and is believed to be true by the agent. In the case of knowledge claims based on parapsychological phenomena, condition (i) must be ignored, since any attempt to deal with it will beg the question. But condition (ii) holds because the knower takes it that he had also fulfilled condition (iii)—he takes it that he has adequate evidence for his belief. Generally speaking, condition (iii) cannot be fulfilled under a paranormal knowledge claim, and if it should happen that condition (ii) is fulfilled, it will not be for reasons that relate to the typical condition (iii) reasons.

Thus we see that a counterargument to the effect that many apparent claims to knowledge fail the standard three criteria and that such failure constitutes no argument against paranormal claims does not go through because paranormal claims do not fail in the same way as the other cases—their failure is much more significant. Even if we were to accept clairvoyance, telepathy, and precognition as possible—despite the objections of the first section—we would still have reason to deny that they constitute means of knowledge.

Conclusion

In this brief critique of certain sorts of parapsychological claims, we have followed two major lines of argument. The first was that, even if it could be shown that some or most of Broad's basic limiting principles were not violated by the claims of precognition, telepathy, or clairvoyance, it seems clear that the first principle—the set of causal conditions—would be violated. This in itself constitutes a major objection to the claim that these types of paranormal phenomena are possible as sources of knowledge.

Second, we have acknowledged that it might be argued that certain claims to knowledge based on paranormal phenomena are somehow analogous or similar to other claims that might fail the ordinary criteria for knowledge. But we have shown that although it is true that both groups might be said to fail the traditional versions of JTB, paranormal claims do so in a significantly different way from, for example, Gettier-type problems. This would also seem to constitute a major objection to allowing such claims to count as knowledge claims.

Clearly, more work needs to be done on how principles (2), (3), and (4) might be violated before any definitive statements can be made with regard to the relations between paranormal claims and those principles. But the large conclusion of our thinking has been that, even if principles (2), (3), and (4) are not violated, the apparent violation of principle (1) combined with the violation of the relevant two portions of the JTB model should count heavily, if not decisively, against paranormal experience as actual knowledge.

Notes

1. C. D. Broad, "The Relevance of Psychical Research to Philosophy," in *Philosophy and Parapsychology*, ed. Jan Ludwig (New York: Prometheus Books, 1978), pp. 43–64; principles cited on pp. 45–49.

2. I exclude from discussion other phenomena that are frequently referred to in an exhaustive categorization of the paranormal, such as psychokinesis, survival after death, and so on, for the sake of manageability. It should be obvious that they merit similar treatment.

3. Broad, pp. 43–44.

4. C. J. Ducasse, *Causation and the Types of Necessity* (New York: Dover, 1969), p. 42.

5. C. W. K. Mundle, "Does the Concept of Precognition Make Sense?" *International Journal of Parapsychology* 6 (1964): 179–198, 182. Reprinted in Ludwig, *Philosophy and Parapsychology*, pp. 327–341.

6. Daniel C. Dennett, *Brainstorms* (White River, Vermont: Bradford Books, 1978), p. xii.

7. John W. Godbey, Jr., in *Analysis* (1975): 23.

8. C. J. Ducasse, "The Philosophical Importance of Psychic Phenomena," in Ludwig, p. 131.

9. Edmund Gettier, "Is Justified True Belief Knowledge?" *Analysis* 23 (1963): 121–123.

10. Ducasse, "Psychic Phenomena."

11. Keith Lehrer, "How Reasons Give Us Knowledge, or the Case of the Gypsy Lawyer," *Journal of Philosophy* 68 (1971): 311–313.

12. Gettier, pp. 122–123.

Precognition and the Paradoxes of Causality

BOB BRIER AND MAITHILI SCHMIDT-RAGHAVAN

The logical impossibility of a future event causing an event in the past is a well-known and widely accepted thesis. We propose to challenge the soundness of this dominant belief in the light of the results of an experiment on precognition conducted at the Institute for Parapsychology (formerly the Duke Parapsychology Laboratory) in Durham, North Carolina.[1] In what follows we shall argue that an explanation of the phenomenon of precognition involves a revision of the concept of cause. Once a surprising aspect of that concept is admitted, then precognition will be more easily assimilated into our world view. We shall first give an account of the experiment and its results. An atemporal notion of cause will then be defended with a view to accommodating precognition. Finally, we shall turn to particle physics and show the relevance of our conclusion regarding backward causation to that field.

There had always been talk around the lab of possible bizarre effects in ESP experiments. In the experiment we shall describe, we tested one of these possibilities, the effect of the checker (the person who scores a test) in precognition experiments. We had always wondered what the target was in a precognition experiment. Did the subject "see" the final record sheet? Did he "see" the cards in experiments where cards were used? Did he "see" the person who was scoring his record sheet, as well as the correct answers? If the subject perceived not only the answers, but the entire surroundings, then the surroundings could affect the scoring. If he did not like the person checking his guesses he might score negatively, and if he liked the person he might score positively.

In our experiment we tested groups of students by the standard precognition technique using random numbers. After the subjects had made their guesses, an entry point into the random number tables was found and the numbers were translated by a code into the five ESP symbols. Finally, a

243

checker entered the correct answers on the record sheet and tabulated the hits. Each subject did four runs of twenty-five guesses. In our experiment, the "experimenter" checked two of the runs and an "assistant" checked the other two. The subject never knew who would check which runs because this was randomly determined after the guesses were made and after the entry point into the table of random numbers was obtained.

The point is that, in this type of experiment, once the entry point is obtained, the scores are completely determined and it would seem that nothing done after can influence the scores. But in our experiment this was *not* the case. It mattered who scored the runs! There was a significant difference in scoring levels between the runs scored by the "experimenter" and those scored by the "assistant." The experiment was repeated a second time and similar results were obtained.

The phenomenon of precognition (as distinct from other forms of ESP such as clairvoyance or psychokinesis) appears to contradict certain widely accepted "a priori truths." For example, in this experiment it would seem that once an event (the hits being determined by the entry point into the table of random numbers) is fixed, nothing later (the selection of the checker) can affect it because the event is in the past. This self-evident truth seems to be contradicted by our results.

Backward Causation

We believe that it is possible for causes to come after their effects and that this backward causation may help explain how precognition works. Consider a standard precognition test in which a subject predicts which of five ESP symbols (star, waves, circle, square, or plus) will be selected the following day by a random procedure. If the subject repeatedly scores significantly above chance, we may be justified in believing that there is a causal connection between the guesses on one day and the actual order of the symbols on the following day. Which causes which? It is possible that the guess causes the selection of the symbols in some mysertious way. It is also possible that by backward causation the selection of the symbol causes the subject to have guessed that particular symbol. We could check to see if the process of selection is in fact random. (We could do this by checking the symbols generated for patterns or indications of non-randomness.) If it *is* random, a possible conclusion is that it was unaffected by the guesses, and it is the guesses that are caused by the symbols selected. This is not airtight, but it is the kind of evidence that might help to decide which of the two options to select. We might also consider spontaneous cases of precognition.

If a psychic precognizes a plane crash which in fact occurs on the following day, we again might wish to say that there is a causal connection between the plane's crashing and the precognitive experience—but which causes which? Here it seems more likely that the plane crashing caused the precognitive experience rather than the other way around. Even if we knew of the precognition, we would still look for the effective cause of the crash. We would search the wreckage for signs of metal fatigue, an improperly hatched door, and so on. Thus we might locate the cause of the crash; but we must still locate the cause of precognition. It would seem to be the crash. So we would have some basis for choosing backward causation as our description of this pair of related events, as opposed to the equally mysterious forward causation. But this choice might be spurious.

Here it might be argued that backward causation is logically impossible. We are aware that most philosophers [2,3,4] who have considered the question of the logical possibility of backward causation have concluded that it is not logically possible, that in every instance where backward causation is assumed some contradiction can be derived. Elsewhere [5] the argument has been made that most philosophers who attempt to derive a contradiction from backward causation make some error. Indeed, they usually derive their contradictions by assuming that if backward causation were possible, then one could change the past. If something I do now can have its effect yesterday, or last year, why can we not change the past? Perhaps we are great fans of Napoleon and wish that he had won at Waterloo. Can we change history? The answer is no, we cannot *change* the past, but we can *affect* it. To make this clear, we shall borrow an example made up by Richard Gale.

Imagine that you are at a party on Saturday night and someone who works with your good friend Smith mentions that yesterday, Friday, he saw someone place a bomb in Smith's desk drawer. The bomb was set to go off in one hour. He also tells you that urgent business called him out of the office and he forgot to warn Smith. As a matter of fact, he has forgotten about the incident until now and does not know what happened to Smith. Is there anything you can do to save your friend? Seemingly not. Either he was blown up by the bomb, in which case he is dead, or he or someone else may have discovered the bomb and safely removed it (or the bomb did not go off, etc.), and he is now alive. In any case, there is nothing you can do about it. *If*, however, backward causation is possible, there may be something you *can* do to save your friend. While at the party you might call out, "Smith, this is me, warning you from the future. There is a bomb in your desk drawer. Get rid of it!"

So far there is nothing impossible about the situation we have described; it is merely somewhat bizarre. But now let us imagine that you meet Smith on the street the day after the party. He tells you of his strange experience on Friday. He was sitting at his desk and he had a vision of you at a party saying, "Smith, this is me, warning you from the future. There is a bomb in your desk drawer. Get rid of it!" He then found the bomb and safely disposed of it. What do we say of such a situation? We would say it is an example of backward causation. Something you did on Saturday (the warning called out at the party) caused something to have happened on the previous day (Smith's precognitive experience which in turn caused him to have been saved). Note, we do not say that you *changed* the past. It is not that Smith was blown up and then your warning changed this. Such a situation does involve a con-tradiction of the form $p \cdot \sim p$: Smith was blown up and Smith was not blown up. Had Smith been blown up, then you still could have tried to save him, but it simply would not have worked. What is the case is that if Smith was saved, it could have been because of something you did.

Once it is realized that backward causation involves merely affecting the past, and not changing it, then the contradictions dissolve. If this conclusion is correct, then it may have some relevance to particle physics where a rather similar controversy still continues.

Atemporal Causality

Our defense of an atemporal view of causality will consist in arguing for the soundness of the following theses:

1. Causal attribution imposes no requirement that the cause precede the effect; in other words, the condition of the temporal priority of cause need not be met in order to warrant the conferring of causality on a verified relationship between facts.

2. If no contradiction can be drawn by entertaining the idea of backward causation then it is a logical possibility.

A definition of causation might be approached by listing the conditions that many philosophers of science consider warrant the conferring of causality on a verified relationship between facts. Within science these causal guarantees are confined to the *necessity* and *sufficiency* of an association. Briefly, the criterion of "necessity" usually refers to the warrant for causality given to condition or event c if effect e never occurs without it. "Sufficiency" refers to one kind of invariance or "uniformity" criterion: whenever c obtains then e obtains, under conditions$_{1,2,3 \ldots}$. The relation between cause and effect is thus a relation of *existential dependence*, the cause of an event being the condition or conditions "without which the event would not have occurred, or whose non-existence or non-occurrence would have made some difference to it."[6]

And here we find that the scientific view of causality differs from the popular view. In the latter that which is the cause is apparently antecedent in time to the effect and the causal relationship is held to be the production of an effect by some power residing in a temporally prior substance. But when, as in the scientific treatment of causality, we treat the cause of an event as the condition or the set of conditions that determine the event we find that these conditions are often contemporary with the effect. The cause of a disease, for example, is not antecedent to the course of the disease, if by antecedent we mean that it produced the disease and then ceased to determine it. The presence of certain microorganisms that may be spoken of as a determining condition (i.e., cause) of the disease is contemporaneous with the disease itself. The scientific analysis of determining conditions does not make the causal relationship one of temporal sequence. There is no reason why we should not call an event a determining condition of another event even if it did not actually antedate the event.

This brings us to an essential point of distinction between the consideration of causal relations as efficient causes and the consideration of them as sufficient reasons. In the latter case we are concerned with establishing conditions under which causal status is conferred on c or conditions under which it is withheld from c; whereas in the former case our inquiry is into the conditions under which c *becomes* e. It is this that marks the distinction between the popular and the scientific view of causality. It also marks the distinction between the direct and the indirect method of difference in Mill's theory. The direct method of difference was formulated by Mill in these words:

> If an instance in which the phenomenon under investigation occurs, and an instance in which it does not occur, have every circumstance in common save one, that one occurring only in the former; the circumstance in which alone the two instances differ is the effect, or the cause, or an indispensable part of the cause of the phenomenon.[7]

The use of this method depends on finding a positive and a negative instance agreeing in every circumstance but one. It is almost impossible to observe two instances agreeing in every circumstance except one, unless it is a case of experimentation. In the experiment a factor is introduced into the situation and a certain change ensues; when the same factor is left out, there is no change. This method, which is essentially a method of experiment, gives to "cause" its popular imprint of "force" or intervention. Mill's canon for the "joint method of agreement and difference" (also called the "indirect method of difference" by Mill) is as follows:

> If two or more instances in which the phenomenon occurs have only one circumstance in common, while two or more instances in which it does not occur have nothing in common save the absence of that circumstance, the circumstance in which alone the two sets of instances differ is the effect, or the cause, or an indispensable part of the cause of the phenomenon.[8]

In this method we examine a number of positive instances to see if they agree in the presence of one common circumstance; we also examine a number of negative instances to see if these agree in the absence of that same circumstance. This method, as Mill puts it, "consists in a double employment of the Method of Agreement, each proof being independent of the other and corroborating it." [9] In the joint method we have only observation of the joint presence and joint absence of two properties, but no indication of the temporal priority of either to the other. The rate of their joint occurrence (or non-occurrence) is the essential piece of information. For so long as the rate is sufficiently high to validate judgments of cause–effect connections, the requirement of temporal priority of the cause can safely be ignored in the attribution of causal relationship to an observed connection between events. The joint method thereby allows us to incorporate the possibility of backward causation within the framework of scientific use. If you decry our scruples in preferring the scientific view of causation to the popular or "interventionist" view, we need only point out that the *model of causation* is open and that as with any explanation the choice of model is in the service of present purposes.

The causal relation is not reducible by definition to the temporal one. However, the language is also characterized by criteria for the application of the causal predicate. In some causal explanations temporal sequence is mentioned among the criteria. The notion of a criterion is still a vague one in philosophy, but this much is clear about it, that a statement of a criterion is not aptly described as empirical, analytic, or necessary. Since it is not analytic or necessary it is both possible for a property to be present when a (selected) criterion for its predicate is not satisfied and conversely possible for the criterion to be satisfied in the absence of the property. Thus it is possible for x to be the cause of y, even though x does not precede y in time, but rather succeeds y in time.

Indeed it is doubtful if the cause in its entirety is ever prior to the effect. The analysis of "cause" as that which is temporally prior to the effect derives much of its strength from the fact that the cause has been considered as separable in time from the effect. And this notion of the temporal discreteness of cause and effect (on which the Humean analysis is based) leads to the absurd conclusion that if the set of conditions we call "cause" had ceased to be before the "effect" obtained then these conditions could not be held to be determining—and if they are not determining in any way then they cannot be said to "produce" the effect. Hence the criterion of the temporal priority of cause, far from being a necessary condition for the attribution of causality, militates against the popular view of causation. And any attempt to have this condition met must end in a fraud.

The demand for the employment of the temporal-priority criterion in causal attributions is based on a misunderstanding; one deriving from the implicit acknowledgment of the dispersal-of-order argument. This argument makes use of a specific kind of phenomenon such as entropic processes. There is the well-known example of a stone being thrown into a placid pool of water, and circular waves spreading out from the center being observed. Although the reverse process—concentric waves contracting toward a center—is a logical possibility, it is never observed. If it were, it would be difficult to explain. One is used to explaining a number of coherent separated items (the waves) by referring to a single event (the dropping of the stone) causing them. If concentric waves were seen as converging at a center, it would seem that this would call for a third event that causes these waves as an explanation. Thus there seems to be a direction of influence, from a central event to coherent separated events. Thus it might be said that explanation runs to events that involve dispersal of order from events that do not. The foregoing is a sketch of a general argument showing why there is a direction of explanation with respect to the dispersal of order and suggesting why one takes the arrow of time as going in the traditional direction. It must now be shown how this relates to the direction of causation.

It is generally accepted that a cause explains its effect in a way an effect does not explain its cause. This is so even when the effect can be used to retrodict the cause. Given this, it is not unreasonable to hope that by considering how the explanatory powers of causes and effects differ, insight might be gained into the nature of causal priority. It must first be noted that when a question is asked by saying "Why?" the answer generally begins with "Because." That is, when one is asking the why of things one is asking for the cause; it is the cause that explains why the effect is. Clearly the effect does not explain why its cause is. All effect can do, if it is a sufficient condition for its cause, is guarantee that the cause occurred. This is not an explanation.

It is this unidirectional and irreversible nature of causal relationships that has been mistaken for temporal priority. We observe and call "causal" the ordering: "Ample rain: large wheat crop." We also allow an inference in reverse: "Small crop, therefore there must have been inadequate rain." But although the first sequence is "causal" we do not accord causality to the second. The asymmetry criterion is thus more important than the sequential. And the asymmetry criterion imposes no requirement that the cause precede the effect. It moves us away from thinking about temporal sequence and toward thinking about the "direction of influence."

Consider the case of a heavy metal ball resting on a soft cushion. Here is a case in which a cause and its effect are simultaneous, and temporal priority will not help to decide which is the cause and which is the effect. It would almost certainly be argued that the downward force of the ball causes the depression in the cushion, as opposed to arguing that when the ball is placed on the cushion, a depression appears and causes the ball to fall. The reason the first alternative is chosen is that taking the downward force exerted by the ball as the cause explains the depression. If the depression were taken as the cause, there would be need of an explanation as to how the depression came about. This example suggests that the kind of explanatory power possessed by a cause cannot be attributed to an effect. The causal relationship has been shown to be asymmetrical but atemporal.

The two models of causal explanations (which we have called the "popular" and the "scientific") are profoundly different, but our analysis has shown that they have the same logical texture: they are both adaptations of the principle of *sufficient reason*. Inasmuch as the backward causal explanation is an explication of this principle, it can be accommodated with confidence within the framework of causal explanations. Each kind of causal explanation is sought and achieved on a special level of inquiry, with its own data and its distinctive perspective; although in each case the mind works on the same logical principle, the principle of *sufficient reason*. As backward causal explanation does not militate against this principle, it cannot be ruled out a priori as a causal explanation.

This brings us to the second thesis, namely, the logical possibility of backward causation. The term "logically possible" is sometimes used in the sense of "logically conceivable" (i.e., it is self-consistent or implies no contradiction) and sometimes in the narrow sense of "imaginable" (i.e., it implies no absurdity). Take, for example, the expression "pitch higher than the highest pitch audible to the human ear." The response of dogs to whistles that produce air vibrations of such high frequency that the corresponding pitch is inaudible to the human ear proves that the statement, which is empirically confirmable, describes a literally unimaginable state of affairs. Although unimaginable it is logically possible in the sense that no contradiction is formally deducible from it. But that p is possible in the sense of being self-consistent does not entail that p is imaginable, as we have demonstrated. It is not enough to point out that backward causation is logically possible in the highly attenuated sense of "possible," namely, that it is self-consistent. For if we wish to make a case for backward causation we need to show that no absurdity is implied in imaging that "x causes y but x succeeds y in time." Our examples of precognitive experiences point clearly to the fact that "backward causal explanation" is not an inexplicable concept, and indicate that there are instances of the explicandum. "Causality" then can be explicated without reference to temporal priority.

Backward Causation and Particle Physics

In a now famous paper [10] Feinberg argued for the possibility of faster-than-light particles. These "tachyons" have spacelike four-momentum and consequently have velocities greater than c. Feinberg demonstrated that such particles do not violate the special theory of relativity, and he presented a quantum field theory for noninteracting tachyons. Arons and Sudarshan [11] rejected Feinberg's schema of quantization but still argued that such particles are possibilities and presented their own reinterpretation. Although tachyons have been empirically searched for, [12] physicists have questioned the logical possibility of there being such particles. [13,14]

The grounds on which the possibility of tachyons is questioned involve what have been called the *paradoxes of causality*. There are various versions of the paradoxes, but essentially they all involve the description of a particle going backward in time and the claim that this entails the paradox that one could communicate with one's past, change the past, describe the same event in contradictory ways, and so on. This has led physicists such as Arons and Sudarshan to reinterpret negative-energy particles going backward in time as positive-energy particles traveling forward in time (but in the opposite direction). It is important to note that when discussing the paradoxes of causality none of the authors ever derives a statement having the form $p \cdot \sim p$. They merely feel there is something peculiar involved and call it a paradox. If what we have suggested above is correct, then there is no reason why there could not be particles going backward in time, and therefore physicists should search for such entities. In support of the position that such a thing is possible, we might mention the work of physicists such as Csonka [15] who have constructed theoretical schemata in which particles with arbitrary quantum numbers interact with normal particles and there is full causality; that is, there is both retarded and advanced causality.

Notes

1. Sara R. Feather and Robert Brier, "The Possible Effect of the Checker in Precognition Tests," *Journal of Parapsychology* 32 (1968): 167–175.

2. Antony Flew, "Can an Effect Precede Its Cause?" *Proceedings of the Aristotelian Society*, supp. vol. 28 (1954): 45–62.

3. Richard M. Gale, "Why a Cause Cannot Be Later than Its Effect," *Review of Metaphysics* 19 (1965): 209–234.

4. Richard G. Swinburne, "Affecting the Past," *Philosophical Quarterly* 16 (1965): 341–347.

5. Bob Brier, *Precognition and the Philosophy of Science* (New York: Humanities Press, 1974).

6. H. W. B. Joseph, *An Introduction to Logic* (Oxford: Oxford University Press, 1916), p. 401.

7. John Stuart Mill, "On the Four Methods of Experimental Inquiry," in *Essays in Logic from Aristotle to Russell*, ed. Ronald Jager (Englewood Cliffs, N.J.: Prentice-Hall, 1963), p. 71.

8. Ibid., p. 74.

9. Ibid., p. 74.

10. G. Feinberg, "Possibility of Faster-than-Light Particles," *Physical Review* 157 (1967): 1089–1105.

11. M. E. Arons and E. C. G. Sudarshan, "Lorentz Invariance, Local Field Theory, and Faster-than-Light Particles," *Physical Review* 173 (1968): 1622–1628.

12. Torsten Alvager and Michael N. Kreisler, "Quest for Faster-than-Light Matter," *Physical Review* 183 (1969): 1357–1361.

13. William B. Rolnick, "Implications of Causality for Faster-than-Light Matter," *Physical Review* 183 (1969): 1105–1108.

14. G. A. Benford et al., "The Tachyonic Antitelephone," *Physical Review D* 2 (1970): 263–265.

15. Paul L. Csonka, "Advanced Effects in Particle Physics: I," *Physical Review* 180 (1969): 1266–1281.

Second Report on a Case
of Experimenter Fraud

J. B. RHINE

It has been about fifteen months since Dr. Walter J. Levy, Jr.,[1] (W.J.L.) was discovered to be falsifying his test results in a PK experiment with rats at the Institute for Parapsychology, of which he was then Director. He was caught by three fellow staff members, James W. David, Jim Kennedy, and Jerry Levin, all of them members of his own research team. As soon as they considered the evidence conclusive, the three men who discovered the deception reported it to me; and after a short but decisive interview with W.J.L., I received his resignation.

Because W.J.L. had published many research papers and his work was widely known here and abroad, it seemed of first importance to warn readers against reliance on any of his research reports. The known evidence against him at the time of the exposé (June 12, 1974) concerned only one of his several lines of research, that in which rats were tested for PK, or psychokinetic ability. The experiment involved the implantation of an electrode in the pleasure center of the animal's brain to provide a means by which random electrical stimulation might make possible the arousal of the animal's desire to increase gratification by influencing (via PK) the rate of stimulation. W.J.L. himself would admit to no other falsification of results, but it seemed advisable to regard as "not acceptable" all reports of his research and all those in which he had played any part (Rhine, 1974b).

In my first report on the W.J.L. affair, I said that there should be a further report on later developments of the case. It is in compliance with that commitment that I now review what has happend thus far. First, a statement will be made about what is known of the extent of the dishonesty involved. Second, an evaluative word will be offered as to the consequences to the field.

I. The Known Extent of the W.J.L. Falsification

It was a natural question in the minds of all concerned as to how far the cheating exposed in the one experiment with rats extended throughout the rest of W.J.L's work and what, if anything, was left unaffected. Consideration was given at once to the possibility of making such a discriminative evaluation. It was reasonable to think that, with all the precautions that had been developed—especially since the elimination of the risk of fraud had long been on our minds—there might be some experiments by W.J.L. in which the conditions ruled out all possibility of dishonesty. It was soon realized, however, that although the precautionary test conditions had been intended to be the best that had been developed in psi research, there was always in actual practice some possibility still open to W.J.L. in his position as Director of the Institute and as a person who thoroughly involved himself in all that was going on. In fact, no one could say that the double-blind conditions, even with a design involving two or more experimenters in each experiment, had been so thoroughly observed that they had completely eliminated all possibility of dishonesty.

It was a unique situation. W.J.L. had been an extremely enterprising and industrious worker. As the chief administrator of the laboratory and an active leader in almost every aspect of its activities, he well knew all the limits and loopholes in a system he was masterminding seven days of the week—often sleeping in the attic while overseeing the 24-hour automatic testing of the animals. In view of these circumstances, there remained none of his work of which it could be said that he himself could not *possibly* have had any opportunity to manipulate the results that had been published. They all had to be put on the same level of uncertainty and unacceptability.

I think it can readily be seen that in the concern over an adequate reappraisal of his research there was nothing to do but to write it all off; and this I have done, at first only warning against any reliance on the W.J.L. reports for the time being. But on further reflection it became increasingly clear that the whole of his work was irretrievably lost, even without knowledge of the exact extent of the deception. In any case, such knowledge is not completely obtainable.

However, among individual members of the staff, especially those whose vigilance led to the discovery of the fraud in the first place, there remained unsatisfied questions. These led to some developments indicating that W.J.L's deception was much more extensive than had first been discovered. In one of these instances, first encountered by Jerry Levin, some indications turned up as to how W.J.L. could have manufactured results in the gerbil experiments on precognition. He could have done it by shorting one or the other of two wires that would have allowed him to manipulate the random target sequence as he watched the recording of the animal's behavior. It was W.J.L's uncalled-for presence at the apparatus with the wires in hand that raised the first question for Levin and led to the exposé. Levin later on observed by accident the presence of fine scratches such as the wires would make if shorted to the aluminum panel.

Another such discovery was made by Jim Kennedy in an examination of the records of W.J.L's automated maze tests of clairvoyance with human subjects, the results of which were stored by the computer. This experiment, as devised by W.J.L., had first been conducted with paper-and-pen methods but later two series had been automated with the computer. Kennedy found that W.J.L's published results were not at all those recorded for the same experiments on the computer. (The original data were not significant.) Apparently a completely different set of results had been invented.

Still another discovery was made by Kennedy who had earlier done some PK testing under W.J.L's supervision using chicken embryos as "subjects." The aim had been to test for possible PK influence exerted on randomly operated heat lamps by the eggs (located as they were in a cool chamber). It had been noticed that there tended to be long strings of hits and an increased trial rate when the scoring was significant. The easiest way to manufacture high scoring in the egg work would produce such effects as these. When the strings were deleted from the data, the scoring rate was at the chance level. Similar stringing was produced as an artifact of W.J.L's admitted manipulation of the rat implantation work. Also, the records of egg research which W.J.L. conducted before Kennedy started working with him showed the presence of astronomically significant strings (of about 100 hits where the probability of a hit was 1/2), yet W.J.L. never mentioned them.

With all the grounds we have for withholding acceptance of W.J.L's work, we cannot—and certainly need not—draw a conclusion about the actual extent of the deception in the whole of his five years of reported research. But, as I have already indicated, no salvageable series or section of sound and significant results has thus far been isolated in which W.J.L. was not around somewhere within reach of an unprotected link in the chain of reliable controls. As will be seen further on, the safest and best type of evidence of psi emerges in the form of incidental hidden "signs of psi" (Rhine 1974a) which the experimenter could not himself have anticipated at the time. However, none of these exceptional fraud-proof signs has so far been found in W.J.L's work.

II. Consequences to Parapsychology of the W.J.L. Affair

The long-term effects of the W.J.L. case on parapsychology will probably depend mainly on the unity of response among colleagues in the field. It should therefore help in the long run in this unification if the fraud problem in general is kept in the focus of attention in the laboratories and the journals. This is not likely, I think, to be overdone and to create a morbid distraction to normal research. It will probably not arouse that much interest.

However, the fraud issue, even in its most generalized form, is but one of several major problems in keeping psi testing on a high level of security. The multiple aspect of research reliability was outlined at the beginning of an article on "Security versus Deception" (Rhine, 1974a). At the moment, however, the primary topic is that of fraud, and the specific issue is that of experimenter dishonesty. Yet, even now, it is still not just the deception of the W.J.L. type that we need to consider here. The very timing of this affair at the FRNM ties it up with the broader problem of possible experimenter dis- honesty beyond this individual case and further complicates the search for adequate solutions. In fact, we can best regard the W.J.L. case as one that pointedly draws attention to a larger insecurity situation which, as I have already indicated (Rhine, 1975), was emerging well before the W.J.L. exposé took place.

Whatever the consequences of this fact are to be and whatever we are to do about them will likely be better understood if we take something of the earlier background into account. When we do so, it is much more under- standable that we find ourselves in our present state of special concern over fraud in parapsychology. It is well known that Henry Sidgwick, the first President of the S.P.R., warned that when critics found no other points open to attack in psychical research, they would charge the investigators with fraud. That this would be the final recourse was, of course, logical. I can recall that in the 1930's experimenter fraud was something we had to force ourselves to take seriously. It was only after we had arrived at a relatively conclusive stage of the ESP tests that McDougall advised me to introduce the two-experimenter test design to meet the anticipated charge of fraud. The experimenter deception which this precaution was designed to exclude was not nearly so immediately urgent as had been the exclusion of sensory cues, the proper mathematical evaluations, the safeguarding of records, and the interpretations of the results. Even the critics did not begin to hammer effectively on the experimenter-fraud question until George Price's article appeared in *Science* (1955), and Price himself argued in effect that since no other criticism was crucial, the conclusion had to be fraud. So this was the predicted last resort of the skeptic.

I am reviewing this familiar situation as background for a closely related point that needs attention here: it is that the very same development that finally closed in on the critics' options, cornered the experimenter himself. At earlier stages of method, the experimenter could alternatively obtain spuri- ous results under test conditions that allowed several uncontrolled factors, but one by one these gaps in method had been closed.

Those who understand science can see that this has been a progressive development in parapsychology in which not only is the unyielding critic driven by elimination into a last-ditch alternative of fraud, but also the experimenter who is unsuccessful in obtaining evidence of psi in his tests is left with the single choice of accepting his failure and trying again or of falsifying his results.

I must interrupt at this point to comment that a few colleagues insist we are *making it too hard* for the experimenter by this gradual improvement of our methods so that at the end he has no alternative to the final evidence we are seeking. They hint that we are thus helping to make him cheat by our very measurement of success. Several spokesmen of this viewpoint are proposing that we lower the standards and abandon our reliance on the measurements of significance. They forget that in present psi testing this requirement of significance is the only way we *can* measure success. Might we not just as well say, "Let us freely allow sensory leakage"; or "Let us use no safeguards against recording errors." Surely we need not be frightened off our course of methodological advance by the fraud issue any more than we have been at other stages of the development of psi methodology. The way to help the less resolute experimenter is not to weaken the safeguards that make conclusions possible, but to strengthen the experimental program against error in every way.

My special point in drawing attention to the fact that dishonesty was the *only* option W.J.L. had was that, with the psi test methods what they are today, this alternative of deception was more easily and conclusively identified as such when the real evidence he had expected failed him. It was not hidden in a loose methodology that would have left the matter inconclusive and confusing. Bad as the situation was, therefore, we can still appreciate the advance in methods in parapsychology that facilitated the expose; we have therefore reached the crossroads which Price, Hansel, and others have long since recognized—the point where, from the view of extreme skepticism, "it had to be fraud." Very well; if that is where we are, it is to our advantage that now we have a clearly drawn contest, and I think we are better prepared to face it than ever.

But again, it is not just the critics' last stand; it is not only the experimenters' either. It involves the whole field quite as much. In the past the trouble has been that no more attention was paid to the question of experimenter fraud than the current urgency of a given case demanded. The principal lesson of the W.J.L. incident is to remind us of the lukewarm concern that has been felt generally throughout our field over the question of experimenter dishonesty. Now we must be ready, more than ever before, to give the attention this issue requires. It is mainly now a matter of how much we can accomplish toward that end while the urgency is still widely and keenly felt.

It is true, much should be accomplished in time by the kind of increasing safeguards that I have been discussing in recent years, and especially as others join the endeavor. Recognizedly, these past advances in methods, even while they do make fraud the only alternative to successful psi demonstration in a well-designed experiment, will also make it equally more difficult for an experimenter to take that option and falsify his results. Better still, these advances in safeguarding methods should make the temptation to cheat less appealing and still more unrewarding as psi-testing designs improve. But I am coming to question, as I ponder the rections to the W.J.L. incident, whether we can afford to wait for this rate of advance. I find it easier to ask: "Why *should* we wait at all in this 'last-ditch' stage of indecisiveness? Is it really necessary?"

I have heard it stated at times in the last year that the suspicion of experimenter fraud could be expected to take an indefinitely long time to vanish completely from psi research. Today I do not think it has to be that way at all if we really take the problem seriously and all come to know the literature of the field adequately. In fact, I even think it should be possible to bring parapsychology through this final challenge to its security, not only with success but with reasonable dispatch. In making this proposal, however, I am assuming we will still go resolutely on with a vigorous development of the methods we already know, giving full attention to the safeguards we now have at our disposal for reducing the risk of experimenter deception. But it is time for a more conclusive step.

A Roundup of Fraud-proof Evidence is Essential

Since circumstances have made a mountain of apprehension over the fraud issue, an equally mountainous formation is called for to represent all the many types of fraud-proof evidence the field of parapsychology has to offer, most of which are little known as such even within the field. It is fair to say, too, that much more may yet be discovered that is still hidden from us all. This will call for a determined search; first, among the stores of existing records available for reexamination; and second, in new types of researches which lend themselves well to the purpose, as (judging by the past) many may be expected to do. It will suffice now, however, merely to indicate some of the principal kinds of evidence already known, to which the term "fraud-proof evidence of psi" can be applied. The category itself and its qualifications and limits can be refined later. The following few types will, I think, serve the purpose of introducing the wealth of this kind of material available without delay.

First, I would propose an examination of all available reports by scientific authors who for some reason wished not to publish their work, but who conducted acceptable experiments for their own satisfaction. As an example, I will cite the case of a report by an eminent scientist who was at the time the head of an internationally known institution. Many years ago he showed me his paper but felt he had to withhold it from publication in order to avoid embarrassment to his colleagues. He was, however, persuaded to publish the report of the experiment if all marks of identification of the author were removed; and this was done. The results were significant and were even repeated later.

This type of incident has other variations. Sometimes the fear of losing his own status has kept an author from publishing, although he was willing to circulate a report privately. Or again, the author may earlier have become known as a critic of psi research and was reluctant to admit that he had later conducted an experiment and that significant results had been obtained. Still other varieties of this kind of situation, one in which it would not be reasonable to suspect the fraud motive, can be assembled. In many cases the actual names could not safely be used. The point these situations have in common is that it would not make sense even to suggest that there might have been any will or intention to fabricate the data. To be sure, such researchers have to be of good quality in other respects as well.

Second, I would draw attention to another type of incident that has occurred from time to time in the editorial offices of the *Journal of Parapsychology* or in the correspondence of the Institute for Parapsychology. In a typical case, a report of a psi experiment is received for publication from an author who is rather new to parapsychology. The results may or may not be reported as significant. In the course of editorial consideration of the paper it becomes evident that the results lend themselves to a type of analysis which is unknown to the author but which may be independent of the analysis he has already made. Or the author's attention may be drawn to a finding reported by another experimenter which was then found by the editorial staff to be confirmed by the data of this newly submitted report. When such an application is clearly new to the author and is based upon an earlier finding by someone else, it can not only be given full acceptance, but it is all the better for having been brought to the author's attention by an independent analyst. The author can, of course, then make an independent check on the analysis from his own records, i.e., to confirm the editor's findings. The fraud-proof value of these cases covers a considerable range of security, often well beyond that of experimenter deception; some are of the ideal type in which one effect is found to be a completely independent replication of an experiment made under a very safe set of conditions, as, for example, when independent analyses are made on duplicate records.

Third, even better than the preceding types, although somewhat more rare, are examples in which already published work may be reexamined long years after publication for a significant distribution of hit patterns in the data, free of any conceivable motivation on the part of either subjects or experimenters. Take, for example, the familiar telepathy tests by G.H. Estabrooks at Harvard in 1925 (Estabrooks, 1961). It will be recalled that he obtained positive results in card guessing in three out of four series of tests. In the fourth, in which the conditions were changed to a greater distance between subject and target, the group of subjects scored well *below* mean chance expectation.

Moreover, in all four series there was a marked top-bottom decline of scoring rate in the 20-trial runs. But Estabrooks himself had been aiming at high-scoring total results, and these were the sole basis of his conclusion. He made no evaluation of the significance of the other features although he noted them; there was no precedent then known to him for doing so. Many years later, when I looked into his data for the type of results we had been getting in my lab at Duke in somewhat similar ESP tests (e.g., psi missing and declines), I found the negative deviation of hits in his fourth series to be significant. Also, the decline in the runs throughout all four of his series was quite definitely significant in the top-bottom difference in scoring rate. This had become a standard way of evaluating decline effects, but that was long after the date of the experiment. It seems quite reasonable to think that neither Estabrooks nor his subjects could have been motivated to fabricate these peculiar results; rather they were safely attributable to the ESP process itself and in time came to be recognized as typical signs of psi.

Other variations, too, have been found that depend on peculiarities of the psi process overlooked by the original experimenter. One of the most complex examples of this third type is that of the PK record book of M.P.R., a Guilford College student of psychology (Rhine & Reeves, 1943). This experimenter's original tests of dice throwing were self-tests, not intended for publication: but because they were faithfully recorded in a very systematic way and were strictly routine, they were reexamined years after the experiment was over when a search was being made for position effects at the Parapsychology Laboratory at Duke. A hit-distribution analysis conducted by Betty Humphrey was applied to six breakdowns of the data in which decline effects could be expected on the basis of other, previously analyzed work. Of these six breakdowns, five showed the significant differences expected (for example, the top-bottom difference of the hits in the run). The consistency of the patterning was unquestionably a remarkably evidential effect which, under the circumstances, ruled out any conceivable question of dishonesty in the original tests. An independent analysis followed Humphrey's, and others could still be repeated. (The nature of the records was such that trickery in the analysis would almost surely have been discovered. For example, erasures and other alterations are detectable by the experienced analyst, and the records were in a bound composition book.)

Fourth, by far the most imposing body of evidence of the fraud-proof type also occurred in the PK branch of the field. By 1944–45 we had accumulated a large number of research reports from our own laboratories and were examining them for methods of appraisal that would get beyond the great variety of conditions under which they were conducted. The most common feature among all 24 of these reports was the record sheet, a more-or-less standard form on which the results of dice throwing by one means or another could be recorded for the most part in regularly structured rows or columns on the page. In many experiments the page was divided into four, six, or eight subunits or sets. Eighteen of these reports had pages that were uniform enough to be subdivided into quarters, sometimes omitting median runs, columns, or sets to get uniformity. Twelve of the 18 had these regular sets on the record page.

This afforded a substantial common base of hit distributions with enough uniformity for generalization, and the expected declines in the column or row from the first to the last trial could be measured both ways by the quartering of the page (and the set). The quarter distribution (QD) of the page then became the basis of appraising hit distributions in the 18 series, and an *independent QD analysis* could be applied to the sets in 12 out of the 18. The test of significance to be applied was the difference between the upper left and the lower right quarters, both in the page QD's and the set QD's. These diagonal declines, as they were called, proved to be highly significant and remarkably consistent. Sixteen of the 18 series showed the decline and the *declines were even more significant in the set* than on the page. The point here, of course, is that none of these original experimenters back in the earlier years when the tests were conducted had any idea such an analysis would ever be made. Most of the series had shown significant total scoring above the chance mean. About those results, in some cases more than others, we could have raised questions of accuracy of recording, of bias in the dice, and of unreliability in the experimenters. But after these QD analyses were made, no human weakness or bias in the apparatus could be found that could have produced these declines. Again, Dr. Humphrey did most of the analyses; but prior to issuing an invitation to qualified outside observers to repeat them, I invited Dr. J.G. Pratt to the Laboratory to make a complete independent analysis. Pratt's analysis confirmed Humphrey's very closely.

At the Duke lab we recognized the fraud-proof conclusiveness of the QD analyses; but in the decade that passed before the fraud issue was first raised in a big way by G. Price (1955), the passage of time had allowed the QD's to be largely forgotten. They were not taken seriously by the critics. Yet, meanwhile, they have lost nothing of their antifraud potential.

These four kinds of special evidence against psi fraud do not exhaust the list of possibilities by any means; but they will, I think, suffice to show that this approach to a final settlement of the establishment of psi occurrence has a conclusiveness of its own. We need not leave the contest to the experimental attempts at fraud control where possibilities of the experimenter's motivation to deceive will also have to be dealt with. I am reasonably sure these fraud-proof types of evidence, if now given precedence for a showdown on the establishment of psi, will fill the need for the scientific mind that considers it. But it will take a thorough effort to draw that issue forcefully. Again, however, it must be repeated, this is not to abandon the effort to build up the control system that will eliminate experimenter deception along standard experimental lines. The evidence of hidden signs of psi, while more conclusive when it applies, is less adaptable (at least as yet) to all test situations. So for the present, we need both lines of control against fraud.

Once the case for psi is conclusive on the basis of the hidden type of evidence, there should be changes welcome to all. Beyond doubt the revolutionary character of psi makes decisive proof exceedingly hard to produce, in contrast to that in the physical sciences. Since there is already some evidence that the subject's belief in psi occurrence affects his test performance, this should be important to the scoring levels. Perhaps experimenters, too, would be affected by a generally relaxed acceptance of the fact of psi ability. We might anticipate that in time there would be little more need for concern about honesty than in any of the other psychological branches of testing and research. Most helpful, perhaps, would be the reduction of skepticism and indifference on the part of those who should be looking into the bearing of psi on their own fields. The future of psi research depends more on this sharing of interest from other disciplines and sciences than upon anything else.

The whole program of parapsychology should be expected to respond to the confidence springing from the assurance that the claim of psi has passed all its tests and that there are no last details which can provide the scientific mind with an excuse even for suspended judgment. More confident steps should then be possible to provide parapsychological education of the highest quality, to obtain adequate economic support, and even to consider practical application over a wide range of possibilities. All this may come in time; but the timing itself may endanger the outcome if left to the unguided circumstances of evolution. So I would say that, with a full appreciation of the major impact of the W.J.L. exposé, it would be best not to forget it until it has spurred us all to clear the terminal barrier to the goal of ultimate establishment.

Notes

1. The first report (Rhine, 1974b) used "Dr. W." as the public identification of Dr. Levy in the slender hope of protecting the interests of innocent people in Levy's nonparapsychological circles. The wide general publicity anticipated at that time can be assumed to have passed. The use of Levy's full name is necessary now, of course, in order to identify his work, which has all been disqualified.

References

Estabrooks, G.H. "A Contribution to Experimental Telepathy." *Journal of Parapsychology*, 1961, 25, 190–213.

Price, G. "Science and the Supernatural." *Science*, August 26, 1955, p. 359.

Rhine, J.B. "Security Versus Deception in Parapsychology." *Journal of Parapsychology*, 1974, 38, 99–121. (a)

Rhine, J.B. "A New Case of Experimenter Unreliability." *Journal of Parapsychology,* 1974, 38, 215–225. (b)

Rhine, J.B. "Psi Methods Reexamined." *Journal of Parapsychology,* 1975, 39, 38–58.

Rhine, J.B., & Reeves, M.P. "The PK effect: II. A study in declines." *Journal of Parapsychology,* 1943, 7, 76–93.

Suggested Readings

As noted in the general introduction, this is not intended as anything like a complete bibliography. The works mentioned are generally standard sources, for the most part easily available, which may be of help at the next stage of the reader's research.

The case for parapsychology:
Murphy, Gardner. *The Challenge of Psychical Research: A Primer of Psychical Research.* New York: Scribner's, 1966.
Pratt, J.G. *ESP Research Today: A Study of Developments in Parapsychology since 1960.* Metuchen, N.J.: Scarecrow Press, 1973.
Rhine, J.B., and Pratt, J.G. *Parapsychology: Frontier Science of the Mind.* Springfield, Ill.: Charles C. Thomas, 1957.
Wolman, B.B., ed. *Handbook of Parapsychology.* New York: Van Nostrand Reinhold, 1977.
The *Journal of Parapsychology* is also an unequaled source.

Critical works:
Hansel, C.E.M. *ESP: A Scientific Evaluation.* New York: Scribner's, 1966.
———. *ESP and Parapsychology: A Critical Re-Evaluation.* Buffalo, N.Y.: Prometheus Books, 1979.
But see also Martin, Michael, "The Problem of Experimenter Fraud: A Re-Evaluation of Hansel's Critique of ESP Experiments," *Journal of Parapsychology,* 43 (1979), 129-139, for a critique of Hansel's critique.

Philosophical treatments:
Alcock, James E., "Parapsychology: Science of the anomalous or search for the soul?," *Behavioral and Brain Sciences,* 10 (1987), pp. 553-565. See also commentaries and replies, pp. 566-643.
Braude, Stephen, *ESP and Psychokinesis: A Philosophical Examination.* Philadelphia: Temple University Press, 1979.
———. *The Limits of Influence: Psychokinesis and the Philosophy of Science.* New York: Routledge and Kegan Paul, 1986.
Brier, Bob. *Precognition and the Philosophy of Science.* New York: Humanities Press, 1974.

Flew, Antony, "Precognition." In *The Encyclopedia of Philosophy*, ed. Paul Edwards. New York: Macmillan and the Free Press, 1967.

Flew, Antony, ed., *Readings in the Philosophical Problems of Parapsychology*. Buffalo, New York: Prometheus Press, 1987.

French, Peter, ed., *Philosophers in Wonderland*. St. Paul, Minn.: Llewellyn, 1975. An anthology including philosophical classics in the area.

Ludwig, Jan, ed., *Philosophy and Parapsychology*. Buffalo, N.Y.: Prometheus Books, 1978. Another good anthology, overlapping to some extent with Peter French's.

Mundle, C.W.K. "ESP: Philosophical Implications." In *The Encyclopedia of Philosophy*, ed. Paul Edwards. New York: Macmillan and the Free Press, 1967.

Rao, K. Ramakrishna and John Palmer, "The anomaly called PSI; Recent Research and Criticism," *Behavioral and Brain Sciences*, 10 (1987), pp. 539-551. See also commentaries and replies, pp. 566-643.

On reverse causality, see especially:

Chisholm, R.M., and Taylor, Richard. "Making Things to Have Happened." *Analysis* 20 (1960): 73-78.

Craig, William L., "Tachyons, Time Travel, and Divine Omniscience." *Journal of Philosophy* 85 (1988), 135-150.

Dummett, Michael. "Bringing About the Past." *Philosophical Review* 73 (1964): 338-359.

Dummett, Michael, and Flew, Antony. "Can an Effect Precede Its Cause?" *Proceedings of the Aristotelian Society*, supp. vol. 28 (1954): 27-62.

Paul Fitzgerald, "Tachyons, Backwards Causation, and Freedom," in *PSA 1970*, Roger C. Buck and Robert S. Cohen, eds., *Boston Studies in the Philosophy of Science*, 8 (Boston: Reidel, 1971), pp. 415-436.

———. "Retrocausality." *Philosophia* 4 (1974), 513-551.

Taylor, Richard. "Causation." In *The Encyclopedia of Philosophy*, ed. Paul Edwards. New York: Macmillan and the Free Press, 1967.

QUANTUM MYSTICISM

Introduction

A major question regarding astrology and parapsychology raised in earlier sections was the question of demarcation: do these qualify as science or pseudoscience? With regard to quantum mechanics—the central body of theory considered in this section—that question is very easily answered.

Quantum mechanics is an indisputably major component of contemporary subatomic physics, and no one disputes the right of *physics* to be considered science. The theory indeed stands as one of the major scientific achievements of the twentieth century: it replaces Newtonian or classical mechanics much as Einsteinian space-time replaces Newtonian space and time and has been empirically tested and repeatedly confirmed in the widest possible range of applications. Quantum mechanics is, quite simply, one of the most successful scientific theories mankind has ever produced.

Nonetheless it has been claimed that the *implications* of this unimpeachable scientific achievement take us directly into the spirit of Eastern mysticism. The basic elements of Hinduism, Taoism, and Buddhism, says Fritjof Capra, "also seem to be the fundamental features of the world view emerging from modern physics." It has been claimed, in particular, that quantum mechanics shows that by our conscious participation we *create* rather than simply observe the universe. "According to quantum mechanics," says Gary Zukav, "there is no such thing as objectivity. We cannot eliminate ourselves from the picture. . . . Physics has become a branch of psychology, or perhaps the other way around."

The case of quantum mechanics thus raises an intriguingly different question from that addressed in preceding sections. There the question was the familiar problem of demarcation, applied to what might be considered aspects of the 'occult': do astrology and parapsychology, for example, qualify as science or not? Here the question is rather whether one of the *firmest* of our scientific theories nonetheless has *implications* that might be considered 'occult'.

Does quantum mechanics legitimately have such implications, or is this a matter of meta-scientific or pseudoscientific misinterpretation? That is the central question that the reader should try to address in the readings that follow.

We begin with brief conflicting notes from ALBERT EINSTEIN and NIELS BOHR as to whether quantum mechanics can be said to offer a *complete* theory of quantum phenomena.

One of the characteristics of quantum theory is that it offers statistical laws—and *only* statistical laws—regarding, for example, radioactive decay. With quantum mechanics we can predict, and predict very well, how many atoms of a lump of uranium are likely to decay in a certain period. What quantum mechanics does not tell us, however—and indeed *cannot* tell us— is when or why one *particular* uranium atom will decay. In that sense the principles of quantum mechanics are essentially statistical.

Does that indicate an essential incompleteness in the theory—a failure to describe a non-statistical reality 'behind' the statistics? In the quotations with which we begin, Einstein answers in the affirmative: "I am, in fact, fully convinced that the essentially statistical character of contemporary quantum theory is solely to be ascribed to the fact that this [theory] operates with an incomplete description of physical systems." Niels Bohr, on the other hand, emphasizes—as would almost all contemporary physicists— that quantum mechanics is *essentially* incompletable. If it is correct, any consistent additions will still leave us with an indeterminacy in nature.[1] "[I]n quantum mechanics, we are not dealing with an arbitrary renunciation of a more detailed analysis of atomic phenomena, but with a recognition that such an analysis is *in principle* excluded."

FRITJOF CAPRA uses this feature of quantum mechanics among others to argue that contemporary physics is approaching the world view of Eastern mysticism. In a piece excerpted here from *The Tao of Physics*, Capra has us consider for example the following quotations from Werner Heisenberg, a twentieth-century founder of quantum mechanics, and Nagarjuna, the second-century founder of Mādhyamika Buddhism:

> The world thus appears as a complicated tissue of events, in which connections of different kinds alternate or overlap or combine and thereby determine the texture of the whole.
>
> —Heisenberg

> Things derive their being and nature by mutual dependence and are nothing in themselves.
>
> —Nagarjuna

Capra introduces his discussion in terms of the Copenhagen interpretation, which will remain a core issue throughout. He also sounds another theme that is brought up repeatedly in later pieces: the claim that in quantum mechanics the consciousness of the observer determines to some extent the physical properties of the objects observed. This central notion of "participation instead of observation," Capra claims, is also at the core of Eastern mysticism:

> Penetrating into ever-deeper realms of matter, [the physicist] has become aware of the essential unity of all things and events. More than that, he has also learnt that he himself and his consciousness are an integral part of this unity. Thus the mystic and the physicist arrive at the same conclusion; one starting from the inner realm, the other from the outer world. The harmony between their views confirms the ancient Indian wisdom that *Brahman*, the ultimate reality without, is identical to *Atman*, the reality within.

A number of the points Capra raises are developed further by GARY ZUKAV in an excerpt from *The Dancing Wu Li Masters*. Zukav outlines some of the strange features of quantum mechanics in more detail—in particular, the Heisenberg uncertainty principle—and concludes that "the implications of quantum mechanics are psychedelic. Not only do we influence our reality, but, in some degree, we actually create it." Zukav echoes Capra in claiming that "the languages of eastern mystics and western physicists are becoming very similar" and, if anything, is even more emphatic in claiming that quantum mechanics entails a metaphysically creative role for consciousness:

> Because it is the nature of things that we can know either the momentum of a particle or its position, but not both, *we must choose* which of these two properties we want to determine. Metaphysically, this is very close to saying that we *create* certain properties because we choose to measure those properties.
>
> . . . physics is the study of the structure of consciousness.

In the piece that follows, however, ROBERT P. CREASE and CHARLES C. MANN take Capra and Zukav as primary targets in an unstintingly critical appraisal of quantum mysticism. "Things have got a little unfastened in the world of scientific popularization," they say, "and quantum mechanics is the culprit."

In particular, Crease and Mann attack the claim of parallelism between contemporary physics and Eastern mysticism evident in both Capra and Zukav. What Crease and Mann point out, following the work of Sal Restivo,[2] is that any comparison between physics and mysticism will inevitably rest on debatable issues of the proper selection and translation of representative texts from both mystics and physicists. Crease and Mann claim, moreover, that the parallels that *have* been drawn ultimately prove to have weak foundations; the paradoxes Capra touts in both physics and mysticism, for example, prove to be 'paradoxes' of quite different types that serve quite different ends in quite different human endeavors.

> Quantum mystics wind up distorting both science and Eastern mysticism. In general, they distort science by deemphasizing the role of mathematics, mathematical reasoning, and technology, claiming that what is essential about science can be cleanly separated from these. They distort Eastern mysticism by overlooking the social and historical roots of the various Eastern religions . . .

In the course of their essay Crease and Mann also pose the following intriguing challenge for parallelism: If we *do* take Eastern mysticism to correspond to contemporary physical theory, will Eastern mysticism then be subject—along with current theory—to empirical test and potential refutation?

In "Quantum Mysteries for Anyone," N. DAVID MERMIN takes us deeper into the perplexities of quantum mechanics. Mermin leads us through the construction of a device foreshadowed by the Einstein-Podolsky-Rosen thought experiment of the 1930s which in effect instantiates Bell's theorem and exemplifies a number of recent actual experiments as well. The philosophical point of the exercise is simply to emphasize how boggling the behavior of such a machine really is. "The device behaves as it behaves, and no mention of wave-functions, reduction hypotheses, measurement theory, superposition principles, wave-particle duality, incompatible observables, complementarity, or the uncertainty principle, is needed to bring home its peculiarity. It is not the Copenhagen interpretation of quantum mechanics that is strange, but the world itself."

Though Mermin claims to be trying to avoid philosophical interpretations regarding the phenomena at issue, he does use a line which echoes earlier pieces by Capra and Zukav: "We now know that the moon is demonstrably not there when nobody looks."

It is this repeated claim that there is a special role for consciousness in quantum mechanics that MARSHALL SPECTOR takes to task in the final piece of the section. Spector begins with a sampler of quotations from earlier pieces. Much of this, he says, "is reminiscent of the subjective idealism of Bishop Berkeley. . . . 'To be is to be perceived' . . ." But does the genuine *physics* of quantum mechanics really have these *metaphysical* implications?

In the end, on the basis of a careful review of five aspects that genuinely distinguish quantum mechanics from classical mechanics—wave-particle duality, the statistical character of nature, essential incompleteness, nonseparability, and the role of observation—Spector concludes that "there is nothing about the new *physical* aspects of quantum mechanics—astounding as they may be *in physics*—that has any implications for the mind-matter distinction of the western metaphysical tradition . . ."

[Quantum mechanics] does not imply that matter does not exist, or that it exists only at our sufferance; it does not imply subjective idealism or anything like it. Physics has not become 'a branch of psychology' (Zukav); it is not 'the study of the structure of consciousness' (Zukav). There is nothing about quantum mechanics . . . that implies or supports these claims in a manner in which classical physics does not.

The lines of controversy regarding the purported implications of quantum mechanics have thus been clearly drawn. Here the reader will want to try to decide just what conclusion is called for: what the essential features of quantum mechanics really are—at least as much as possible from an informal presentation such as this one—and what implications those features genuinely have.

But there are also a number of more general questions that arise here. Consider again, for example, the essentially statistical character of laws within quantum mechanics. Does the claim that such laws are to be taken as *fundamental* violate our *a priori* expectations of what the universe and its explainability must be like? And if so, does that show that there must be something incomplete about such a theory—as Einstein thought—or merely that our expectations even of what the universe *must* be like are subject to empirical disconfirmation? This may lead to an even more general question: Are there ultimate principles regarding what things must be like that shape theory formation and selection but are *not* themselves open to empirical disconfirmation?

Quantum mechanics also raises questions as to what, precisely, a scientific theory is supposed to do. By all accounts, quantum mechanics offers a mathematical formalism that allows precise predictions of subatomic phenomena. But by all accounts it also fails miserably to offer any accessible or intuitively intelligible picture of such phenomena. Is the former enough to establish quantum mechanics as a satisfactory scientific theory, or does the explanatory purpose of science demand something like the second as well? What in fact does scientific explanation demand?

Fritjof Capra begins this section with a claim of convergence between Western physics and Eastern mysticism—a claim roundly attacked, we've seen, by both Crease and Mann on the one hand and Marshall Spector on the other. In the course of their essay Crease and Mann note that part of the difficulty is that two very different human enterprises are being compared: "Eastern mysticism . . . is less like a body of truths than a program for a spiritual quest, an attitude adopted toward the world . . ." That line of thought can be seen as leading us to the other approaches considered in section V.

Notes

1. This point is perhaps most carefully handled in section 2 C of Marshall Spector's "Mind, Matter, and Quantum Mechanics."

2. See Sal Restivo, *The Social Relations of Physics, Mysticism, and Mathematics* (Dordrecht, Holland: Reidel, 1985). It should be noted that Crease and Mann also offer a critique of certain aspects of Restivo's treatment, however.

Conflicting Notes from Einstein and Bohr

Einstein

Born, Pauli, Heitler, Bohr, and Margenau . . . are all firmly convinced that the riddle of the double nature of all corpuscles (corpuscular and undulatory character) has in essence found its final solution in the statistical quantum theory. On the strength of the successes of this theory they consider it proved that a theoretically complete description of a system can, in essence, involve only statistical assertions concerning the measurable quantities of this system. They are apparently all of the opinion that Heisenberg's indeterminacy-relation (the correctness of which is, from my own point of view, rightfully regarded as finally demonstrated) is essentially prejudicial in favor of the character of all thinkable reasonable physical theories in the mentioned sense. . . . I am, in fact, firmly convinced that the essentially statistical character of contemporary quantum theory is solely to be ascribed to the fact that this [theory] operates with an incomplete description of physical systems.
—Albert Einstein, in his reply to criticisms, P. A. Schilpp, ed., *Albert Einstein: Philosopher-Scientist*, p. 666.

Bohr

[I]n quantum mechanics, we are not dealing with an arbitrary renunciation of a more detailed analysis of atomic phenomena, but with a recognition that such an analysis is *in principle* excluded. The peculiar individuality of the quantum effects presents us, as regards the comprehension of well-defined evidence, with a novel situation unforeseen in classical physics and irreconcilable with conventional ideas suited for our orientation and adjustment to ordinary experience. It is in this respect that quantum theory has called for a renewed revision of the foundation for the unambiguous use of elementary concepts, as a further step in the development which, since the advent of relativity theory, has been so characteristic of modern science.

—Niels Bohr, "Discussion with Einstein," in P. A. Schilpp, ed., *Albert Einstein: Philosopher-Scientist*, p. 235.

The Unity of All Things

Fritjof Capra

... A Hindu and a Taoist may stress different aspects of the experience; a Japanese Buddhist may interpret his or her experience in terms which are very different from those used by an Indian Buddhist; but the basic elements of the world view which has been developed in all these traditions are the same. These elements also seem to be the fundamental features of the world view emerging from modern physics.

The most important characteristic of the Eastern world view—one could almost say the essence of it—is the awareness of the unity and mutual interrelation of all things and events, the experience of all phenomena in the world as manifestations of a basic oneness. All things are seen as interdependent and inseparable parts of this cosmic whole; as different manifestations of the same ultimate reality. The Eastern traditions constantly refer to this ultimate, indivisible reality which manifests itself in all things, and of which all things are parts. It is called *Brahman* in Hinduism, *Dharmakaya* in Buddhism, *Tao* in Taoism. Because it transcends all concepts and categories, Buddhists also call it *Tathata*, or Suchness:

> What is meant by the soul as suchness, is the oneness of the totality of all things, the great all-including whole.[1]

In ordinary life, we are not aware of this unity of all things, but divide the world into separate objects and events. This division is, of course, useful and necessary to cope with our everyday environment, but it is not a fundamental feature of reality. It is an abstraction devised by our discriminating and categorizing intellect. To believe that our abstract concepts of sepa-

Excerpted with permission from Fritjof Capra, *The Tao of Physics*. New York: Bantam Books, 1984. Slight changes have been made in the style of bibliographical references.

rate 'things' and 'events' are realities of nature is an illusion. Hindus and Buddhists tell us that this illusion is based on *avidya*, or ignorance, produced by a mind under the spell of *maya*. The principal aim of the Eastern mystical traditions is therefore to readjust the mind by centering and quietening it through meditation. The Sanskrit term for meditation—*samadhi* —means literally 'mental equilibrium.' It refers to the balanced and tranquil state of mind in which the basic unity of the universe is experienced:

> Entering into the *samadhi* of purity, [one obtains] all-penetrating insight that enables one to become conscious of the absolute oneness of the universe.[2]

The basic oneness of the universe is not only the central characteristic of the mystical experience, but is also one of the most important revelations of modern physics. It becomes apparent at the atomic level and manifests itself more and more as one penetrates deeper into matter, down into the realm of subatomic particles. The unity of all things and events will be a recurring theme throughout our comparison of modern physics and Eastern philosophy. As we study the various models of subatomic physics we shall see that they express again and again, in different ways, the same insight—that the constituents of matter and the basic phenomena involving them are all interconnected, interrelated and interdependent; that they cannot be understood as isolated entities, but only as integrated parts of the whole.

... I shall discuss how the notion of the basic interconnectedness of nature arises in quantum theory, the theory of atomic phenomena, through a careful analysis of the process of observation.[3] Before entering this discussion, I have to return to the distinction between the mathematical framework of a theory and its verbal interpretation. The mathematical framework of quantum theory has passed countless successful tests and is now universally accepted as a consistent and accurate description of all atomic phenomena. The verbal interpretation, on the other hand—i.e., the metaphysics of quantum theory—is on far less solid ground. In fact, in more than forty years, physicists have not been able to provide a clear metaphysical model.

The following discussion is based on the so-called Copenhagen interpretation of quantum theory which was developed by Bohr and Heisenberg in the late 1920s and is still the most widely accepted model. In my discussion I shall follow the presentation given by Henry Stapp of the University of California[4] which concentrates on certain aspects of the theory and on a certain type of experimental situation that is frequently encountered in subatomic physics. ...

The starting point of the Copenhagen interpretation is the division of the physical world into an observed system ("object") and an observing system.

The observed system can be an atom, a subatomic particle, an atomic process, etc. The observing system consists of the experimental apparatus and will include one or several human observers. A serious difficulty now arises from the fact that the two systems are treated in different ways. The observing system is described in the terms of classical physics, but these terms cannot be used consistently for the description of the observed "object." We know that classical concepts are inadequate at the atomic level, yet we have to use them to describe our experiments and to state the results. There is no way we can escape this paradox. The technical language of classical physics is just a refinement of our everyday language and it is the only language we have to communicate our experimental results.

The observed systems are described in quantum theory in terms of probabilities. This means that we can never predict with certainty where a subatomic particle will be at a certain time, or how an atomic process will occur. All we can do is predict the odds. For example, most of the subatomic particles known today are unstable, that is, they disintegrate—or "decay"—into other particles after a certain time. It is not possible, however, to predict this time exactly. We can only predict the probability of decay after a certain time or, in other words, the average lifetime of a great number of particles of the same kind. The same applies to the "mode" of decay. In general, an unstable particle can decay into various combinations of other particles, and again we cannot predict which combination a particular particle will choose. All we can predict is that out of a large number of particles 60 percent, say, will decay in one way, 30 percent in another way, and 10 percent in a third way. It is clear that such statistical predictions need many measurements to be verified. Indeed, in the collision experiments of high-energy physics, tens of thousands of particle collisions are recorded and analyzed to determine the probability for a particular process.

It is important to realize that the statistical formulation of the laws of atomic and subatomic physics does not reflect our ignorance of the physical situation, like the use of probabilities by insurance companies or gamblers. In quantum theory, we have come to recognize probability as a fundamental feature of the atomic reality which governs all processes, and even the existence of matter. Subatomic particles do not exist with certainty at definite places, but rather show "tendencies to exist," and atomic events do not occur with certainty at definite times and in definite ways, but rather show "tendencies to occur."

It is not possible, for example, to say with certainty where an electron will be in an atom at a certain time. Its position depends on the attractive force binding it to the atomic nucleus and on the influence of the other electrons in the atom. These conditions determine a probability pattern which represents the electron's tendencies to be in various regions of the atom. The picture [below] shows some visual models of such probability

visual models of probability patterns

patterns. The electron is likely to be found where the patterns are bright and unlikely to be present where they are dark. The important point is that the entire pattern represents the electron at a given time. Within the pattern, we cannot speak about the electron's position, but only about its tendencies to be in certain regions. In the mathematical formalism of quantum theory, these tendencies, or probabilities, are represented by the so-called probability function, a mathematical quantity which is related to the probabilities of finding the electron in various places at various times.

The contrast between the two kinds of description—classical terms for the experimental arrangement and probability functions for the observed objects—leads to deep metaphysical problems which have not yet been resolved. In practice, however, these problems are circumvented by describing the observing system in operational terms; that is, in terms of instructions which permit scientists to set up and carry out their experiments. In this way, the measuring devices and the scientists are effectively joined into one complex system which has no distinct, well-defined parts, and the experimental apparatus does not have to be described as an isolated physical entity.

For the further discussion of the process of observation it will be useful to take a definite example, and the simplest physical entity that can be used is a subatomic particle, such as the electron. If we want to observe and measure such a particle, we must first isolate it, or even create it, in a process which can be called the preparation process. Once the particle has been prepared for observation, its properties can be measured, and this constitutes the process of measurement. The situation can be represented symbolically as follows. A particle is prepared in the region A, travels from

A to B, and is measured in the region B. In practice, both the preparation and the measurement of the particle may consist of a whole series of quite complicated processes. In the collision experiments of high-energy physics, for example, the preparation of the particles used as projectiles consists in sending them around a circular track and accelerating them until their energy is sufficiently high. This process takes place in the particle accelerator. When the desired energy is reached, they are made to leave the accelerator (A) and travel to the target area (B), where they collide with other particles. These collisions take place in a bubble chamber where the particles produce visible tracks which are photographed. The properties of the particles are then deduced from a mathematical analysis of their tracks; such an analysis can be quite complex and is often carried out with the help of computers. All these processes and activities constitute the act of measurement.

The important point in this analysis of observation is that the particle constitutes an intermediate system connecting the processes at A and B. It exists and has meaning only in this context; not as an isolated entity, but as an interconnection between the processes of preparation and measurement. The properties of the particle cannot be defined independently of these processes. If the preparation or the measurement is modified, the properties of the particle will change, too.

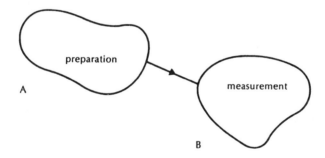

observation of a particle in atomic physics

On the other hand, the fact that we speak about "the particle," or any other observed system, shows that we have some independent physical entity in mind which is first prepared and then measured. The basic problem with observation in atomic physics is, then—in the words of Henry Stapp—that "the observed system is required to be isolated in order to be defined, yet interacting in order to be observed."[5] This problem is resolved in quantum theory in a pragmatic way by requiring that the observed system be free from the external disturbances caused by the process of observation during some interval between its preparation and subsequent measurement. Such a condition can be expected if the preparing and measuring devices are physically separated by a large distance, so that the observed object can travel from the region of preparation to the region of measurement.

How large, then, does this distance have to be? In principle, it must be infinite. In the framework of quantum theory, the concept of a distinct physical entity can be defined precisely only if this entity is infinitely far away from the agencies of observation. In practice, this is of course not possible; neither is it necessary. We have to remember, here, the basic attitude of modern science—that all its concepts and theories are approximate. In the present case, this means that the concept of a distinct physical entity need not have a precise definition, but can be defined approximately. This is done in the following way.

The observed object is a manifestation of the interaction between the processes of preparation and measurement. This interaction is generally complex and involves various effects extending over different distances; it has various "ranges," as we say in physics. Now, if the dominant part of the interaction has a long range, the manifestation of this long-range effect will travel over a large distance. It will then be free from external disturbances and can be referred to as a distinct physical entity. In the framework of quantum theory, distinct physical entities are therefore idealizations which are meaningful only to the extent that the main part of the interaction has a long range. Such a situation can be defined mathematically in a precise way. Physically, it means that the measuring devices are placed so far apart that their main interaction occurs through the exchange of a particle or, in more complicated cases, of a network of particles. There will always be other effects present as well, but as long as the separation of the measuring devices is large enough these effects can be neglected. Only when the devices are not placed far enough apart will the short-range effects become dominant. In such a case, the whole macroscopic system forms a unified whole, and the notion of an observed object breaks down.

Quantum theory thus reveals an essential interconnectedness of the universe. It shows that we cannot decompose the world into independently existing smallest units. As we penetrate into matter, we find that it is made of particles, but these are not the "basic building blocks" in the sense of Democritus and Newton. They are merely idealizations which are useful from a practical point of view, but have no fundamental significance. In the words of Niels Bohr, "Isolated material particles are abstractions, their properties being definable and observable only through their interaction with other systems."[6]

The Copenhagen interpretation of quantum theory is not universally accepted. There are several counterproposals, and the philosophical problems involved are far from being settled. The universal interconnectedness of things and events, however, seems to be a fundamental feature of the atomic reality which does not depend on a particular interpretation of the mathematical theory. The following passage from a recent article by David Bohm, one of the main opponents of the Copenhagen interpretation, confirms this fact most eloquently:

> One is led to a new notion of unbroken wholeness which denies the classical idea of analyzability of the world into separately and independently existing parts. . . . We have reversed the usual classical notion that the independent "elementary parts" of the world are the fundamental reality, and that the various systems are merely particular contingent forms and arrangements of these parts. Rather, we say that inseparable quantum interconnectedness of the whole universe is the fundamental reality, and that relatively independently behaving parts are merely particular and contingent forms within this whole.[7]

At the atomic level, then, the solid material objects of classical physics dissolve into patterns of probabilities, and these patterns do not represent probabilities of things, but rather probabilities of interconnections. Quantum theory forces us to see the universe not as a collection of physical objects, but rather as a complicated web of relations between the various parts of a unified whole. This, however, is the way in which Eastern mystics have experienced the world, and some of them have expressed their experience in words which are almost identical with those used by atomic physicists. Here are two examples:

> The material object becomes . . . something different from what we now see, not a separate object on the background or in the environment of the rest of nature but an indivisible part and even in a subtle way an expression of the unity of all that we see.[8]

> Things derive their being and nature by mutual dependence and are nothing in themselves.[9]

If these statements could be taken as an account of how nature appears in atomic physics, the following two statements from atomic physicists could, in turn, be read as a description of the mystical experience of nature:

> An elementary particle is not an independently existing unanalyzable entity. It is, in essence, a set of relationships that reach outward to other things.[10]

> The world thus appears as a complicated tissue of events, in which connections of different kinds alternate or overlap or combine and thereby determine the texture of the whole.[11]

The picture of an interconnected cosmic web which emerges from modern atomic physics has been used extensively in the East to convey the mystical experience of nature. For the Hindus, *Brahman* is the unifying thread in the cosmic web, the ultimate ground of all being:

He on whom the sky, the earth, and the atmosphere
Are woven, and the wind, together with all life-breaths,
Him alone know as the one Soul.[12]

In Buddhism, the image of the cosmic web plays an even greater role. The core of the *Avatamsaka Sutra*, one of the main scriptures of Mahayana Buddhism, is the description of the world as a perfect network of mutual relations where all things and events interact with each other in an infinitely complicated way. Mahayana Buddhists have developed many parables and similes to illustrate this universal interrelatedness, some of which will be discussed later on, in connection with the relativistic version of the 'web philosophy' in modern physics. The cosmic web, finally, plays a central role in Tantric Buddhism, a branch of the Mahayana which originated in India around the third century A.D. and constitutes today the main school of Tibetan Buddhism. The scriptures of this school are called the *Tantras*, a word whose Sanskrit root means "to weave" and which refers to the interwovenness and interdependence of all things and events.

In Eastern mysticism, this universal interwovenness always includes the human observer and his or her consciousness, and this is also true in atomic physics. At the atomic level, "objects" can be understood only in terms of the interaction between the processes of preparation and measurement. The end of this chain of processes lies always in the consciousness of the human observer. Measurements are interactions which create "sensations" in our consciousness—for example, the visual sensation of a flash of light, or of a dark spot on a photographic plate—and the laws of atomic physics tell us with what probability an atomic object will give rise to a certain sensation if we let it interact with us. "Natural science," says Heisenberg, "does not simply describe and explain nature; it is part of the interplay between nature and ourselves."[13]

The crucial feature of atomic physics is that the human observer is not only necessary to observe the properties of an object, but is necessary even to define these properties. In atomic physics, we cannot talk about the properties of an object as such. They are meaningful only in the context of the object's interaction with the observer. In the words of Heisenberg, "What we observe is not nature itself, but nature exposed to our method of questioning."[14] The observer decides how he is going to set up the measurement and this arrangement will determine, to some extent, the properties of the observed object. If the experimental arrangement is modified, the properties of the observed object will change in turn.

This can be illustrated with the simple case of a subatomic particle. When observing such a particle, one may choose to measure—among other quantities—the particle's position and its momentum (a quantity defined as the particle's mass times its velocity). ... [A]n important law of quantum theory—Heisenberg's uncertainty principle—says that these two quantities can never be measured simultaneously with precision. We can either obtain a precise knowledge about the particle's position and remain completely ignorant about its momentum (and thus about its velocity), or vice versa; or we can have a rough and imprecise knowledge about both quantities. The important point now is that this limitation has nothing to do with the imperfection of our measuring techniques. It is a principle limitation which is inherent in the atomic reality. If we decide to measure the particle's position precisely, the particle simply *does not have* a well-defined momentum, and if we decide to measure the momentum, it does not have a well-defined position.

In atomic physics, then, the scientist cannot play the role of a detached objective observer, but becomes involved in the world he observes to the extent that he influences the properties of the observed objects. John Wheeler sees this involvement of the observer as the most important feature of quantum theory and he has therefore suggested replacing the word "observer" by the word "participator." In Wheeler's own words,

> Nothing is more important about the quantum principle than this, that it destroys the concept of the world as "sitting out there," with the observer, safely separated from it by a 20-centimeter slab of plate glass. Even to observe so minuscule an object as an electron, he must shatter the glass. He must reach in. He must install his chosen measuring equipment. It is up to him to decide whether he shall measure position or momentum. To install the equipment to measure the one prevents and excludes his installing the equipment to measure the other. Moreover, the measurement changes the state of the electron. The universe will never afterward be the same. To describe what has happened, one has to cross out that old word "observer" and put in its place the new word "participator." In some strange sense, the universe is a participatory universe.[15]

The idea of "participation instead of observation" has been formulated in modern physics only recently, but it is an idea which is well known to any student of mysticism. Mystical knowledge can never be obtained just by observation, but only by full participation with one's whole being. The notion of the participator is thus crucial to the Eastern world view, and the Eastern mystics have pushed this notion to the extreme, to a point where observer and observed, subject and object, are not only inseparable but also become indistinguishable. The mystics are not satisfied with a situation analogous to atomic physics, where the observer and the observed cannot be separated, but can still be distinguished. They go much further, and in deep meditation they arrive at a point where the distinction between observer and observed breaks down completely, where subject and object fuse into a unified undifferentiated whole. Thus the *Upanishads* say:

Where there is a duality, as it were, there one sees another; there one smells another; there one tastes another. . . . But where everything has become just one's own self, then whereby and whom would one see? Then whereby and whom would one smell? Then whereby and whom would one taste?[16]

This, then, is the final apprehension of the unity of all things. It is reached— so the mystics tell us—in a state of consciousness where one's individuality dissolves into an undifferentiated oneness, where the world of the senses is transcended and the notion of "things" is left behind. In the words of Chuang Tzu:

My connection with the body and its parts is dissolved. My perceptive organs are discarded. Thus leaving my material form and bidding farewell to my knowledge, I become one with the Great Pervader. This I call sitting and forgetting all things.[17]

Modern physics, of course, works in a very different framework and cannot go that far in the experience of the unity of all things. But it has made a great step toward the world view of the Eastern mystics in atomic theory. Quantum theory has abolished the notion of fundamentally separated objects, has introduced the concept of the participator to replace that of the observer, and may even find it necessary to include the human consciousness in its description of the world. It has come to see the universe as an interconnected web of physical and mental relations whose parts are defined only through their connections to the whole. To summarize the world view emerging from atomic physics, the words of a Tantric Buddhist, Lama Anagarika Govinda, seem to be perfectly apropos:

The Buddhist does not believe in an independent or separately existing external world, into whose dynamic forces he could insert himself. The external world and his inner world are for him only two sides of the same fabric, in which the threads of all forces and of all events, of all forms of consciousness and of their objects, are woven into an inseparable net of endless, mutually conditioned relations.[18]. . .

In contrast to the mystic, the physicist begins his inquiry into the essential nature of things by studying the material world. Penetrating into ever-deeper realms of matter, he has become aware of the essential unity of all things and events. More than that, he has also learnt that he himself and his consciousness are an integral part of this unity. Thus the mystic and the physicist arrive at the same conclusion; one starting from the inner realm, the other from the outer world. The harmony between their views confirms the ancient Indian wisdom that *Brahman*, the ultimate reality without, is identical to *Atman*, the reality within.

Notes

1. Ashvagosha. *The Awakening of Faith*, transl. D.T. Suzuki, Chicago: Open Court, 1900, p. 55.

2. *Ibid.*, p. 93.

3. Although I have suppressed all the mathematics and simplified the analysis considerably, the following discussion may nevertheless appear to be rather dry and technical. It should perhaps be taken as "yogic" exercise which—like many exercises in the spiritual training of the Eastern traditions—may not be much fun, but may lead to a profound and beautiful insight into the essential nature of things.

4. H.P. Stapp, "S-Matrix Interpretation of Quantum Theory," *Physical Review*, Vol. D3 (March 15th, 1971), pp. 1303-20.

5. *Ibid.*, p. 1303.

6. N. Bohr, *Atomic Physics and the Description of Nature*, Cambridge: Cambridge University Press, 1934.

7. D. Bohm and B. Hiley, "On the Intuitive Understanding of Nonlocality as Implied by Quantum Theory," *Foundations of Physics*, 5 (1975), pp. 96, 102.

8. S. Aurobindo, *The Synthesis of Yoga*, Pondicherry, India: Aurobindo Ashram Press, 1957, p. 993.

9. Nagarjuna, quoted in T.R.V. Murti, *The Central Philosophy of Buddhism*, London: Allen & Unwin, 1955, p. 138.

10. H.P. Stapp, *op. cit.*, p. 1310.

11. W. Heisenberg, *Physics and Philosophy*, New York: Harper Torchbooks, 1958, p. 107.

12. *Mundaka Upanishad*, 2.2.5.

13. W. Heisenberg, *op. cit.*, p. 81.

14. *Ibid.*, p. 58.

15. J.A. Wheeler, in J. Mehra (ed.), *The Physicist's Conception of Nature*, Dordrecht, Holland: D. Reidel, 1973, p. 244.

16. *Brihad-aranyaka Upanishad*, 4.5.15.

17. Chuang Tzu, transl. James Legge, arranged by Clae Waltham, New York: Ace Books, 1971, ch. 6.

18. Lama Anagarika Govinda, *Foundations of Tibetan Mysticism*, New York: Samuel Weiser, 1974, p. 93.

Einstein Doesn't Like It

GARY ZUKAV

Quantum mechanics are not the fellows who repair automobiles in Mr. Quantum's garage. Quantum mechanics is a branch of physics. There are several branches of physics. Most physicists believe that sooner or later they will construct an overview large enough to incorporate them all.

According to this point of view, we eventually will develop, in principle, a theory which is capable of explaining everything so well that there will be nothing left to explain. This does not mean, of course, that our explanation necessarily will reflect the way that things actually are. We still will not be able to open the watch, as Einstein put it, but every occurrence in the *real* world (inside the watch) will be accounted for by a corresponding element of our final supertheory. We will have, at last, a theory that is consistent within itself and which explains all observable phenomena. Einstein called this state the "ideal limit of knowledge."[1]

This way of thinking runs into quantum mechanics the same way that the car runs into the proverbial brick wall. Einstein spent a large portion of his career arguing against quantum mechanics, even though he himself made major contributions to its development. Why did he do this? To ask this question is to stand at the edge of an abyss, still on the solid ground of Newtonian physics, but looking into the void. To answer it is to leap boldly into the new physics.

* * *

Quantum mechanics forced itself upon the scene at the beginning of this century. No convention of physicists voted to start a new branch of physics

Excerpted with permission from Gary Zukav, *The Dancing Wu Li Masters,* New York: William & Morrow Co., 1979. Slight changes have been made in the style of some bibliographical references.

called "quantum mechanics". No one had any choice in the matter, except, perhaps, what to call it.

A "quantum" is a quantity of something, a specific amount. "Mechanics" is the study of motion. Therefore, "quantum mechanics" is the study of the motion of quantities. Quantum theory says that nature comes in bits and pieces (quanta), and quantum mechanics is the study of this phenomenon.

Quantum mechanics does not replace Newtonian physics, it includes it. The physics of Newton remains valid within its limits. To say that we have made a major new discovery about nature is one side of a coin. The other side of the coin is to say that we have found the limits of our previous theories. What we actually discover is that the way that we have been looking at nature is no longer comprehensive enough to explain all that we can observe, and we are forced to develop a more inclusive view. In Einstein's words:

> ... creating a new theory is not like destroying an old barn and erecting a skyscraper in its place. It is rather like climbing a mountain, gaining new and wider views, discovering unexpected connections between our starting point and its rich environment. But the point from which we started out still exists and can be seen, although it appears smaller and forms a tiny part of our broad view gained by the mastery of obstacles on our adventurous way up.[2]

Newtonian physics still is applicable to the large-scale world, but it does not work in the subatomic realm. Quantum mechanics resulted from the study of the subatomic realm, that invisible universe underlying, embedded in, and forming the fabric of everything around us.

In Newton's age (late 1600s), this realm was entirely speculation. The idea that the atom is the indivisible building block of nature was proposed about four hundred years before Christ, but until the late 1800s it remained just an idea. Then physicists developed the technology to observe the effects of atomic phenomena, thereby "proving" that atoms exist. Of course, what they really proved was that the theoretical existence of atoms was the best explanation of the experimental data that anyone could invent at the time. They also proved that atoms are not indivisible, but themselves are made of particles smaller yet, such as electrons, protons, and neutrons. These new particles were labeled "elementary particles" because physicists believed that, at last, they really had discovered the ultimate building blocks of the universe.

The elementary particle theory is a recent version of an old Greek idea. To understand the theory of elementary particles, imagine a large city made entirely of bricks. This city is filled with buildings of all shapes and sizes. Every one of them, and the streets as well, have been constructed with only a few different types of brick. If we substitute "universe" for "city" and "particle" for "brick," we have the theory of elementary particles.

It was the study of elementary particles that brought physicists nose to nose with the most devastating (to a physicist) discovery: *Newtonian phys-*

ics does not work in the realm of the very small! The impact of that earthshaking discovery still is reshaping our world view. Quantum mechanical experiments repeatedly produced results which the physics of Newton could neither predict nor explain. Yet, although Newton's physics could not account for phenomena in the microscopic realm, it continued to explain macroscopic phenomena very well (even though the macroscopic is made of the microscopic)! This was perhaps the most profound discovery of science.

Newton's laws are based upon observations of the everyday world. They predict events. These events pertain to real things like baseballs and bicycles. Quantum mechanics is based upon experiments conducted in the subatomic realm. It predicts probabilities. These probabilities pertain to subatomic phenomena. Subatomic phenomena cannot be observed directly. None of our senses can detect them.[3] Not only has no one ever seen an atom (much less an electron), no one has ever tasted, touched, heard, or smelled one either.

Newton's laws depict events which are simple to understand and easy to picture. Quantum mechanics depicts the probabilities of phenomena which defy conceptualization and are impossible to visualize. Therefore, these phenomena must be understood in a way that is not more difficult than our usual way of understanding, but different from it. Do not try to make a complete mental picture of quantum mechanical events. (Physicists make partial pictures of quantum phenomena, but even these pictures have a questionable value.) Instead, allow yourself to be open without making an effort to visualize anything. Werner Heisenberg, one of the founders of quantum physics, wrote:

> The mathematically formulated laws of quantum theory show clearly that our ordinary intuitive concepts cannot be unambiguously applied to the smallest particles. All the words or concepts we use to describe ordinary physical objects, such as position, velocity, color, size, and so on, become indefinite and problematic if we try to use them of elementary particles.[4]

The idea that we do not understand something until we have a picture of it in our heads is a by-product of the Newtonian way of looking at the world. If we want to get past Newton, we have to get past that . . .

Newton's laws of motion describe what happens to a moving object. Once we know the laws of motion we can predict the future of a moving object provided that we know certain things about it initially. The more initial information that we have, the more accurate our predictions will be. We also can retrodict (predict backward in time) the past history of a given object. For example, if we know the present position and velocity of the earth, the moon, and the sun, we can predict where the earth will be in relation to the moon and the sun at any particular time in the future, giving us a foreknowledge of eclipses, seasons, and so on. In like manner, we can calculate where the earth has been in relation to the moon and the sun, and when similar phenomena occurred in the past.

Without Newtonian physics the space program would not be possible. Moon probes are launched at the precise moment when the launch site on the earth (which simultaneously is rotating around its axis and moving forward through space) is in a position, relative to the landing zone on the moon (which also is rotating and moving) such that the path traversed by the spacecraft is the shortest possible. The calculations of the earth, moon, and spacecraft movements are done by computer, but the mechanics used are the same ones that are described in Newton's *Philosophiae Naturalis Principia Mathematica*.

In practice, it is very difficult to know all the initial circumstances pertaining to an event. Even a simple action such as bouncing a ball off a wall is surprisingly complex. The shape, size, elasticity and momentum of the ball, the angle at which it was thrown, the density, pressure, humidity and temperature of the air, the shape, hardness and position of the wall, to name a few of the essential elements, are all required to know where and when the ball will land. It is increasingly difficult to obtain all of the data necessary for accurate predictions when more complex actions are involved. According to the old physics, however, it is possible, in principle, to predict *exactly* how a given event is going to unfold if we have enough information about it. In practice, it is only the enormity of the task that prevents us from accomplishing it.

The ability to predict the future based on a knowledge of the present and the laws of motion gave our ancestors a power they had never known. However, these concepts carry within them a very dispiriting logic. If the laws of nature determine the future of an event, then, given enough information, we could have predicted our present at some time in the past. That time in the past also could have been predicted at a time still earlier. In short, if we are to accept the mechanistic determination of Newtonian physics—if the universe really is a great machine—then from the moment that the universe was created and set into motion, everything that was to happen in it already was determined.

According to this philosophy, we may seem to have a will of our own and the ability to alter the course of events in our lives, but we do not. Everything, from the beginning of time, has been predetermined, including our illusion of having a free will. The universe is a prerecorded tape playing itself out in the only way that it can. The status of men is immeasurably more dismal than it was before the advent of science. The Great Machine runs blindly on, and all things in it are but cogs.

According to quantum mechanics, however, it is not possible, *even in principle*, to know enough about the present to make a complete prediction about the future. Even if we have the time and the determination, it is not possible. Even if we have the best possible measuring devices, it is not possible. It is not a matter of the size of the task or the inefficiency of detectors. The very nature of things is such that we must choose which aspect of them we wish to know best, for we can know only one of them with precision.

As Niels Bohr, another founder of quantum mechanics, put it:

> ... in quantum mechanics, we are not dealing with an arbitrary renunciation of a more detailed analysis of atomic phenomena, but with a recognition that such an analysis is *in principle* excluded.[5] (Italics in the original)

For example, imagine an object moving through space. It has both a position and a momentum which we can measure. This is an example of the old (Newtonian) physics. (Momentum is a combination of how big an object is, how fast it is going, and the direction that it is moving.) Since we can determine both the position and the momentum of the object at a particular time, it is not a very difficult affair to calculate where it will be at some point in the future. If we see an airplane flying north at two hundred miles per hour, we know that in one hour it will be two hundred miles farther north if it does not change its course or speed.

The mind-expanding discovery of quantum mechanics is that Newtonian physics does not apply to subatomic phenomena. In the subatomic realm, we cannot know both the position *and* the momentum of a particle with absolute precision. We can know both, approximately, but the more we know about one, the less we know about the other. We can know either of them precisely, but in that case, we can know nothing about the other. This is Werner Heisenberg's uncertainty principle. As incredible as it seems, it has been verified repeatedly by experiment.

Of course, if we picture a moving particle, it is very difficult to imagine not being able to measure both its position and momentum. Not to be able to do so defies our "common sense." This is not the only quantum mechanical phenomenon which contradicts common sense. Commonsense contradictions, in fact, are at the heart of the new physics. They tell us again and again that the world may not be what we think it is. It may be much, much more.

Since we cannot determine both the position and momentum of subatomic particles, we cannot predict much about them. Accordingly, quantum mechanics does not and cannot predict specific events. It does, however, predict *probabilities*. Probabilities are the odds that something is going to happen, or that it is not going to happen. Quantum theory can predict the probability of a microscopic event with the same precision that Newtonian physics can predict the actual occurrence of a macroscopic event.

Newtonian physics says, "If such and such is the case now, then such and such is going to happen next." Quantum mechanics says, "If such and such is the case now, then the *probability* that such and such is going to happen next is . . . (whatever it is calculated to be)." We never can know with certainty what will happen to the particle that we are "observing." All that we can know for sure are the probabilities for it to behave in certain ways. This is the most that we can know because the two data which must be included in a Newtonian calculation, position and momentum, cannot both be known with precision. *We must choose*, by the selection of our experiment, which one we want to measure most accurately.

The lesson of Newtonian physics is that the universe is governed by laws that are susceptible to rational understanding. By applying these laws we extend our knowledge of, and therefore our influence over, our environment. Newton was a religious person. He saw his laws as manifestations of God's perfection. Nonetheless, Newton's laws served man's cause well. They enhanced his dignity and vindicated his importance in the universe. Following the Middle Ages, the new field of science ("Natural Philosophy") came like a fresh breeze to revitalize the spirit. It is ironic that, in the end, Natural Philosophy reduced the status of men to that of helpless cogs in a machine whose functioning had been preordained from the day of its creation.

Contrary to Newtonian physics, quantum mechanics tells us that our knowledge of what governs events on the subatomic level is not nearly what we assumed it would be. It tells us that we cannot predict subatomic phenomena with any certainty. We only can predict their probabilities.

Philosophically, however, the implications of quantum mechanics are psychedelic. Not only do we influence our reality, but, in some degree, we actually *create* it. Because it is the nature of things that we can know either the momentum of a particle or its position, but not both, *we must choose* which of these two properties we want to determine. Metaphysically, this is very close to saying that we *create* certain properties because we choose to measure those properties. Said another way, it is possible that we create something that has position, for example, like a particle, because we are intent on determining position and it is impossible to determine position without having some *thing* occupying the position that we want to determine.

Quantum physicists ponder questions like, "Did a particle with momentum exist before we conducted an experiment to measure its momentum?"; "Did a particle with position exist before we conducted an experiment to measure its position?"; and "Did any particles exist at all before we thought about them and measured them?" *"Did we create the particles that we are experimenting with?"* Incredible as it sounds, this is a possibility that many physicists recognize.

John Wheeler, a well-known physicist at Princeton, wrote:

May the universe in some strange sense be "brought into being" by the participation of those who participate? . . . The vital act is the act of participation. "Participator" is the incontrovertible new concept given by quantum mechanics. It strikes down the term "observer" of classical theory, the man who stands safely behind the thick glass wall and watches what goes on without taking part. It can't be done, quantum mechanics says.[6]

The languages of eastern mystics and western physicists are becoming very similar.

Newtonian physics and quantum mechanics are partners in a double irony. Newtonian physics is based upon the ideas of laws which govern phenomena and the power inherent in understanding them, but it leads to impotence in the face of a Great Machine which is the universe. Quantum mechanics is based upon the idea of minimal knowledge of future phenomena (we are limited to knowing probabilities) but it leads to the possibility that our reality is what we choose to make it.

There is another fundamental difference between the old physics and the new physics. The old physics assumes that there is an external world which exists apart from us. It further assumes that we can observe, measure, and speculate about the external world without changing it. According to the old physics, the external world is indifferent to us and to our needs.

Galileo's historical stature stems from his tireless (and successful) efforts to quantify (measure) the phenomena of the external world. There is great power inherent in the process of quantification. For example, once a relationship is discovered, like the rate of acceleration of a falling object, it matters not who drops the object, what object is dropped, or where the dropping takes place. The results are always the same. An experimenter in Italy gets the same results as a Russian experimenter who repeats the experiment a century later. The results are the same whether the experiment is done by a skeptic, a believer, or a curious bystander.

Facts like these convinced philosophers that the physical universe goes unheedingly on its way, doing what it must, without regard for its inhabitants. For example, if we simultaneously drop two people from the same height, it is a verifiable (repeatable) fact that they both will hit the ground at the same time, regardless of their weights. We can measure their fall, acceleration, and impact the same way that we measure the fall, acceleration, and impact of stones. In fact, the results will be the same as if they *were* stones.

"But there is a difference between people and stones!" you might say. "Stones have no opinions or emotions. People have both. One of these dropped people, for example, might be frightened by his experience and the other might be angry. Don't their feelings have any importance in this scheme?"

No. The feelings of our subjects matter not in the least. When we take them up the tower again (struggling this time) and drop them off again, they fall with the same acceleration and duration that they did the first time, even though now, of course, they are both fighting mad. The Great Machine is impersonal. In fact, it was precisely this impersonality that inspired scientists to strive for "absolute objectivity."

The concept of scientific objectivity rests upon the assumption of an external world which is "out there" as opposed to "I" which is "in here." (This way of perceiving, which puts other people "out there," makes it very lonely "in here.") According to this view, Nature, in all her diversity, is "out there." The task of the scientist is to observe the "out there" as objectively as possible. To observe something objectively means to see it as it would appear to an observer who has no prejudices about what he observes.

The problem that went unnoticed for three centuries is that a person who carries such an attitude certainly is prejudiced. His prejudice is to be "objective," that is, to be without a preformed opinion. In fact, it is impossible to be without an opinion. An opinion is a point of view. The point of view that we can be without a point of view is a point of view. The decision itself to study one segment of reality instead of another is a subjective expression of the researcher who makes it. It affects his perceptions of reality, if nothing else. Since reality is what we are studying, the matter gets very sticky here.

The new physics, quantum mechanics, tells us clearly that it is not possible to observe reality without changing it. If we observe a certain particle collision experiment, not only do we have no way of proving that the result would have been the same if we had not been watching it, all that we know indicates that it would not have been the same, because the result that we got was affected by the fact that we were looking for it.

Some experiments show that light is wave-like. Other experiments show equally well that light is particle-like. If we want to demonstrate that light is a particle-like phenomenon or that light is a wave-like phenomenon, we only need to select the appropriate experiment.

According to quantum mechanics there is no such thing as objectivity. We cannot eliminate ourselves from the picture. We are a part of nature, and when we study nature there is no way around the fact that nature is studying itself. Physics has become a branch of psychology, or perhaps the other way round.

Carl Jung, the Swiss psychologist, wrote:

> The psychological rule says that when an inner situation is not made conscious, it happens outside, as fate. That is to say, when the individual remains undivided and does not become conscious of his inner contradictions, the world must perforce act out the conflict and be torn into opposite halves.[7]

Jung's friend, the Nobel Prize-winning physicist, Wolfgang Pauli, put it this way:

> From an inner center the psyche seems to move outward, in the sense of an extraversion, into the physical world . . .[8]

If these men are correct, then physics is the study of the structure of consciousness.

*　*　*

The descent downward from the macroscopic level to the microscopic level, which we have been calling the realm of the very small, is a two-step process. The first step downward is to the atomic level. The second step downward is to the subatomic level.

The smallest object that we can see, even under a microscope, contains millions of atoms. To see the atoms in a baseball, we would have to make the baseball the size of the earth. If a baseball were the size of the earth, its atoms would be about the size of grapes. If you can picture the earth as a huge glass ball filled with grapes, that is approximately how a baseball full of atoms would look.

The step downward from the atomic level takes us to the subatomic level. Here we find the particles that make up atoms. The difference between the atomic level and the subatomic level is as great as the difference between the atomic level and the world of sticks and rocks. It would be impossible to see the nucleus of an atom the size of a grape. In fact, it would be impossible to see the nucleus of an atom the size of a room. To see the nucleus of an atom, the atom would have to be as high as a fourteen-story building! The nucleus of an atom as high as a fourteen-story building would be about the size of a grain of salt. Since a nuclear particle has about 2,000 times more mass than an electron, the electrons revolving around this nucleus would be about as massive as dust particles!

The dome of Saint Peter's basilica in the Vatican has a diameter of about fourteen stories. Imagine a grain of salt in the middle of the dome of Saint Peter's with a few dust particles revolving around it at the outer edges of the dome. This gives us the scale of subatomic particles. It is in this realm, the subatomic realm, that Newtonian physics has proven inadequate, and that quantum mechanics is required to explain particle behavior.

A subatomic particle is not a "particle" like a dust particle. There is more than a difference in size between a dust particle and a subatomic particle. A dust particle is a *thing*, an object. A subatomic particle cannot be pictured as a thing. Therefore, we must abandon the idea of a subatomic particle as an object.

Quantum mechanics views subatomic particles as "tendencies to exist" or "tendencies to happen." How strong these tendencies are is expressed in terms of probabilities. A subatomic particle is a "quantum," which means a quantity of something. What that something is, however, is a matter of speculation. Many physicists feel that it is not meaningful even to pose the question. It may be that the search for the ultimate "stuff" of the universe is a crusade for an illusion. At the subatomic level, mass and energy change unceasingly into each other. Particle physicists are so familiar with the phenomena of mass becoming energy and energy becoming mass that they routinely measure the mass of particles in energy units.[9] Since the tendencies of subatomic phenomena to become manifest under certain conditions are probabilities, this brings us to the matter (no pun) of statistics.

Because there are millions of millions of subatomic particles in the smallest space that we can see, it is convenient to deal with them statistically. Statistical descriptions are pictures of crowd behavior. Statistics cannot tell us how one individual in a crowd will behave, but they can give us a fairly accurate description, based on repeated observations, of how a group as a whole behaves.

For example, a statistical study of population growth may tell us how many children were born in each of several years and how many are predicted to be born in years to come. However, the statistics cannot tell us which families will have the new children and which ones will not. If we want to know the behavior of traffic at an intersection, we can install devices there to gather data. The statistics that these devices provide may tell us how many cars, for instance, turn left during certain hours, but not *which* cars.

Statistics is used in Newtonian physics. It is used, for example, to explain the relationship between gas volume and pressure. This relation is named Boyle's Law after its discoverer, Robert Boyle, who lived in Newton's time. It could as easily be known as the Bicycle Pump Law, as we shall see. Boyle's Law says that if the volume of a container holding a given amount of gas at a constant temperature is reduced by one half, the pressure exerted by the gas in the container doubles.

Imagine a person with a bicycle pump. He has pulled the plunger fully upward, and is about ready to push it down. The hose of the pump is connected to a pressure gauge instead of to a bicycle tire, so that we can see how much pressure is in the pump. Since there is no pressure on the plunger, there is no pressure in the pump cylinder and the gauge reads zero. However, the pressure inside the pump is not actually zero. We live at the bottom of an ocean of air (our atmosphere). The weight of the several miles of air above us exerts a pressure at sea level of 14.7 pounds on every square inch of our bodies. Our bodies do not collapse because they are exerting 14.7 pounds per square inch outward. This is the state that we usually read as zero on a bicycle pressure gauge. To be accurate, suppose that we set our gauge to read 14.7 pounds per square inch before we push down on the pump handle.

Now we push the piston down halfway. The interior volume of the pump cylinder is now one half of its original size, and no air has been allowed to escape, because the hose is connected to a pressure gauge. The gauge now reads 29.4 pounds per square inch, or twice the original pressure. Next we push the plunger two thirds of the way down. The interior volume of the pump cylinder is now one third of its original size, and the pressure gauge reads three times the original pressure (44.1 pounds per square inch). This is Boyle's Law: At a constant temperature the pressure of a quantity of gas is inversely proportional to its volume. If the volume is reduced to one half, the pressure doubles; if the volume is reduced to one third, the pressure triples, etc. To explain why this is so, we come to classical statistics.

The air (a gas) in our pump is composed of millions of molecules (molecules are made of atoms). These molecules are in constant motion, and at any given time, millions of them are banging into the pump walls. Although we do not detect each single collision, the macroscopic effect of these millions of impacts on a square inch of the pump wall produces the phenomenon of "pressure" on it. If we reduce the volume of the pump cylinder by one half, we crowd the gas molecules into a space twice as small as the original one, thereby causing twice as many impacts on the same square inch of pump wall. The macroscopic effect of this is a doubling of the "pressure." By crowding the molecules into one third of the original space, we cause three times as many molecules to bang into the same square inch of pump wall, and the "pressure" on it triples. This is the kinetic theory of gases.

In other words, "pressure" results from the group behavior of a large number of molecules in motion. It is a collection of individual events. Each individual event can be analyzed because, according to Newtonian physics, each individual event is theoretically subject to deterministic laws. In principle, we can calculate the path of each molecule in the pump chamber. This is how statistics is used in the old physics.

Quantum mechanics also uses statistics, but there is a very big difference between quantum mechanics and Newtonian physics. In quantum mechanics, there is no way to predict individual events. This is the startling lesson that experiments in the subatomic realm have taught us.

Therefore, quantum mechanics concerns itself only with group behavior. It intentionally leaves vague the relation between group behavior and individual events because individual subatomic events cannot be determined accurately (the uncertainty principle) and, as we shall see in high-energy particles, they constantly are changing. Quantum physics abandons the laws which govern individual events and states *directly* the statistical laws which govern collections of events. Quantum mechanics can tell us how a group of particles will behave, but the only thing that it can say about an inidvidual particle is how it *probably* will behave. Probability is one of the major characteristics of quantum mechanics.

This makes quantum mechanics an ideal tool for dealing with subatomic phenomena. For example, take the phenomenon of common radioactive decay (luminous watch dials). Radioactive decay is a phenomenon of predictable overall behavior consisting of unpredictable individual events.

Suppose that we put one gram of radium in a time vault and leave it there for sixteen hundred years. When we return, do we find one gram of radium? No! We find only half a gram. This is because radium atoms naturally disintegrate at a rate such that every sixteen hundred years half of them are gone. Therefore, physicists say that radium has a "half life" of sixteen hundred years. If we put the radium back in the vault for another sixteen hundred years, only one fourth of the original gram would remain when we opened the vault again. Every sixteen hundred years one half of all the radium atoms in the world disappear. How do we know which radium atoms are going to disintegrate and which radium atoms are not going to disintegrate?

We don't. We can predict how many atoms in a piece of radium are going to disintegrate in the next hour, but we have no way of determining *which* ones are going to disintegrate. There is no physical law that we know of which governs this selection. Which atoms decay is purely a matter of chance. Nonetheless, radium continues to decay, on schedule, as it were, with a precise and unvarying half life of sixteen hundred years. Quantum theory dispenses with the laws governing the disintegration of individual radium atoms and proceeds directly to the statistical laws governing the disintegration of radium atoms as a group. This is how statistics is used in the new physics.

Another good example of predictable overall (statistical) behavior consisting of unpredictable individual events is the constant variation of intensity among spectral lines. . . . [A]ccording to Bohr's theory, the electrons of an atom are located only in shells which are specific distances from the nucleus. . . . Normally, the single electron of a hydrogen atom remains in the shell closest to the nucleus (the ground state). If we excite it (add energy to it) we cause it to jump to a shell farther out. The more energy we give it, the farther out it jumps. If we stop exciting it, the electron jumps inward to a shell closer to the nucleus, eventually returning all the way to the innermost shell. With each jump from an outer shell to an inner shell, the electron emits an energy amount equal to the energy amount that it absorbed when we caused it to jump outward. These emitted energy packets (photons) constitute the light which, when dispersed through a prism, forms the spectrum of one hundred or so colored lines that is peculiar to hydrogen. Each colored line in the hydrogen spectrum is made from light emitted from hydrogen electrons as they jump from a particular outer shell to a particular inner shell.

. . . Some of the lines in the hydrogen spectrum are more pronounced than others. The lines that are more pronounced are always more pronounced and the lines that are faint are always faint. The intensity of the lines in the hydrogen spectrum varies because hydrogen electrons returning to the ground state do not always take the same route.

Shell five, for example, may be a more popular stopover than shell three. In that case, the spectrum produced by millions of excited hydrogen atoms will show a more pronounced spectral line corresponding to electron jumps from shell five to shell one and a less pronounced spectral line corresponding to electron jumps from, say, shell three to shell one. That is because, in this example, more electrons stop over at shell five before jumping to shell one than stop over at shell three before jumping to shell one.

In other words, the probability is very high, in this example, that the electrons of excited hydrogen atoms will stop at shell five on their way back to shell one, and the probability is lower that they will stop at shell three. Said another way, we know that a certain number of electrons probably will stop at shell five and that a certain lesser number of electrons probably will stop at shell three. Still, we have no way of knowing *which* electrons will stop where. As before, we can describe precisely an overall behavior without being able to predict a single one of the individual events which comprise it.

This brings us to the central philosophical issue of quantum mechanics, namely, "What is *it* that quantum mechanics describes?" Put another way, quantum mechanics statistically describes the overall behavior and/or predicts the probabilities of the individual behavior of what?

In the autumn of 1927, physicists working with the new physics met in Brussels, Belgium, to ask themselves this question, among others. What they decided there became known as the Copenhagen Interpretation of Quantum Mechanics.[10] Other interpretations developed later, but the Copenhagen Interpretation marks the emergence of the new physics as a consistent way of viewing the world. It is still the most prevalent interpretation of the mathematical formalism of quantum mechanics. The upheaval in physics following the discovery of the inadequacies of Newtonian physics was all but complete. The question among the physicists at Brussels was not whether Newtonian mechanics could be adapted to subatomic phenomena (it was clear that it could not be), but rather, what was to replace it.

The Copenhagen Interpretation was the first consistent formulation of quantum mechanics. Einstein opposed it in 1927 and he argued against it until his death, although he, like all physicists, was forced to acknowledge its advantages in explaining subatomic phenomena.

The Copenhagen Interpretation says, in effect, that *it does not matter* what quantum mechanics is about![11] The important thing is that it works in all possible experimental situations. This is one of the most important statements in the history of science. The Copenhagen Interpretation of Quantum Mechanics began a monumental reunion which was all but unnoticed at the time. The rational part of our psyche, typified by science, began to merge again with that other part of us which we had ignored since the 1700s, our irrational side.

The scientific idea of truth traditionally had been anchored in an absolute truth somewhere "out there"—that is, an absolute truth with an independent existence. The closer that we came in our approximations to the absolute truth, the truer our theories were said to be. Although we might never be able to perceive the absolute truth directly—or to open the watch, as Einstein put it—still we tried to construct theories such that for every facet of absolute truth, there was a corresponding element in our theories.

The Copenhagen Interpretation does away with this idea of a one-to-one correspondence between reality and theory. This is another way of saying what we have said before. Quantum mechanics discards the laws governing individual events and states directly the laws governing aggregations. It is very pragmatic.

The philosophy of pragmatism goes something like this. The mind is such that it deals only with ideas. It is not possible for the mind to relate to anything other than ideas. Therefore, it is not correct to think that the mind actually can ponder reality. All that the mind can ponder is its *ideas* about reality. (Whether or not that is the way reality actually is, is a metaphysical issue.) Therefore, whether or not something is true is not a matter of how closely it corresponds to the absolute truth, but of how consistent it is with our experience.[12]

The extraordinary importance of the Copenhagen Interpretation lies in the fact that for the first time, scientists attempting to formulate a consistent physics were forced by their own findings to acknowledge that a complete understanding of reality lies beyond the capabilities of rational thought. It was this that Einstein could not accept. "The most incomprehensible thing about the world," he wrote, "is that it is comprehensible."[13] But the deed was done. The new physics was based not upon "absolute truth," but upon *us*.

Henry Pierce Stapp, a physicist at the Lawrence Berkeley Laboratory, expressed this eloquently:

[The Copenhagen Interpretation of Quantum Mechanics] was essentially a rejection of the presumption that nature could be understood in terms of elementary space-time realities. According to the new view, the complete description of nature at the atomic level was given by probability functions that referred, not to underlying microscopic space-time realities, but rather to the macroscopic objects of sense experience. The theoretical structure did not extend down and anchor itself on fundamental microscopic space-time realities. Instead it turned back and anchored itself in the concrete sense realities that form the basis of social life. . . . This pragmatic description is to be contrasted with descriptions that attempt to peer "behind the scenes" and tell us what is "really happeining."[14]

Another way of understanding the Copenhagen Interpretation (in retrospect) is in terms of split-brain analysis. The human brain is divided into two halves which are connected at the center of the cerebral cavity by a tissue. To treat certain conditions, such as epilepsy, the two halves of the brain sometimes are separated surgically. From the experiences reported by and the observations made of persons who have undergone this surgery, we have discovered a remarkable fact. Generally speaking, the left side of our brain functions in a different manner than the right side. Each of our two brains sees the world in a different way.

The left side of our brain perceives the world in a linear manner. It tends to organize sensory input into the form of points on a line, with some points coming before others. For example, language, which is linear (the words which you are reading flow along a line from left to right), is a function of the left hemisphere. The left hemisphere functions logically and rationally. It is the left side of the brain which creates the concept of causality, the image that one thing causes another because it always precedes it. The right hemisphere, by comparison, perceives whole patterns.

Persons who have split-brain operations actually have two separate brains. When each hemisphere is tested separately, it is found that the left brain remembers how to speak and use words, while the right brain generally cannot. However, the right brain remembers the lyrics of songs! The left side of our brain tends to ask certain questions of its sensory input. The right side of our brain tends to accept what it is given more freely. Roughly speaking, the left hemisphere is "rational" and the right hemisphere is "irrational."[15]

Physiologically, the left hemisphere controls the right side of the body and the right hemisphere controls the left side of the body. In view of this, it is no coincidence that both literature and mythology associate the right hand (left hemisphere) with rational, male, and assertive characteristics and the left hand (right hemisphere) with mystical, female, and receptive characteristics. The Chinese wrote about the same phenomena thousands of years ago (yin and yang) although they were not known for their split-brain surgery.

Our entire society reflects a left hemispheric bias (it is rational, masculine, and assertive). It gives very little reinforcement to those characteristics representative of the right hemisphere (intuitive, feminine, and receptive). The advent of "science" marks the beginning of the ascent of left hemispheric thinking into the dominant mode of western cognition and the descent of right hemispheric thinking into the underground (underpsyche) status from which it did not emerge (with scientific recognition) until Freud's discovery of the "unconscious" which, of course, he labeled dark, mysterious, and irrational (because that is how the left hemisphere views the right hemisphere).

The Copenhagen Interpretation was, in effect, a recognition of the limitations of left hemispheric thought, although the physicists at Brussels in 1927 could not have thought in those terms. It was also a *re-cognition* of those psychic aspects which long had been ignored in a rationalistic society. After all, physicists are essentially people who wonder at the universe. To stand in awe and wonder is to understand in a very specific way, even if that understanding cannot be described. The subjective experience of wonder is a message to the rational mind that the object of wonder is being perceived and understood in ways other than the rational.

The next time you are awed by something, let the feeling flow freely through you and do not try to "understand" it. You will find that you *do understand*, but in a way that you will not be able to put into words. You are perceiving intuitively through your right hemisphere. It has not atrophied from lack of use, but our skill in listening to it has been dulled by three centuries of neglect.

Wu Li Masters perceive in both ways, the rational and the irrational, the assertive and the receptive, the masculine and the feminine. They reject neither one nor the other. They only dance.

DANCING LESSON FOR NEWTONIAN PHYSICS	DANCING LESSON FOR QUANTUM MECHANICS
Can picture it.	Cannot picture it.
Based on ordinary sense perceptions.	Based on behavior of subatomic particles and systems not directly observable.
Describes *things*; individual objects in space and their changes in time.	Describes statistical behavior of *systems*.
Predicts events.	Predicts probabilities.
Assumes an objective reality "out there."	Does not assume an objective reality apart from our experience.
We can observe something without changing it.	We cannot observe something without changing it.
Claims to be based on "absolute truth"; the way that nature really is "behind the scenes."	Claims only to correlate experience correctly.

Notes

1. Albert Einstein and Leopold Infeld, *The Evolution of Physics,* New York: Simon and Schuster, 1938, p. 31.

2. *Ibid.,* p. 152.

3. The dark-adapted eye can detect a single photon. Otherwise, only the *effects* of subatomic phenomena are available to our senses (a track on a photographic plate, a pointer movement on a meter, etc.).

4. Werner Heisenberg, *Across the Frontiers,* New York: Harper & Row, 1974, p. 114.

5. Niels Bohr, *Atomic Theory and Human Knowledge,* New York: John Wiley, 1958, p. 62.

6. J. A. Wheeler, K. S. Thorne, and C. Misner, *Gravitation,* San Francisco: Freeman, p. 1273.

7. Carl G. Jung, *Collected Works,* vol. 9, Bollingen Series XX, Princeton: Princeton University Press, 1969, pp. 70-71.

8. Carl G. Jung and Wolfgang Pauli, *The Interpretation of Nature and the Psyche,* Bollingen Series LI, Princeton, Princeton Unviersity Press, 1955, p. 175.

9. Strictly speaking, mass, according to Einstein's special theory of relativity, *is* energy and energy *is* mass. Where there is one, there is the other.

10. This was the 5th Solvay Congress at which Bohr and Einstein conducted their now-famous debates. The term "Copenhagen Interpretation" reflects the dominant influence of Niels Bohr (from Copenhagen) and his school of thought.

11. The Copenhagen Interpretation says that quantum theory is about correlations in our experiences. It is about what will be observed under specified conditions.

12. The philosophy of pragmatism was created by the American psychologist, William James. Recently, the pragmatic aspects of the Copenhagen Interpretation of Quantum Mechanics have been emphasized by Henry Pierce Stapp, a theoretical physicist at the Lawrence Berkeley Laboratory in Berkeley, California. The Copenhagen Interpretation, in addition to the pragmatic part, has the claim that quantum theory is in some sense complete; that no theory can explain subatomic phenomena in any more detail.

An essential feature of the Copenhagen Interpretation is Bohr's principle of complementarity ... Some historians practically equate the Copenhagen Interpretation and complementarity. Complementarity is subsumed in a general way in Stapp's pragmatic interpretation of quantum mechanics, but the special emphasis on complementarity is characteristic of the Copenhagen Interpretation.

13. Albert Einstein, "On Physical Reality," *Franklin Institute Journal,* 221, 1936, 349ff.

14. Henry Stapp, "The Copenhagen Interpretation and the Nature of Space-Time," *American Journal of Physics,* 40, 1972, 1098ff.

15. Robert Ornstein, ed., *The Nature of Human Consciousness,* New York: Viking, 1974, pp. 61-149.

The Yogi and the Quantum

ROBERT P. CREASE
CHARLES C. MANN

> The new physics, they called it. It sounded very much like ancient Eastern mysticism.
>
> — Shirley MacLaine, *Dancing in the Light*

In 1935, Austrian physicist Erwin Schrödinger invented a "diabolical device" to illustrate the bizarre implications of quantum mechanics, the theory of the subatomic domain that he had helped to establish a decade earlier. Imagine, Schrödinger said, that a cat is sealed in a steel box, along with a Geiger counter, a small amount of radioactive material, a hammer, and a vial of hydrocyanic acid. The device is arranged so that when an atom of the radioactive substance decays, the Geiger counter discharges and, through a triggering mechanism, causes the hammer to smash the vial and release the poison. The problem is to describe the condition of the cat after one hour has passed—without looking inside the box.

Common sense, of course, decrees that an atom will or will not have decayed, and the cat will be either dead or alive but not both. Radioactive decays, however, take place in the realm of quantum mechanics, where things are different: according to the standard interpretation of the theory, such decays are indeterminate—they neither take place, nor don't take place—until someone performs a measurement, which in this case means opening the box and examining the cat. Until then, Schrödinger wrote, "the living and dead cat are, pardon the expression, blended or smeared out."[1]

Schrödinger's Cat Paradox, as it has come to be known, is often cited as evidence for the existence of an important connection between quantum

Portions of this essay are taken from Robert P. Crease and Charles C. Mann, "The Quantum Cat: Physics and the Search for Mystical Insight," *The Sciences*, July/August 1987. The authors would like to thank that periodical for permission to reprint these passages.

mechanics and Eastern mysticism. Quantum mysticism, as this view might be called, had its origins in some statements by certain of the pioneers of quantum mechanics, blossomed in the 1970s and 1980s, and today appears to be on the verge of becoming as firmly entrenched in popular culture as astrology.

Quantum mysticism raises numerous questions for scientists, sociologists, historians of ideas, and scholars of Eastern religions. To address these questions one must begin with the vigorous decade-long discussion, beginning in the mid-1920s, among the creators of quantum mechanics over whether the comings and goings of the quantum world were *anschaulich*—that is, intuitable or picturable—for Schrödinger's Cat Paradox was a by-product of this debate. Some physicists, including Schrödinger, felt that it was possible and even imperative to develop an intuitive comprehension of quantum mechanics, which would facilitate progress in the field. Others, including Werner Heisenberg, argued that quantum happenings were sufficiently alien from those of our ordinary, macroscopic world that the human mind lacked the appropriate nonmathematical concepts; the attempt to picture these quantum events, they claimed, would in fact confuse and mislead physicists. The urge to visualize proved irresistable, however, and the results were strange indeed. They inevitably raised the question of what "meaning," if any, quantum mechanics had for our world.

The troubles stemmed largely from the role of what is known as the "Schrödinger wave equation" invented by Schrödinger himself in 1926. (It is also called the "Schrödinger wave function"—the terms are interchangeable.) Now an essential tool of physics, Schrödinger's equation accounts for the behavior of the basic components of matter with enormous accuracy. Until its invention these components were thought, more or less, to be tiny particles of some sort—little marbles, albeit marbles with some peculiar features. Schrödinger's equation, however, depicted everything inside the atom entirely in terms of waves. Physicists found this startling: how could insubstantial waves build up the warp and woof of the world? If the denizens of the subatomic world are waves, waves of what?[2]

The first interpretation of this puzzling state of affairs is commonly credited to Heisenberg's adviser, Max Born. A few months after Schrödinger wrote down the wave function, Born argued that the frequency and wavelength of the waves in Schrödinger's equation are linked in a mathematically precise way not to *real* frequencies and wavelengths, but to statistical distributions—the probability, for example, that a given electron might be found at point A with momentum B in time C. Thus the wave function for an individual particle is a sort of catalogue of all its possible states. When a physicist performs a measurement, the wave function "collapses": all but one possibility is excluded, and the experimenter winds up with actual values for A, B, and C.[3]

At first glance there appears to be nothing radical or surprising about Born's interpretation or the inability to specify precisely the outcome of a measurement before performing it. People have bet on coin tosses and the like for centuries without disturbing the verities of physics. The equation describing the result of a coin toss, for instance, says that there is a 50 percent chance of getting heads and a 50 percent chance of getting tails. But the probabilistic description of such a toss simply means that we don't know the precise speed and angle of the toss, the wind direction, and so on. If we knew all of these things, we could know the outcome. The mortality curves used by insurance companies employ statistical distributions in a similar way; if enough about each individual human being and that person's activities were known, we could know that person's lifetime. The presence of statistics in both coin toss equations and mortality tables merely indicates our ignorance of the relevant factors.

But the probabilities in quantum mechanics are different, as Niels Bohr was the first to argue. In September of 1927, at the Volta congress in Como, Italy, Bohr asserted that quantum mechanics is *complete*—that in principle one cannot know more than what the Schrödinger wave equation is able to say. The wave function must represent a single system, and until an observer measures the system—an electron, say—and the wave function collapses, it does not have a definite position, momentum, energy, and so on. This pronouncement seemed to plunge physicists into an ontological quagmire. For if the properties of an object do not have values until they are measured, the object does not exist in any ordinary sense.

Famously, Einstein objected. Quantum mechanics, he said, is *not* complete; there must be some other factors—"hidden variables"—that, when discovered, will allow scientists to describe the behavior of subatomic particles as exactly as that of marbles. The wave function, in effect, must represent an ensemble of systems, and we are simply ignorant of what factors govern which system will prevail. Bohr disagreed, and the two men spent many years in argument, neither convincing the other. The dispute soon sent the antagonists down some peculiar byways, especially when they began to devise contraptions to translate their views of what is happening in the subatomic world into visualizable episodes in ours.

"Suppose you have a ball in one of two closed boxes," Einstein wrote Bohr in June of 1935. "If you say that there's a fifty-fifty chance for the ball being in either box, that's an incomplete description, because the ball is clearly in one or the other box, right? Why is it different when the Schrödinger equation says an electron has an equal chance of being in two places? It is not halfway in both, is it?"[4]

Schrödinger, too, was troubled by the implications of quantum mechanics—and by the fact that the absurdities seemed to stem from his own contribution to the theory. In November of 1935, a few months after

Einstein's letter to Bohr, Schrödinger weighed in with his own image—the now-famous cat—which has become a maddeningly elegant symbol for the perplexities of quantum mechanics. Here the wave function describes the state not only of the radioactive substance but of the entire setup of Geiger counter, hammer, vial, and cat. When a person measures this system—which in this case means opening the box and examining the cat—the wave function for that system, which is a superposition of two quantum states (one in which the atom has not decayed and the acid is still inside the vial, and one in which it has decayed and the acid is sloshing around the bottom of the box) collapses, and the cat is either dead or alive.

Schrödinger's device is thus categorically different from, say, an automated coin-toss mechanism that activates a lethal poison when the coin comes up heads. In theory, the outcome of such a classical process can be predicted if enough is known about the mechanism—the angle of release, the height to which it sends the coin, its sensitivity to such factors as atmospheric pressure and gravity—and any uncertainty over the outcome would be neither scientifically nor philosophically interesting. The cat will be *either* dead or alive. But if quantum theory is complete, the occurrence of the decay *cannot* be predicted; the atom and the rest of the setup exist in theory—and hence in reality—only as a superposition of possibilities, and, truth to tell, the cat is both alive *and* dead until an observer opens the box.

For Schrödinger, the moral of the story was that something was wrong with quantum mechanics. For Bohr and Heisenberg, however, the moral was different. Quantum mechanics was right, they said. If anything was wrong, it was with our cognitive capacity; human beings were simply not equipped to visualize the totally different world "down there." Whereas classical physics depicts objects as having properties all of which could be specified in principle at any given time, quantum mechanics requires us to regard some of these properties as "complementary but exclusive"; that is, some property can be specified precisely at one particular time only at the cost of precision in specifying another property, the choice being left up to the scientist.

This idea was the cornerstone of what was soon called the Copenhagen Interpretation (after the city in which Bohr did most of his work), which is still the most important elaboration of the meaning of quantum mechanics. For Bohr and others, the Copenhagen Interpretation was a kind of philosophical key allowing them to feel comfortable with the peculiarities of the subatomic realm. Nevertheless, it almost immediately forced physicists to face a bewildering situation, for it seemed to say that conventional assumptions about the nature of reality do not apply in the subatomic world. And this is just what Bohr argued. The Schrödinger equation, he said, does not depict actual, independent objects. As Bohr said at Como, "[A]n independent reality in the ordinary physical sense can neither be ascribed to the phenomena nor to the agencies of observations."[5]

The Copenhagen Interpretation couched these views in terms of an "observer," whose intervention in the act of measurement—in opening Schrödinger's box, for instance—collapses the wave function and makes the quantum situation definite, in effect turning electrons back into marbles. Unfortunately, when Bohr and his colleagues tried to decide precisely what constitutes an observer their philosophical discussions were less rigorous than their physics. The use of the word "observer" proved treacherous, for it appeared to introduce subjectivity into physics. After some debate, the Copenhagen group decided that this implication was correct. The observer who opens Schrödinger's box causes the wave function to collapse and, hence, decides the life or death of the cat. It was only a short step from this position to the conclusion that the existence of the world depends on consciousness—that, indeed, reality is our creation.

Few would have guessed that a theory with such practical value (quantum-mechanical predictions are extraordinarily accurate) could open such a Pandora's box of foggy speculation. Over the years the ontological implications of the Copenhagen Interpretation have inspired commentary by philosophically inclined physicists and physically inclined philosophers. Some of their conclusions have been odd enough to inspire a secondary growth of writing by philosophically and physically inclined cranks, who have taken quantum mechanics out of the subatomic domain and endowed it with wide and general meaning for human action and human destiny. In the 1930s some theologians believed the uncertainties of quantum mechanics showed how God could perform acausal miracles in the world; today, half a century later, the Maharishi Mahesh Yogi contends that twentieth-century physics proves the validity of Transcendental Meditation. Other authors have argued, among other things, that twentieth-century physics proves the existence of telepathy, the collective unconscious, and faster-than-light communication. Like so many modern Saint Anselms, Bohr and his *confrères* are even supposed to have created a proof of the existence of God.

Especially popular recently has been quantum mysticism, the view that links exist between quantum mechanics and Eastern mysticism. The groundwork for quantum mysticism was to some extent laid by scientists themselves; when Niels Bohr was knighted in 1947, for instance, he selected for his coat-of-arms design a yin-yang symbol, while Heisenberg and Robert Oppenheimer, among others, sometimes drew connections between their work and ideas of Eastern religions. The founding father of the modern movement, however, is Fritjof Capra, a maverick physicist whose *The Tao of Physics* sold millions of copies as it asserted a connection between physics and Zen Buddhism; indeed, writes Capra, this connection is of critical significance for it demonstrates the inadequacy of our present world view and points to the need for a change so drastic as to amount to a cultural revolution. "The survival of our whole civilization may depend on whether we can bring about such a change."[6] A non-scientist, Gary Zukav, then produced *The Dancing Wu Li Masters*, which made the breezy announcement that "philosophically, ... the implications of [quantum mechanics] are psychedelic," and won an American Book Award. Still more visible and effusive has been actress Shirley MacLaine, who quotes Zukav in the epigraph of *Dancing in the Light*. MacLaine's approach to quantum mechanics is mixed in with lengthy accounts of her affair with a Russian director who was her son in at least four past incarnations and her experiences with a woman named J. Z. Guest, who lives in Yelm, Washington, and is the Charlie McCarthy for a 35,000-year-old superbeing named Ramtha. Despite the odd packaging, MacLaine's quantum mystical views are utterly typical of the genre: modern physicists and Eastern mystics are saying the same things in different languages, and their basic message is that human beings and the world are melded together in a kind of cosmic ragout. "Metaphysical experiences," MacLaine writes,

> have led me to understand even more fully what the new physicists and ancient mystics were attempting to reconcile in their own minds: the reality of consciousness. Aside from suddenly seeming to speak the same language, they seemed to be on the brink of agreeing that even the *cosmos* was nothing but consciousness. That the universe and God itself might just be one giant, collective "thought."[7]

Things have got a little unfastened in the world of scientific popularization, and quantum mechanics is the culprit. From the first, there was little doubt of quantum mechanics' ability to account for the phenomena experimenters saw in laboratories. But the theory was couched in a mathematical structure physicists found bizarre, and it failed completely to provide pictures that would enable scientists to understand intuitively why the equations worked and what they were talking about. Still, quantum mechanics *worked*, which convinced most scientists to put the question of its meaning on the back burner.

The popularity of quantum mysticism as well as its extravagant claims, however, has forced the question of interpretation to the fore. Like advocates of creationism and astrology, quantum mystics misrepresent science to the public, and the result could be detrimental to how money is allocated to the field. Moreover, conceptual confusion in physics, as in any area, is worth trying to clear up for its own sake.

The task has not been easy. To avoid having Schrödinger's cat turn the universe into a collective thought, some scientists have claimed that the cat or the Geiger tube or the atom are the "observers" that collapse the wave function. This has led to arguments about whether an observer must be conscious and whether cats and/or Geiger tubes and/or atoms are conscious enough to count. The discussion has come to have the air of theological lunacy associated with the early Christian disputation over whether Christ is of the same substance (Homoousian) or of like substance (Homoiousian) with the Father. Then the fate of Christianity seemed to hang on a dipthong; now, the meaning of reality seems to hang on a semi-alive cat.

Sal Restivo, a professor in the Science and Technology Studies Division at Rensselaer Polytechnic Institute, has critically examined quantum mysticism (his word for it is "parallelism") in his book *The Social Relations of Physics, Mysticism, and Mathematics*. He observes:

> The basic data for parallelism are common language (for example, English) statements on the nature and implications of physics and mysticism that vary in technical content. The methodology of parallelism is the comparative analysis of such statements. Similar rhetoric, imagery, and metaphoric content in such statements constitute the evidence for parallelism. The basic assumption in this approach is that if the imagery and the rhetorical and metaphoric content of statements on physics and mysticism are similar, the conceptual content must be similar, and the experience of reality must also be similar among physicists and mystics.[8]

But even before comparisons are made, Restivo points out, the success of such a procedure must first depend on proper selection and translation of representative texts from both mystics and physicists. Several perils stand in the way. The first lies in selection; the texts chosen must truly reflect the spirit of Eastern mysticism and physics and neither be idiosyncratic nor overlook conflict and diversity. Restivo finds, however, that Capra and others tend to pick and choose their texts, ignoring the immense variety in mystical experiences and assuming that the world views of Buddhism, Hinduism, and Taoism are essentially the same. It might be added that quantum mystics also tend to pick and choose their physics texts as well, sweeping under the rug arguments about the proper nature and direction of the field. Here we can cite one of Capra's own examples, the bootstrap hypothesis, which denies that there are fundamental particles and asserts that each type of particle is linked with all the others. In the mid-1970s, when Capra's book was written, this model was popular among some physicists. Others, however, embraced the quark model, based on the distinctly unmystical notion of fundamental particles. By promoting the bootstrap hypothesis as typical of the world view of modern physics, and then comparing it to certain Eastern modes of thought, Capra uses an idiosyncratic picture of physics to buttress his quantum mystical claims.

Another peril in the quantum mystical methodology is contamination. Many concepts of modern physics have filtered into common vocabulary, including "space-time," "complementarity," "quantum," "relativity," and so forth, as have many concepts of Eastern religions. This may lead to conscious or unconscious misuse of terminology, suggesting links where there are none. An example of deliberate, playful misuse is Murray Gell-Mann's 1962 system of representing subatomic particles which assigned the baryons to one group of eight; Gell-Mann named his system the "Eightfold Way" in joking homage to the Buddha's teaching. Any resemblence to actual teachings of the Buddha is, of course, superficial.

But far and away the most formidable peril is that of translation. A prerequisite for comparing statements by physicists and mystics is that both be in a similar language. What must be translated, however, is in each case profoundly different. The theories of modern physics, for instance, are couched in a mathematical formalism that requires years of training to understand and are inevitably rendered inadequately when couched in nonmathematical language; indeed, the late Nobel Prizewinning physicist Richard Feynman claimed that it is impossible to explain the meaning of the laws of nature to those unfamiliar with mathematics.[9] Similarly, ancient Eastern mystics wrote not only in a foreign language but about a kind of experience that many say defies any translation into language whatsoever; Restivo quotes one observer of Mahayana Buddhism to the effect that its practitioners undergo intuitive experiences "intrinsically beyond words and symbols."[10]

Language may therefore have entirely different relations to what is addressed in the hands of physicists and mystics. If so, both the conceptual content of what physicists and mystics say and their experiences of reality are likely to be different even when they speak in similar language. Zukav's assertions that "physics is not mathematics" and that most physical ideas are "essentially simple" do not reassure; neither does Capra's statement regarding Hinduism, Buddhism, and Taoism that "the basic features of their world view are the same."[11] Translations on each side may be done in different ways, moreover, some of which may suggest links and others not. Physicist Gerald Feinberg, for instance, has managed to write an excellent book explaining quantum mechanics in nonmathematical terms without recourse to any mystical language at all.[12] Quantum events thus may be *anschaulich*, but with different degrees of adequacy and resemblance to mysticism.

Even if such methodological hurdles are overcome, however, the question of the justice of specific links asserted by comparative analyses remains. Capra, for instance, finds parallels in the following themes: organicism, paradoxes, transcendence, space-time, oneness, and empiricism. In addition, he finds certain "equivalencies" between modern physics and Eastern mysticism: the quantum field and Ch'i (the vital energy animating the Universe), the physical vacuum and the Great Void (the ultimate reality that gives birth to, sustains and reabsorbs everything), S-matrix theory (a mathematical theory describing the interactions between certain kinds of subatomic particles) and the *I Ching* (the *Book of Changes* describing the interplay of yin and yang), complementarity and the Tao (the Way of the Universe), and the bootstrap model and the Buddhist view of nature.[13] These parallels and equivalencies are also often cited by others.

Upon close scrutiny, however, these parallels and equivalencies prove to have weak foundations. To take one example, quantum mystics frequently point to the widespread presence of paradoxes in modern physics (such as Schrödinger's Cat) and Zen (the sound of one hand clapping). Zukav, for instance, asks, "Is it a coincidence that Buddhists exploring 'internal' reality a millenium ago and physicists exploring 'external' reality a millennium later both discovered that 'understanding' involves passing the barrier of paradox?" The answer, in a word, is yes. Paradoxes in mysticism are conundrums that cannot be resolved by reason, and aim to force the mind to jump beyond its natural habits into apprehending a wholly nonrational truth. Such paradoxes are thus eternally helpful and are to be cherished in themselves. Paradoxes in physics, on the other hand, are signposts indicating where rational understanding of the theory is still lacking and must be achieved by the forging of new concepts and techniques. These paradoxes are to be eliminated as quickly as possible. Similar objections could be raised in connection with the other frequently cited parallels and equivalencies.

Eastern mysticism, moreover, is less like a body of truths than a program for a spiritual quest, an attitude adopted toward the world, the gods, the transcendental. The wisdom expressed by this attitude, according to its practitioners, has remained essentially unchanged for thousands of years. Science, however, continually evolves, and even its fundamental concepts are subject to change, as when Einstein overtook Newton. If quantum mystics argue that contemporary scientists are making the *same* claims about the world that ancient Eastern mystics make, is ancient Eastern mysticism refuted when and if new discoveries force scientists to revise their views? Capra's bootstrap hypothesis is again a good example. In the late 1970s, after Capra's book was published, the quark model triumphed quite decisively over the bootstrap model. This triumph was wholly traditional; quarks explained the data better. Must Buddhists now cast their world view out the window?

Moreover, if links between the world views of physics and Eastern mysticism are more than superficial, the truths of the latter ought to lead to discoveries in the former. As physicist John Bell has said, "We'll all go and sit at the feel of the Maharishi if he tells us where the Higgs boson [a long-sought particle] is to be found!"[14] Yet to our knowledge this cross-fertilization has never succeeded, nor even been attempted.

Quantum mystics wind up distorting both science and Eastern mysticism. In general, they distort science by deemphasizing the role of mathematics, mathematical reasoning, and technology, claiming that what is essential about science can be cleanly separated from these. They distort Eastern mysticism by overlooking the social and historical roots of the various Eastern religions and the extent to which the cosmologies of Buddhism, Hinduism, and Taoism not only were individually shaped by the structures of the societies that gave birth to them but also evolved in the light of exposure to changing social conditions. The convenient neglect by quantum mystics of such all-important elements of Eastern religions as rejection of the world, the long period of discipleship, and the preoccupation with death indicate that today's "quantum Zen" is probably as ignorant of the real religions of the East as was the "California Zen" of a generation ago.

Restivo is convinced that an ideological motive lies behind quantum mysticism which works to the benefit of both physics and mysticism. Physics is justified in the face of counter-cultural criticism that it is sterile and inhuman (the boys who brought you the Bomb are shown to be equally in character in sandals and love beads), while Eastern mysticism becomes innoculated against the intentions and methods of critical inquiry (you'd better believe the utterances of the sandal and love bead brigade, for they've got the backing of those who operate the huge particle accelerators).

Here Restivo is less convincing. He begins grinding a Marxist ax as he pursues a "materialist sociology of knowledge in opposition to all forms of transcendental, idealist and idealistic, purist, individualist-heroic, and religious, theological, and spiritual analyses and explanations of social life"; he even turns self-critical and apologizes to the reader if he hasn't purged all traces of non-materialist modes of thought from his work.[15] There are indeed obvious traces of ideology in some quantum mystical writings; Capra, a competent and serious physicist, confesses to a desire to improve "the image of science."[16] But it is unlikely that the enthusiasm of a movie actress like MacLaine, for instance, can be satisfactorally explained in this way; it's far more likely that some sort of religious, theological, or spiritual motive is at work.

Ever since the age of Copernicus, the world no longer revolved around us, and science has steadily whittled us down to size. According to evolutionary theory we are but one species among millions whose existence is accidental and whose demise may be imminent. At best our noblest thoughts, works, and institutions may be adaptive mechanisms whose sole purpose is to ensure the transfer of genetic material from one generation to the next; at worst, they are useless appendages to life. Zukav, MacLaine, and other quantum mystics desperately want to find some evidence for a reversal of this tendency, and suppose that this reversal is heralded by such quantum riddles as Schrödinger's cat, which seems to show that we make the Universe happen. They want to be told that, after centuries of diminishing humanity's significance in the universe, science is about to restore us to our rightful place in the Sun. To paraphrase a recent campaign slogan, they want to hear that "It's morning in the Universe again."

But Restivo's Marxist approach has more serious shortcomings. Science is more than just one tool in a conflict between different social groups for power and prestige; the things that science discloses and makes available to us are different from those things disclosed by other disciplines. Astrology, witchcraft, and alchemy, too, have been viewed as tools in social conflict—what makes their products different from those of science? In addition, Restivo fails to answer the question of why modern physics lends itself to quantum mystical interpretation. True criticism includes an account of why others have been led astray.

The questions at issue here are both extremely important and extremely difficult; anyone attempting to answer them must surely be daunted by Stephen Brush's remark that "no one has yet formulated a consistent world view that incorporates the CI [Copenhagen Interpretation] of QM [quantum mechanics] while excluding what most scientists would call pseudo-sciences—astrology, parapsychology, creationism, Velikovsky's theories, and thousands of other cults and doctrines."[17] We can only venture some guesses as to the elements of that world view here. One reason why quantum mechanics suggests links to Eastern mysticism is the deeply held theological and metaphysical Western conviction that objectivity belongs to those things whose fundamental characteristics can be all present at the same time, at least in principle. The guiding image here is of a divine mind capable of witnessing the spectacle of the world's *eidoi* at a glance; the things of the world must all be *anschaulich* to the Demiurge. The aim of scientific theory, in this traditional view, is to develop a similar picture of the "fundamental furniture of the world." As the early disputes over *Anschaulichkeit* by the founders of quantum mechanics showed, however, modern physics reveals the impossibility of this aim.[18] Some properties of certain quantum objects are fleshed out differently in different contexts, these contexts depending, in the laboratory, on the actions of the experimenter.

Quantum mystics then argue, essentially, as follows: If theory aims to picture the basic things of the world, and if modern physics shows that such a picture is not possible and that a picture emerges only when we engage with the world, then the world as it emerges in our theories is an illusion and reality is our own creation. Put this way, the discoveries of quantum mechanics do indeed suggest some similarities to certain beliefs of Eastern mystics, cursorily examined.

But surely there is an alternative. Suppose, for instance, that theory explains phenomena by providing the "score," as it were, for the phenomena that appear through our instruments in experimental "performances."[19] The point is that musical scores do not represent or picture music, and one needs more than the score to have music; one needs, among other things, an instrument, a skilled performer, an audience, a tradition of playing, and so forth, each of which belongs to the culture of human life and varies depending on the social and historical context. While our theory of electrons, therefore, may tell us something abstract about them, the form they will take in any given experiment depends on our instruments, our techniques, our skill, and so forth. If theory is viewed in this way, then it is an exaggeration to say, as Zukav does, that "Not only do we influence our reality, but, in some degree, we actually *create* it."[20] That would be equivalent to saying that we create electrons when we measure their charge—or create a symphony afresh in performing one.

Quantum mechanics does indeed pose a serious challenge to traditional philosophy of science and its notions of objectivity. But this does not necessarily mean that the tenets of Eastern mysticism meet that challenge—or even that the tenets of mysticism are invalidated if they fail the challenge. It does mean that the philosophy of science will have to work out its notions of theory, objectivity, and scientific phenomena in more detail.[21] Until then, Schrödinger's spectral feline will continue to haunt quantum mechanics with occult implications. The misguided search for mystical insight can be taken as an indication of how much work physicists and philosophers still have before them.

Notes

1. Erwin Schrödinger, "The Current Situation in Quantum Mechanics," *Die Naturwissenschaften*, 29 November 1935, p. 812.

2. For more about quantum mechanics, *Anschaulichkeit*, and the wave equation, see Mara Beller, "The Genesis of Interpretations of Quantum Physics: 1925-1927" (Ph.D. dissertation, University of Maryland, 1983); and Arthur I. Miller, *Imagery in Scientific Thought* (Cambridge, Mass.: MIT Press, 1986), pp. 127-183. For a more narrative account of these events, see Robert P. Crease and Charles C. Mann, *The Second Creation: Makers of the Revolution in Twentieth Century Physics* (Macmillan, 1986), pp. 52-59.

3. Max Born, Z. *Physik* 37 (1926), p. 863.

4. For more about the Einstein-Bohr dispute, cf. Arthur Fine, *The Shaky Game: Einstein, Realism, and the Quantum Theory* (Chicago: University of Chicago Press, 1986).

5. Niels Bohr, *Atomic Theory and the Description of Nature* (Cambridge, England: Cambridge University Press, 1934), p. 54.

6. Fritjof Capra, *The Tao of Physics: An Exploration of the Parallels Between Modern Physics and Eastern Mysticism* (Berkeley: Shambhala Publications, 1976), p. 298. Gary Zukav, *The Dancing Wu Li Masters* (New York: Morrow, 1979). Among the numerous other authors who espouse quantum mysticism is Alex Comfort (of *Joy of Sex* fame) in *Reality & Empathy: Physics, Mind & Science in the 21st Century* (Buffalo: State Univ. of New York Press, 1984). Quantum mysticism has even appeared in movies; the baseball groupie played by Susan Sarandon in *Bull Durham* is a quantum mystic.

7. Shirley MacLaine, *Dancing in the Light* (New York: Bantam, 1985), p. 403; the exergue to this article is from p. 323.

8. Sal Restivo, *The Social Relations of Physics, Mysticism, and Mathematics* (Dordrecht, Holland: Reidel, 1985), p. 22.

9. Richard Feynman, *The Character of Physical Law* (Cambridge, Mass.: MIT Press, 1965), pp. 39-40.

10. Restivo, *op. cit.*, p. 25.

11. Zukav, *op. cit.*, p. 31; Capra, *op. cit.*, p. 5.

12. Gary Feinberg, *What is the World Made Of?* (Garden City, New York: Anchor, 1977).

13. These are listed by Restivo, *op. cit.*, p. 9.

14. Interview, John Bell, Geneva, Switzerland, 20 February 1987.

15. Restivo, *op. cit.*, p. viii.

16. Capra, *op. cit.*, p. 92.

17. Stephen G. Brush, *The History of Modern Science* (Ames: Iowa State University Press, 1988), p. 409.

18. The clearest way to state the reason for this impossibility is that quantum phenomena are represented by infinite-dimensional vectors in Hilbert space, which precludes the possibility of any complete "picture" of a quantum state.

19. The image first appears in Patrick A. Heelan, "Experiment and Theory: Constitution and Reality," *Journal of Philosophy* 85:10 (October 1988), p. 522.

20. Zukav, *op. cit.*, pp. 53-54.

21. See, for instance: Robert P. Crease, "The Rediscovery of Experiment" (*Missouri Review*, 1988); Patrick A. Heelan, *Quantum Mechanics and Objectivity* (The Hague: Nijhoff, 1965), and *Space-Perception and the Philosophy of Science* (Berkeley: University of California Press, 1983).

Quantum Mysteries for Anyone

N. DAVID MERMIN

> We often discussed his notions on objective reality. I recall that during one walk Einstein suddenly stopped, turned to me and asked whether I really believed that the moon exists only when I look at it.
>
> A. Pais[1]

> As O. Stern said recently, one should no more rack one's brain about the problem of whether something one cannot know anything about exists all the same, than about the ancient question of how many angels are able to sit on the point of a needle. But it seems to me that Einstein's questions are ultimately always of this kind.
>
> W. Pauli[2]

Pauli and Einstein were both wrong. The questions with which Einstein attacked the quantum theory do have answers; but they are not the answers Einstein expected them to have. We now know that the moon is demonstrably not there when nobody looks.

The impact of this discovery on philosophy may have been blunted by the way in which it is conventionally stated, which leaves it fully accessible only to those with a working knowledge of quantum mechanics. I hope to remove that barrier by describing this remarkable aspect of nature in a way that presupposes no background whatever in the quantum theory or, for that matter, in classical physics either. I shall describe a piece of machinery that presents without any distortion one of the most strikingly peculiar features of the atomic world. No formal training in physics or mathematics is needed to grasp and ponder the extraordinary behavior of the device; it is only necessary to follow a simple counting argument on the level of a newspaper braintwister.

Being a physicist, and not a philosopher, I aim only to bring home some strange and simple facts which might raise issues philosophers would be

Reprinted with permission from the *Journal of Philosophy* 78 (1981), 397-408.

interested in addressing. I shall try, perhaps without notable success, to avoid raising and addressing such issues myself. What I describe should be regarded as something between a parable and a lecture demonstration. It is less than a lecture demonstration for technical reasons: even if this were a lecture, I lack the time, money, and particular expertise to build the machinery I shall describe. It is more than a parable because the device could in fact be built with an effort almost certainly less than, say, the Manhattan project, and because the conundrum posed by the behavior of the device is no mere analogy, but the atomic world itself, acting at its most perverse.

There are some black boxes within the device whose contents can be described only in highly technical terms. This is of no importance. The wonder of the device lies in what it does, not in how it is put together to do it. One need not understand silicon chips to learn from playing with a pocket calculator that a machine can do arithmetic with superhuman speed and precision; one need not understand electronics or electrodynamics to grasp that a small box can imitate human speech or an orchestra. At the end of the essay I shall give a brief technical description of what is in the black boxes. That description can be skipped. It is there to serve as an existence proof only because you cannot buy the device at the drugstore. It is no more essential to appreciating the conundrum of the device than a circuit diagram is to using a calculator or a radio.

The device has three unconnected parts. The question of connectedness lies near the heart of the conundrum, but I shall set it aside in favor of a few simple practical assertions. There are neither mechanical connections (pipes, rods, strings, wires) nor electromagnetic connections (radio, radar, telephone or light signals) nor any other relevant connections. Irrelevant connections may be hard to avoid. All three parts might, for example, sit atop a single table. There is nothing in the design of the parts, however, that takes advantage of such connections to signal from one to another, for example, by inducing and detecting vibrations in the table top.

By insisting so on the absence of connections I am inevitably suggesting that the wonders to be revealed can be fully appreciated only by experts on connections or their lack. This is not the right attitude to take. Were we together and had I the device at hand, you could pick up the parts, open them up, and poke around as much as you liked. You would find no connections. Neither would an expert on hidden bugs, the Amazing Randi, or any physicists you called in as consultants. The real worry is unknown connections. Who is to say that the parts are not connected by the transmission of unknown Q-rays and their detection by unrecognizable Q-detectors? One can only offer affidavits from the manufacturer testifying to an ignorance of Q-technology and, in any event, no such intent.

Evidently it is impossible to rule out conclusively the possibility of connections. The proper point of view to take, however, is that it is precisely the wonder and glory of the device that it impels one to doubt these assurances from one's own eyes and hands, professional magicians, and technical experts of all kinds. Suffice it to say that there are no connections that suspicious lay people or experts of broad erudition and unimpeachable integrity can discern. If you find yourself questioning this, then you have grasped the mystery of the atomic world.

Two of the three parts of the device (A and B) function as detectors. Each detector has a switch that can be set in one of three positions (1, 2, and 3) and a red and a green light bulb (Fig. 1). When a detector is set off it flashes either its red light or its green. It does this no matter how its switch is set, though whether it flashes red or green may well depend on the setting. The only purpose of the lights is to communicate information to us; marks on a ribbon of tape would serve as well. I mention this only to emphasize that the unconnectedness of the parts prohibits a mechanism in either detector that might modify its behavior according to the color that may have flashed at the other.

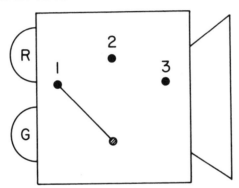

Fig. 1. A detector. Particles enter on the right. The red (R) and green (G) lights are on the left. The switch is set to 1.

The third and last part of the device is a box (C) placed between the detectors. Whenever a button on the box is pushed, shortly thereafter two particles emerge, moving off in opposite directions toward the two detectors (Fig. 2). Each detector flashes either red or green whenever a particle reaches it. Thus within a second or two of every push of the button, each detector flashes one or the other of its two colored lights.

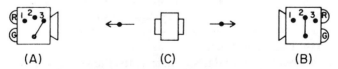

(A)   (C)   (B)

Fig. 2. The complete device. A and B are the two detectors. C is the box from which the two particles emerge.

Because there are no connections between parts of the device, the link between pressing the button on the box and the subsequent flashing of the detectors can be provided only by the passage of the particles from the box to the detectors. This passage could be confirmed by subsidiary detectors between the box and the main detectors A and B, which can be designed so as not to alter the functioning of the device. Additional instruments or shields could also be used to confirm the lack of other communication between the box and the two detectors or between the detectors themselves (Fig. 3).

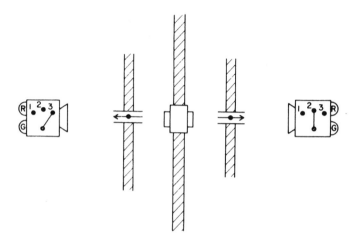

Fig. 3. Possible refinement of the device. The box is embedded in a wall that cuts off one detector from the other. Subsidiary detectors confirm the passage of the particles to the main detectors.

The device is operated repeatedly in the following way. The switch on each detector is set at random to one of its three possible positions, giving nine equally likely settings for the pair of detectors: 11, 12, 13, 21, 22, 23, 31, 32, and 33. The button on the box is then pushed, and somewhat later each detector flashes one of its lights. The flashing of the detectors need not be simultaneous. By changing the distance between the box and the detectors we can arrange that either flashes first. We can also let the switches be given their random settings either before or after the particles leave the box. One could even arrange for the switch on B not to be set until after A had flashed (but, of course, before B flashed).

After both detectors have flashed their lights, the settings of the switches and the colors that flashed are recorded, using the following notation: 31 GR means that detector A was set to 3 and flashed green, while B was set to

1 and flashed red; 12 RR describes a run in which A was at 1, B at 2, and both flashed red; 22 RG describes a run in which both detectors were set to 2, A flashed red and B flashed green; and so on. A typical fragment from a record of many runs is shown in Fig. 4.

```
                  __GG  22GG  1_
          __RR  31RG  13RG  22GG  22k
      _R  21GR  32RG  11GG  32GR  33GG  2_
    22GG  11RR  11GG  23GG  12RR  32GR  11GG
  G  12RG  13RG  33GG  21RG  13GR  31RR  32GR  _
  _GR  13GR  21RG  33RR  13GR  11RR  11GG  13RG  31
  2GG  32GR  33GG  21GR  21GG  33RR  23RG  21GG  21k
 13GR  11GG  32GG  31GR  32RG  33RR  13RR  13RG  12R'
 11GG  31RG  33RR  12RG  21GR  11GG  22GG  33GG  23GI
 11RR  22RR  12RG  22GG  23GR  12GR  33GG  31GG  13GI
 13GR  21RR  33RR  33RR  13RG  23RG  33GG  32RR  12R'
  3RR  32RG  11RR  11RR  11RR  32RG  12RG  21RG  11G
   RG  23RR  21RG  33RR  13GR  12GR  23RG  21RR  32
    R  21GR  12RR  31GR  12RG  13GR  13RG  22RR  ]
     23GR  11RR  12RR  33RR  21RG  13GR  21RR
       _R  12RR  23GG  13RG  21RG  11GG  1?
        _2RG  32RG  32GR  11GG  22R_
          _G_  31RG  21__
```

Fig. 4. Fragment of a page of a volume from the set of notebooks recording a long series of runs.

The accumulated data have a random character, but, like data collected in many tossings of a coin, they reveal certain unmistakable features when enormously many runs are examined. The statistical character of the data should not be a source of concern or suspicion. Blaming the behavior of the device on repeated, systematic, and reproducible accidents, is to offer an explanation even more astonishing than the conundrum it is invoked to dispel.

The data accumulated over millions (or, if you prefer, billions or trillions) of runs can be summarized by distinguishing two cases:

Case a. In those runs in which each switch ends up with the same setting (11, 22, or 33) both detectors always flash the same color. RR and GG occur in a random pattern with equal frequency; RG and GR never occur.

Case b. In the remaining runs, those in which the switches end up with different settings (12, 13, 21, 23, 31, or 32), both detectors flash the same color only a quarter of the time (RR and GR occurring randomly with equal frequency); the other three quarters of the time the detectors flash different colors (RG and GR occurring with equal frequency).

These results are subject to the fluctuations accompanying any statistical predictions, but, as in the case of a coin-tossing experiment, the observed ratios will differ less and less from those predicted, as the number of runs becomes larger and larger.

This is all it is necessary to know about how the device operates. The particular fractions ¼ and ¾ arising in case b are of critical importance. If the smaller of the two were 1/3 or more (and the larger 2/3 or less) there would be nothing wonderful about the device. To produce the conundrum it is necessary to run the experiment sufficiently many times to establish with overwhelming probability that the observed frequencies (which will be close to 25% and 75%) are not chance fluctuations away from expected frequencies of 33 1/3% and 66 2/3%. (A million runs is more than enough for this purpose.)

These statistics may seem harmless enough, but some scrutiny reveals them to be as surprising as anything seen in a magic show, and leads to similar suspicions of hidden wires, mirrors, or confederates under the floor. We begin by seeking to explain why the detectors invariably flash the same colors when the switches are in the same positions (case a). There would be any number of ways to arrange this were the detectors connected, but they are not. Nothing in the construction of either detector is designed to allow its functioning to be affected in any way by the setting of the switch on the other, or by the color of the light flashed by the other.

Given the unconnectedness of the detectors, there is one (and, I would think, only one) extremely simple way to explain the behavior in case a. We need only suppose that some property of each particle (such as its speed, size, or shape) determines the color its detector will flash for each of the three switch positions. What that property happens to be is of no consequence; we require only that the various states or conditions of each particle can be divided into eight types; RRR, RRG, RGR, RGG, GRR, GRG, GGR, and GGG. A particle whose state is of type RGG, for example, will always cause its detector to flash red for setting 1 of the switch, green for setting 2, and green for setting 3; a particle in a state of type GGG will cause its detector to flash green for any setting of the switch; and so on. The eight types of states encompass all possible cases. The detector is sensitive to the state of the particle and responds accordingly; putting it another way, a particle can be regarded as carrying a specific set of flashing instructions to its detector, depending on which of the eight states the particle is in.

The absence of RG or GR when the two switches have the same settings can then be simply explained by assuming that the two particles produced in a given run are both produced in the same state; i.e., they carry identical instruction sets. Thus if both particles in a run are produced in states of type RRG, then both detectors will flash red if both switches are set to 1 or 2, and both will flash green if both switches are set to 3. The detectors flash the same colors when the switches have the same settings because the particles carry the same instructions.

This hypothesis is the obvious way to account for what happens in case a. I cannot prove that it is the only way, but I challenge the reader, given the lack of connections between the detectors, to suggest any other.

The apparent inevitability of this explanation for the perfect correlations in case a forms the basis for the conundrum posed by the device. For the explanation is quite incompatible with what happens in case b.

If the hypothesis of instruction sets were correct, then both particles in any given run would have to carry identical instruction sets whether or not the switches on the detectors were set the same. At the moment the particles are produced there is no way to know how the switches are going to be set. For one thing, there is no communication between the detectors and the particle-emitting box, but in any event the switches need not be set to their random positions until after the particles have gone off in opposite directions from the box. To ensure that the detectors invariably flash the same color every time the switches end up with the same settings, the particles leaving the box in each run must carry the same instructions even in those runs (case b) in which the switches end up with different settings.

Let us now consider the totality of all case b runs. In none of them do we ever learn what the full instruction sets were, since the data reveal only the colors assigned to two of the three settings. (The case a runs are even less informative). Nevertheless we can draw some nontrivial conclusions by examining the implications of each of the eight possible instruction sets for those runs in which the switches end up with different settings. Suppose, for example, that both particles carry the instruction set RRG. Then out of the six possible case b settings, 12 and 21 will result in both detectors flashing the same color (red), and the remaining four settings, 13, 31, 23, and 32, will result in one red flash and one green. Thus both detectors will flash the same color for two of the six possible case b settings. Since the switch settings are completely random, the various case b settings occur with equal frequency. Both detectors will therefore flash the same color in a third of those case b runs in which the particles carry the instruction sets RRG.

The same is true for case b runs where the instruction set is RGR, GRR, GGR, GRG, or RGG, since the conclusion rests only on the fact that one color appears in the instruction set once and the other color, twice. In a third of the case b runs in which the particles carry any of these instruction sets, the detectors will flash the same color.

321

The only remaining instruction sets are RRR and GGG; for these sets both detectors will evidently flash the same color in every case b run.

Thus, regardless of how the instruction sets are distributed among the different runs, in the case b runs *both detectors must flash the same color at least a third of the time.* (This is a bare minimum; the same color will flash more than a third of the time, unless the instruction sets RRR and GGG never occur.) As emphasized earlier, however, when the device actually operates the same color is flashed only a quarter of the time in the case b runs.

Thus the observed facts in case b are incompatible with the only apparent explanation of the observed facts in case a, leaving us with the profound problem of how else to account for the behavior in both cases. This is the conundrum posed by the device, for there is no other obvious explanation of why the same colors always flash when the switches are set the same. It would appear that there must, after all, be connections between the detectors—connections of no known description which serve no purpose other than relieving us of the task of accounting for the behavior of the device in their absence.

I shall not pursue this line of thought, since my aim is only to state the conundrum of the device, not to resolve it. The lecture demonstration is over. I shall only add a few remarks on the device as a parable.

One of the historic exchanges between Einstein and Bohr, [3, 4] which found its surprising denouement in the work of J. S. Bell nearly three decades later,[5] can be stated quite clearly in terms of the device. I stress that the transcription into the context of the device is only to simplify the particular physical arrangement used to raise the issues. The device is a direct descendant of the rather more intricate but conceptually similar *gedanken* experiment proposed in 1935 by Einstein, Podolsky, and Rosen. We are still talking physics, not descending to the level of analogy.

The Einstein, Podolsky, Rosen experiment amounts to running the device under restricted conditions in which both switches are required to have the same setting (case a). Einstein would argue (as was argued above) that the perfect correlations in each run (RR or GG but never RG or GR) can be explained only if instruction sets exist, each particle in a run carrying the same instructions. In the Einstein, Podolsky, Rosen version of the argument the analogue of case b was not evident, and its fatal implications for the hypothesis of instruction sets went unnoticed until Bell's paper.

The *gedanken* experiment was designed to challenge the prevailing interpretation of the quantum theory, which emphatically denied the existence of instruction sets, insisting that certain physical properties (said to be complementary) had no meaning independent of the experimental procedure by which they were measured. Such measurements, far from revealing the value of a preexisting property, had to be regarded as an inseparable part of the very attribute they were designed to measure. Properties of this kind have no independent reality outside the context of a specific experiment arranged to observe them: the moon is *not* there when nobody looks.

In the case of my device, three such properties are involved for each particle. We can call them the 1-color, 2-color, and 3-color of the particle. The *n*-color of a particle is red if a detector with its switch set to *n* flashes red when the particle arrives. The three *n*-colors of a particle are complementary properties. The switch on a detector can be set to only one of the three positions, and the experimental arrangements for measuring the 1-, 2-, or 3-color of a particle are mutually exclusive. (We may assume, to make this point quite firm, that the particle is destroyed by the act of triggering the detector, which is, in fact, the case in many recent experiments probing the principles that underlie the device.)

To assume that instruction sets exist at all, is to assume that a particle has a definite 1-, 2-, and 3-color. Whether or not all three colors are known or knowable is not the point; the mere assumption that all three have values violates a fundamental quantum-theoretic dogma.

No basis for challenging this dogma is evident when only a single particle and detector are considered. The ingenuity of Einstein, Podolsky, and Rosen lay in discovering a situation involving a *pair* of particles and detectors, where the quantum dogma continued to deny the existence of 1-, 2-, and 3-colors, while, at the same time, quantum theory predicted correlations (RR and GG but never RG or GR) that seemed to require their existence.

Einstein concluded that, if the quantum theory were correct, i.e., if the correlations were, as predicted, perfect, then the dogma on the nonexistence of complementary properties—essentially Bohr's doctrine of complementarity—had to be rejected.

Pauli's attitude toward this in his letter to Born is typical of the position taken by many physicists: since there is no known way to determine all three *n*-colors of a particle, why waste your time arguing about whether or not they exist? To deny their existence has a certain powerful economy—why encumber the theory with inaccessible entities? More importantly, the denial is supported by the formal structure of the quantum theory which completely fails to allow for any consideration of the simultaneous 1-, 2-, and 3-colors of a particle. Einstein preferred to conclude that all three *n*-colors did exist, and that the quantum theory was incomplete. I suspect that many physicists, though not challenging the completeness of the quantum theory, managed to live with the Einstein, Podolsky, Rosen argument by observing that though there was no way to establish the existence of all three *n*-colors, there was also no way to establish their nonexistence. Let the angels sit, even if they can't be counted.

Bell changed all this, by bringing into consideration the case b runs, and pointing out that the quantitative numerical predictions of the quantum theory (1/4 vs. 1/3) unambiguously ruled out the existence of all three *n*-colors. Experiments done since Bell's paper confirm the quantum-theoretic predictions.[6] Einstein's attack, were he to maintain it today, would be more than an attack on the metaphysical underpinnings of the quantum theory—more, even, than an attack on the quantitative numerical predictions of the quantum theory. Einstein's position now appears to be contradicted by nature itself. The device behaves as it behaves, and no mention of wave-functions, reduction hypotheses, measurement theory, superposition principles, wave-particle duality, incompatible observables, complementarity, or the uncertainty principle, is needed to bring home its peculiarity. It is not the Copenhagen interpretation of quantum mechanics that is strange, but the world itself.

As far as I can tell, physicists live with the existence of the device by implicitly (or even explicitly) denying the absence of connections between its pieces. References are made to the "wholeness" of nature: particles, detectors, and box can be considered only in their totality; the triggering and flashing of detector A cannot be considered in isolation from the triggering and flashing of detector B—both are part of a single indivisible process. This attitude is sometimes tinged with Eastern mysticism, sometimes with Western know-nothingism, but, common to either point of view, as well as to the less trivial but considerably more obscure position of Bohr, is the sense that strange connections are there. The connections are strange because they play no explicit role in the theory: they are associated with no particles or fields and cannot be used to send any kinds of signals. They are there for one and only one reason: to relieve the perplexity engendered by the insistence that there are no connections.

Whether or not this is a satisfactory state of affairs is, I suspect, a question better addressed by philosophers than by physicists.

I conclude with the recipe for making the device, which, I emphasize again, can be ignored:

The device exploits Bohm's version[7] of the Einstein, Podolsky, Rosen experiment. The two particles emerging from the box are spin ½ particles in the singlet state. The two detectors contain Stern-Gerlach magnets, and the three switch positions determine whether the orientations of the magnets are vertical or at $\pm 120°$ to the vertical in the plane perpendicular to the line of flight of the particles. When the switches have the same settings the magnets have the same orientation. One detector flashes red or green according to whether the measured spin is along or opposite to the field; the other uses the opposite color convention. Thus when the same colors flash the measured spin components are different.

It is a well-known elementary result that, when the orientations of the magnets differ by an angle θ, then the probability of spin measurements on each particle yielding opposite values is $\cos^2(\theta/2)$. This probability is unity when $\theta = 0$ (case a) and ¼ when $\theta = \pm 120°$ (case b).

If the subsidiary detectors verifying the passage of the particles from the box to the magnets are entirely nonmagnetic they will not interfere with this behavior.

Notes

1. *Reviews of Modern Physics,* LI, 863 (1979): 907.
2. From a 1954 letter to M. Born, in *The Born-Einstein Letters* (New York: Walker, 1971), p. 223.
3. A. Einstein, B. Podolsky, and N. Rosen, *Physical Review*, XI.VII, 777 (1935).
4. N. Bohr, *Physical Review*, XI.VIII, 696 (1935).
5. J. S. Bell, *Physics*, I, 195 (1964).
6. Theoretical and experimental aspects of the subject are reviewed by J. F. Clauser and A. Shimony, *Reports on Progress in Physics*, XLI, 1991 (1978). For a less technical survey see B. d'Espagnat, *Scientific American*, ccXL, 5 (November 1979): 158.
7. D. Bohm, *Quantum Theory* (Englewood Cliffs, N.J.: Prentice-Hall, 1951), pp. 614-619.

Mind, Matter, and Quantum Mechanics

MARSHALL SPECTOR

Introduction: Some Startling Claims

The popular literature about quantum mechanics bristles with startling claims regarding the implications of this theory for a new understanding of the role of mind or consciousness in the physical world. Even some of the more technical literature, from the founders of the theory through contemporary physicists, contains eyebrow-raising assertions in this regard.

Here are just a few examples, chosen entirely from the earlier selections in this book (the emphases and bracketed insertions are mine):

> In Eastern mysticism, this universal interwovenness always includes the *human observer and his or her consciousness*, and *this is also true in atomic physics*. At the atomic level, "objects" can be understood only in terms of the interaction between the processes of preparation and measurement. The end of this chain of processes lies always in the *consciousness of the human observer*. Measurements are interactions which *create "sensations" in our consciousness*—for example, the visual sensation of a flash of light, or a dark spot on a photographic plate—and the laws of atomic physics tell us with what probability an *atomic object will give rise to a certain sensation* if we let it interact with us. "Natural science," [not just quantum mechanics] says Heisenberg, "does not simply describe and explain nature; it is part of the interplay between nature and ourselves" [our consciousness—our minds].
>
> *The crucial feature of atomic physics is that the human observer is not only necessary to observe the properties of an object, but is necessary even to define these properties.*
>
> —Fritjof Capra, *The Tao of Physics*, p. 126; p. 231 above.
>
> Quantum theory . . . may even find it necessary to include human consciousness in its description of the world.
>
> —Fritjof Capra, *The Tao of Physics*, p. 129; p. 283 above.

The distinction in classical physics between observer and observed or between mind and matter is no longer valid, we are told. Here the physicist

326

John Wheeler is quoted for support:

> May the universe in some strange sense be "brought into being" by the partic-
> ipation of those who participate? . . . "Participator" is the incontrovertible
> new concept given by quantum mechanics. It strikes down the term "observer"
> of classical theory, the man who stands safely behind the thick glass wall and
> watches what goes on without taking part. It can't be done, quantum mechan-
> ics says.
> —Gary Zukav, *The Dancing Wu Li Masters*, p. 54; p. 291 above.

Zukav comments, "The languages of eastern mystics and western physicists are becoming very similar" (*Ibid*). Later on the same page Zukav writes:

> There is another fundamental difference between the old physics and the new
> physics. The old physics assumes that there is an external world [matter] which
> exists apart from us [minds].

Moreover,

> According to quantum mechanics there is no such thing as objectivity. . . . We
> [our minds?] are a part of nature, and when we study nature . . . nature is
> studying itself. Physics has become a branch of psychology, or perhaps the
> other way around.
> —Gary Zukav, *The Dancing Wu Li Masters*, p. 56; p. 292 above.

Then, after quoting the psychologist Jung and the physicist Pauli:

> If these men are correct, then *physics is the study of the structure of
> consciousness.*
> —Gary Zukav, *The Dancing Wu Li Masters*, p. 56; p. 293 above.

Quantum mechanics, Zukav claims,

> does not assume an objective reality [matter] apart from our experience [mind].

but instead,

> claims only to correlate experience correctly.
> —Gary Zukav, *The Dancing Wu Li Masters*, p. 66; p. 300 above.

Finally, and perhaps most startling of all for its directness and specificity,
there is the physicist N. David Mermin's claim that

> We now know that the moon is demonstrably not there when nobody looks.
> —N. David Mermin, "Quantum Mysteries for Anyone," p. 397; p. 315 above.

Much of this, and in particular the substance of the last two quotations, is reminiscent of the subjective idealism of Bishop Berkeley. Writing over two hundred years ago, Berkeley emphatically denied the existence of matter and argued that all talk ostensibly about matter, insofar as it is intelligible at all, must be understood as shorthand talk about the patterns of our experience—the patterns of "ideas" in our minds. "To be is to be perceived" was his own encapsulation of his view, and he would have embraced the last two quotations from Zukav as characteristic of all physical theories.

Berkeley's theory was one response to an ancient metaphysical problem, and if indeed the *physical* theory that is quantum mechanics shows that one particular *metaphysical* theory is correct—subjective idealism, that is certainly news worth knowing. In this paper I want to assess the correctness of the view, encapsulated in the above quotations, that quantum mechanics does indeed have this deep metaphysical implication.

My strategy will *not* be to examine directly the various arguments used to support the quotations above, however; I shall instead proceed in a more general manner.

First, in part 1, I will attempt a brief and elementary sketch of the general metaphysical framework of Western philosophy in which issues regarding matter and mind and their relations have been discussed, with an eye in particular to what the role of physics has been in this discussion.

Then, in part 2, I will state (somewhat dogmatically) what is *physically* new about quantum mechanics that seems to generate the radical claims exemplified by the above quotations.

In part 3, I will examine what these new characteristics of quantum mechanics *qua physics* have to say about the *metaphysical* issues of mind and matter laid out in part 1. In particular, I want to examine whether the ways in which quantum mechanics differs from *classical* mechanics are sufficient to imply a new central role for consciousness—mind—in describing and understanding nature, as claimed in the above quotations. Section 3 will be the central part of this paper, and my conclusion will be an attempt to throw cold water on the alleged idealist implications of quantum mechanics.

Finally, in part 4, I want to say some things of a more general nature about what would be required for there to be an acceptable account of mind-matter interaction in which quantum mechanics played a major role in integrating these two realms. In summary I shall also offer a few speculative observations and suggestions regarding the history and future of theories of the relations between mind and matter and what physics can tell us about these metaphysical issues.

1. Matter, Mind, and Metaphysics

Briefly, 'metaphysics' is one label for the perennial quest to determine the ultimate nature of reality. It is the search for the basic entities or concepts in terms of which all else is constituted or understood. A metaphysical system or theory is one which uses or constructs a small set of concepts and principles on the basis of which it tries to show how all else falls into place as "merely" the (perhaps complex) elaboration of these concepts and principles and the interplay of the entities to which the concepts refer.

We today refer to thinkers who construct or assess such systems as *philosophers*, as opposed to *scientists*, but in the history of this endeavor in Western thought the distinction is not as clear as we often take it to be. Names such as Aristotle, Descartes, Leibniz, Galileo, Newton, Mach, Einstein, and Whitehead come to mind as examples of the fact that the distinction is, at the very least, not hard-edged. And if those who gave birth to and/or thought deeply about quantum mechanics—Bohr, Heisenberg, Planck, Einstein, Schrödinger, Bohm—are considered physicists rather than philosophers (or, heaven forbid, "metaphysicians"), it need only be pointed out that they were also concerned with issues usually referred to as "metaphysical." Indeed, it is that concern, shown by these physicists and by their popularizers, which has generated the discussion of which this paper is a part.

In one central strand of the metaphysical tradition in Western thought there is a cluster of metaphysical views which center on a distinction between matter and mind. Although the roots of the mind-matter distinction go back much further, it is useful for our purposes to pick up the distinction with Descartes, for two reasons. First, he set the stage for much of the ensuing discussion; even those who found his metaphysical views objectionable often felt the need to refute him before elaborating their own views (a common way in which philosophers use their past masters). Second, the rapid growth of science as we now know it—physics in particular—began at about the time Descartes wrote. Indeed, the controversies under consideration in this paper grow out of a dissatisfaction with the metaphysical and scientific frameworks generated by Descartes and his approximate contemporaries, thinkers such as Galileo and Newton: the frameworks of the mind-matter distinction and classical physics.

Descartes formulated (or reformulated in clear and stark terms) a mind-matter distinction as the foundation of the system of entities which he took to be the ultimate constituents of reality. Reality, for Descartes, consists of two utterly distinct "substances" (kinds of things which could exist independently): *matter*, which he identified (unlike Newton) with space or "extension," and *mind* (unextended thinking or conscious substance). The question of how two such radically distinct kinds of entity could possibly interact and thus constitute one coherent world was a problem with which later philosophers wrestled. It is also, as we shall see, part of the problem of making sense of some of the positions expressed by the quotations in my introduction above. Interactions among parts or aspects of the realm of *matter* or "material substance," at any rate, came to be handled by what we today call *physics*. It is the mathematical physics of Descartes, Galileo, and Newton, which came to full flower in the late 19th century, which is now referred to as "classical physics." Classical physics provided the "mere details" of what goes on in one of the two main realms constituting reality. Physics was the science of matter.

So, the 19th century classical world view or metaphysics said, the universe consists of two distinct kinds of entity: matter, and mind(s). Each exists and unfolds within and upon the double stage of time and space. Physics provides the details of the spatial-material realm. But what about the mental realm? What is to be the "science of mind"? And if there is a science of mind, with its own distinct set of concepts and principles, how could these disparate realms *interact* or indeed be related at all?

These are difficult questions, and the difficulties were seen by some of Descartes' critics. Let me offer just one odd anecdote, by way of example:

There was a (physical? metaphysical?) controversy between Leibniz and certain followers of Descartes during the late 17th century about the proper formulation of a theorem in mechanics which we today refer to as the "conservation of momentum" (mass times velocity). Cartesians believed that *direction* of motion need not and *should not* be taken into account in adding up the mv' s of the bodies of a system. Leibniz held that direction was essential (viewing velocity as a vector, not just "speed"). One of the considerations in the controversy was this: If *direction* as well as speed was *not* part of the law of conservation, then the motions of a system of material bodies, for example a human body, is left somewhat undetermined by the operative mechanical laws, and this openness is where mind, in the form of volition, can enter.

You *mentally* determine what your body does—you raise your arm, for example by (somehow) changing the *direction* of material motion. Thus on the Cartesian approach to this issue *mind* could affect *matter* without interfering with physical laws. (And, of course, matter affects mind by causing sensations in it. Does this sound a bit like some of the quotations in my introduction?) Unfortunately, the mechanism—the detailed "how" of this effect—is left both unexplained and unexplainable.

If direction *were* part of the conservation law, however, then it seems that the unfolding motion of a system of bodies—for example, a human body—is fully determined, and mind is then totally left out of the picture. Your mind becomes an ineffectual ghost in the bodily mechanism. You must be fooling yourself when you believe your *willing* to raise your arm to be the *cause* of its motion, for its motion is fully accounted for by physical material causes in accordance with mechanical laws.

This difficulty and others like it (does matter lose energy when it produces sensations in minds?) which arise out of a radical mind-matter distinction, with physics conceived as the science of the latter, may strike one as quaint and simplistic—naive puzzles from the childhood of modern physics. But I would suggest (without argument at this point) that updated versions of this problem and others are still with us, haunting the metaphysical system known as mind-matter dualism (*interacting* dualism in Descartes' version, parallelism in Leibniz's).

Dualism is not the only game in town, however. Assuming a basic conceptual distinction between mind and matter, one can deny the existence of either one of these categories of entities while understanding the other in roughly its original sense.

Thus one can be a "materialist" and deny the existence of mind(s). Of course one must then say something about why mind *seems* to exist as a distinct category; one must convincingly *reinterpret* those phenomena which were taken to be paradigmatically mental (and thus not material) in such a way as to show that they are really material after all.

Conversely, one might deny the separate existence of matter and offer arguments reinterpreting those phenomena previously taken to be paradigmatically material in nature as really at bottom, in some complex and clever sense, mental after all—trying to avoid the kinds of difficulties of dualism mentioned above, of course. But this latter alternative is just the "idealism" discussed earlier, where it appeared that physicists (such as Mermin) and others quoted in the introduction were espousing a Berkeleyian subjective idealism—a subjective idealism supposedly based on the new *science* of matter, quantum mechanics.

Does that argument work? Does modern physics have this metaphysical implication? Does the new science of matter—quantum mechanics—imply the nonexistence of its own substrate, *matter*, as a category distinct from or independent of mind—the other major metaphysical category of (at least) the last three hundred years? To put it paradoxically: Does the new science of matter imply that only mind exists?

More generally, and less startlingly, we can ask: What, if anything, does quantum mechanics, our current *physics*, have to say about this nest of issues usually considered *metaphysical*? What is new about quantum mechanics, as compared to the older classical mechanics, which appears to speak to metaphysical issues in a manner in which classical mechanics was allegedly mute?

I will suggest answers to these questions in part 3. But before one can responsibly make claims in response to these questions one must first lay out what indeed *is* new about quantum mechanics *qua physics*. This I will try to do in part 2.

2. What's New About Quantum Mechanics?

What are the essential new features of quantum mechanics which render it physically different in important respects from classical physics? It is essential to have a clear idea of the *physical* differences between classical physics and quantum mechanics before one can validly draw any conclusions about what the new physics has to say about the *metaphysical* issues of mind and matter that the old physics did not say.

It is difficult to say something at once non-technical, correct, and non-controversial about the differences between quantum mechanics and classical mechanics. But I think it may be generally agreed that the five features I shall highlight are indeed the relevant ones. Here I should also say that I won't claim to provide an exhaustive list of *all* the differences, nor claim that the features which I shall discuss are logically independent of one another (indeed, they are not). Let me emphasize again that my purpose in presenting this list is to eventually answer the question of whether quantum mechanics has anything *different* to say about the *metaphysical* issues involved in the mind-matter distinction—whether quantum mechanics has any major metaphysical implications which classical physics does not. Here then are the key differences:

A. *Wave-particle duality.* Classical physics contained two distinct strands: *particle theories* and *continuum theories*. Both of these sets of theories use the conceptual and mathematical apparatus of Newton's laws of motion, which state the manner in which the motions of a given distribution of masses will unfold under the actions of given forces. But particle and continuum theories apply these laws to different types of postulated mass distributions and make different assumptions about allowable kinds of force laws.[1] Briefly, particle theories allow for action-at-a-distance force laws to account for the interactions of regions of mass which are otherwise physically unconnected. They assume the physical possibility of independent lumps of matter with wholly empty space in between them influencing each other instantaneously, without the benefit of any material medium of interaction. Continuum theories assume that there can be contact action only, and apply these same Newtonian laws of motion to postulated continuous distributions of matter. (The notion of a *wave* enters simply because of the ubiquity of spatio-temporal sinusoidal oscillations of various physical parameters when one has time-consuming contact action only, in a continuous medium—hence the usual distinction between *particles* and *waves* which one sees in the popular literature.)

After a history of controversy in classical physics, it came to be accepted that *light* consisted of waves rather than particles. By the end of the 19th century, in fact, it seemed that the physical world—the "matter" of the matter-mind distinction—consisted of two major sorts of entities: material *particles* and electromagnetic *waves*. Newtonian mechanics dealt with the former and Maxwellian electrodynamics with the latter.[2]

One of the surprising new phenomena of nature accounted for by quantum mechanics (or at least describable within the formalism of the theory) is that *particles* of matter have wave-like characteristics, and that electromagnetic *waves*—light, for example—display *particle*-like characteristics. Thus, a stream of electrons (classically understood as particles) beamed at a screen through two parallel intervening slits will indeed be detected at the screen as distinct scintillations or spots characteristic of *particle impacts*, but the pattern of distribution of these spots on the screen will take the form of an *interference pattern* which one would obtain from a similar experiment using continuous light waves rather than electron particles.

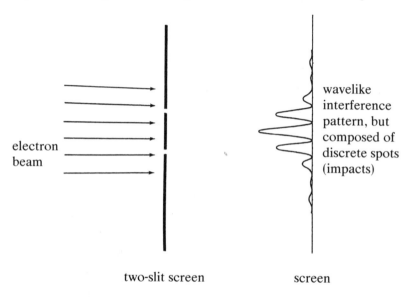

electron beam

two-slit screen

screen

wavelike interference pattern, but composed of discrete spots (impacts)

Rather than obtaining a distribution of spots which is simply the sum of what would have occurred by opening one slit at a time, one gets an interference pattern that suggests an interaction of the electrons (or of the electrons with the slits) as if there were *an* electron-wave, rather than a swarm of discrete electrons. But this pattern occurs even if one sends one electron through at a time, so *simple interaction* cannot be invoked to refute the apparent wave-like feature of the electron beam. Nor can the particle-like feature of discrete spots (impacts) on the screen be challenged. Electrons are *not* simply classical discrete particles; nor are "they" simply classical waves (or a classical wave), undulations in an underlying continuous physical medium. What are they? They are—well, entities which display both wave-like and particle-like characteristics in a manner which classical physical theory cannot explain (or even describe), but which finds a place within the mathematical formalism of quantum mechanics.

All of the above is true of *light* as well. In this case one would *expect* an interference pattern at the screen, but one also finds that this pattern is composed of discrete spots, as if produced by the impact of light particles (quanta). Light too, then (and electromagnetic "waves" in general) is composed neither of classical continuous waves (undulations of an underlying electromagnetic ether) nor of classical discrete particles—as Newton believed, by the way—but of an entity or entities displaying aspects of both classical categories. The attempt to apply the classical wave-particle distinction leads to unintelligible claims.

It is this startling new phenomenon, unanticipated and *unanticipatable* by classical physics, which is known as "wave-particle duality." It is indeed genuinely new in quantum mechanics. Or, perhaps more accurately, this is a genuinely newly discovered feature of nature—the material world—which falls into place in quantum mechanics.

Later, in part 3, we shall see whether this new feature of *physics* has any consequences for the mind-matter distinction of *metaphysics*.

B. The statistics of quantum mechanics are a characteristic of the world, not just an expression of our (contingent) ignorance of underlying non-statistical processes. This revolutionary feature of quantum mechanics has been discussed in some of the earlier selections in this book. In classical physics, statistical laws and probabilities enter only with our ignorance. They enter only when we are ignorant regarding the exact numerical magnitudes of the parameters of a very complex system—the molecules of a gas, for example—or when we do not have fully accurate knowledge of some of the features of simple systems—the precise manner in which a coin is flipped, for example. In such cases we can in principle find out more, get more complete knowledge, and thereby get rid of the probabilistic statements. Their use reflects our contingently incomplete *knowledge* of the world, not the *world itself*, which is assumed to be fully determinate and determinable, completely describable by non-statistical laws.

But statistical laws and probabilities appear in quantum mechanics in cases where *there is simply nothing further to be known* about some of the systems dealt with. Where on the screen the next impact from the next electron will occur, for example, is not accurately determin*able*: one can state *only* the probability of the next scintillation being within this or that specified region of the screen. The "state function" and the "wave equation" which characterize the electron and its behavior in quantum mechanics are taken *not* as describing an actual continuous physical "electron wave." For if they did, one would have to say that this space-filling wave somehow collapsed to a point (how?) at the instant of the appearance of the scintillation on the screen (and more rapidly than the speed of light—indeed instantaneously!). The value of the state function at a given point is therefore taken rather as expressing the *probability* of finding the electron-particle at that point.

A bit more technically: the wave function Ψ (x, t) is interpreted as a probability-density function which, when integrated over a given region of space and time, gives the probability of finding an electron in that spatio-temporal region if one were to look. This, by the way, is the heart of what has been called the "Copenhagen Interpretation of quantum mechanics," due originally to Max Born. Note that this does not mean that there is a physical theory—quantum mechanics—which is completely given by its mathematical equations, and to which is added some extra (philosophical?) "interpretation" of the theory. The *mathematical formalism* is just that—a set of mathematical symbols and formulas. This formalism becomes an (essential) ingredient of a *physical theory*—a set of statements about the physical world—when one gives physical meaning to the symbols of the formalism. The physical theory, then, is the symbolic equations *plus* the interpretation of the symbols. In the so-called "Copenhagen Interpretation of quantum mechanics" the Schrödinger wave equation as a *physical statement or law* uses the wave function Ψ (x, t) to express a probability-density function rather than an actual physical wave.

When in this paper I use the term "quantum mechanics" I am referring to that *physical theory*—not just the mathematical formalism—which some authors refer to as the "Copenhagen Interpretation of quantum mechanics." A different but related theory might take the wave function as designating an actual physical wave. But such an "interpretation" would constitute a *different hypothesis* about the world, not just a different "interpretation" of some underlying physical theory. This is in fact a hypothesis that can be—and has been—refuted physically. It was seriously entertained by some in the early days of quantum theory but is not accepted today.

According to quantum mechanics, then, the world is fundamentally indeterminate. It is indeterminate in precisely specified respects, by the way—there is nothing sloppy here. But if the world itself is indeterminate, it is not just that we happen not to know the classical details of what is "really going on." As I said above, it is rather that there is simply *nothing further to be known.*

"Nothing further to be known"? That's a rather strong claim. How can a physical theory make such a meta-claim about itself—about what we may or may not later find out? How can it judge itself and the future of physics in this way? This genuinely new feature of physics brings us to the next distinctive feature of quantum mechanics.

C. No complete (non-probabilistic) description of physical systems—in terms of classical concepts, say—is possible in principle. It must be emphasized that the claim of "essential incompleteness" is *not* that it is logically impossible that we will eventually discard quantum mechanics in favor of some new *complete* and *deterministic* theory reminiscent of classical physics. No physical theory can validly say of itself that it is in some sense "necessarily" true or never to be replaced.

Given the vast sweep and power of quantum mechanics, however—its explanatory and predictive power is indeed prodigious in ways which are difficult for lay people to fully appreciate—it is easy to understand how most physicists today are thoroughly convinced that no matter what future research brings, the essentially statistical character of the *world* will not be overthrown. It is true that Einstein could never bring himself to believe that "God plays dice with the universe." To the end, he believed that, *because of its incompleteness*, quantum mechanics could not be the final theory of the material world. Physicists today generally "forgive" the master for his "error." (In my view, the question is still open.)

The correct way of taking the essential incompleteness claim is that present quantum mechanics is logically incompatible with any additions which would attempt to patch in "hidden variables"—variables which, if known, would lead us out of the probabilistic wilderness back to the classical land of determinism. In brief, quantum mechanics is not completeble. If quantum mechanics is correct, even additions to it will leave us with indeterminacy *in nature*. Thus not only is there nothing further now known according to the theory, but *in principle* nothing further that *can* be known.

These three new characteristics of quantum mechanics which render it essentially different from classical mechanics as a physical theory are not independent of one another. There are conceptual, logical, and mathematical interrelations among *wave-particle duality*, the *statistical character of nature*, and the *essential incompleteness* of quantum mechanics—interrelations which I shall not attempt to set out here, however. These are, after all, three characteristics of one unified set of concepts and laws constituting one coherent theory: quantum mechanics, our currently accepted theory of the material world. In a sense, it is only in comparison with another (fundamentally incorrect) theory of nature—classical mechanics—that these features appear as *distinct* properties of the new physics. With this caveat in mind, however, let me sketch two more features of quantum mechanics which are truly new in comparison with classical mechanics. The first is a feature sometimes referred to as *non-separability* and is related to a consequence of the theory known as "Bell's theorem."

D. The world is not a multitude of radically distinct entities. Nature is essentially "non-separable." In classical physics, the material world is believed to consist of independent entities whose mutual influence falls off "rapidly" with distance. Here I shall leave "rapidly" unexplained. The details do not matter: whether it's Newtonian atoms, portions of a Cartesian plenum, Rutherford atoms, tables and chairs, or planets, stars, and galaxies that are at issue, the farther these distinct entities are from one another the less they influence one another. In the classical world of Newtonian mechanics and Maxwellian electrodynamics, nature is composed of a large number of distinct simple components of a few kinds, in space, and acting only locally. Influence becomes vanishingly small as a function of distance. Near things affect far things only via intermediate things, and this takes time, the effect diminishing with distance. In brief, *there is no non-local causality*.

This *separability condition*, which seems so very obvious, was used as a premise by Einstein, Podolsky, and Rosen in a famous thought experiment over fifty years ago designed to show that quantum mechanics was false (and incomplete). For as they realized even then, *non-separability* is an essential feature of the theory, which says in effect that *if* one wishes to conceive of nature as an array of distinct spatially localizable parts, then *if any pair of these parts have once interacted they will always interact*.[3] Various *actual* experiments testing this startling consequence have more recently been devised, and they seem to bear out this conclusion. These experiments can be summed up by the following "generic experiment" or *experiment-schema* related to the Einstein-Podolsky-Rosen thought experiment:

Emit two entities (electrons, photons, or whatever—it does not matter) in opposite directions from a central source. Wait as long as you like. Let them travel as far from one another as you like (or as far as you are able to arrange given the degree of cleverness with which you can devise an experimental set-up—ah, there's the rub!). Now, *do* something to *one* of them: measure its spin or its polarization, for example. This act of measurement, which affects the entity being measured (we'll get to that in section E below) will also affect the *other* entity and subsequent measurements made upon *it*.[4] This is a truly astounding feature of the world according to quantum mechanics.[5]

This brings us finally to a new feature of quantum mechanics which is usually described not as a property of the theory but as a relation between the theory and *our* role as investigators: the manner in which the theory is related to the ways in which we *observe* its predictions, *measure* its parameters, and *detect* its entities. The next section begins with one way of stating this feature of the theory.

E. The particular manner in which one chooses to measure, detect or observe[6] a subatomic parameter, entity, or event determines (to some extent) the nature of that parameter, entity, or event. Therefore, quantum *entities* (let me use this term generically for electrons, photons, etc., and their spins, positions, etc.), unlike classical entities, do not pre-exist independently of our coming to observe, detect, or measure them. The metaphysical implications of this will probably seem obvious—but let's save that for part 3.

At first blush, this may seem *not* to be new in principle. After all, the act of observing, measuring, or detecting a *classical* entity in principle affects or "disturbs" it, but we can of course take this into account and calculate back to the nature of the undisturbed entity. In classical physics this does not generate any real difficulties for we usually deal with "large" entities, and our observation, detection, or measurement of them constitutes at worst only minor disturbances which are easy to correct for. In quantum mechanics, one might assume, the difference is only one of degree. Electrons are much smaller than chairs and planets, and the means we use to observe, measure, or detect their properties—position, momentum, spin, and the like—have a much larger effect on them. The corrections are just more difficult to calculate . . . or so it might be thought.

But the difference is more radical than that. It is not just a matter of, say, trying to observe or measure the mass of a fly by bouncing a bowling ball off of it.

In quantum mechanics, one gets situations such as described by the following imaginary "generic experiment" (for actual experiments, see the earlier selections in this book):

Prepare a beam of particles, each of which has a spin given by the angle of its axis of rotation appropriately defined. We want to know what percentage of them have spin-A, and what percentage have spin-B (where A and B are distinct angles). Set up a measuring apparatus that will observe or detect spin-A particles. We find, say, 60% of them have spin-A. Does this mean that at most 40% have spin-B? Surprisingly—if your expectations are classical—the answer can be 'no' for certain systems. For if we now set up a different apparatus, one that observes or detects spin-B particles, we find that, say 80% of the particles have spin-B!

In such a situation, one cannot say that spin is an antecedently existing, independent characteristic which we look in upon without effect or with easily calculable effect. *The nature of the observation or measurement, in a radical sense quite unlike the classical situation, "determines" the properties being observed, measured, or detected.*

For reasons which will become more apparent in part 3, I would prefer not to describe this new situation as a new *relation* between the *theory* and the ways in which *we* (humans) *observe* its entities. I would describe it rather as a difference *within the new theory* which brings the modes of "observation," "measurement," or "detection"—the *measuring apparatus*—*within* its descriptive and explanatory net in a manner radically different from classical mechanics. We shall see in part 3 that this more accurate characterization of the situation will block the unwarranted metaphysical implications characteristic of the quotations with which we began.

In classical mechanics, the "theory of the measuring instrument" was always, in principle, distinct in a clear way. Thus for example we observe the motions of the bodies of the solar system by capturing and analyzing the light bouncing off the planets and their satellites. We look at them with telescopes. The optics involved in these processes of observation, detection, or measurement could be handled in a manner *theoretically disjoint* from the theory of the dynamics of the moving masses which is under investigation with the aid of the telescope. In principle, *what* we observe, measure, or detect takes place in a manner theoretically independent of the *way* we observe, measure, or detect.

In quantum mechanics, however, the interacting system consisting of spinning particle *and* spin-"observing" apparatus is *one system*. It cannot be conceptually and mathematically decomposed into two independent and independently describable sub-systems which could then somehow merely be glued together so that *one* could observe the *other*. A *particle-plus-spin-A detector* is *one* quantum mechanical system; a *particle-plus-spin-B detector* is a *distinct* (though related) *single* system. (This description will remove the paradoxical nature of the "observation *creating* the entity" claim—but more on that in part 3.)

It is this radical inclusion of the "observation (or detection, or measurement) process" within the conceptual ambit of the theory which distinguishes quantum mechanics from classical mechanics. It should then come as no major surprise that paradoxes result if one begins a description or analysis of a *subatomic-entity-plus-complex-measuring-apparatus* system by simply cutting it in two. To use an analogy: a dog is not its front half, plus its rear half, simply glued back together after cutting it in two.

This completes my sketch of the ways in which quantum mechanics, *qua physics*, is radically different from classical mechanics. To turn my description into more than a sketch—to see the underlying unity of what I have arbitrarily set down as five strange new things about the world—one would have to set out the theory itself in at least some mathematical detail.

But what I have presented is I think sufficient for our purpose, which was after all a metaphysical one. We are now in a position to systematically respond to the main question of this paper: What does quantum mechanics, our current *physics*, have to say in virtue of the new *physical* features of the theory about the nest of *metaphysical* issues outlined in part 1? What *metaphysical* implications does quantum mechanics have *differentially*—i.e., that classical physics did not have? In particular, does quantum mechanics have anything to say about the mind-matter distinction? Does quantum mechanics refute a Cartesian dualism of mind and matter, or imply a Berkeleyian subjective idealism or anything like it, as the quotations with which we began seem to claim?

The answer to this question, or nest of related questions, can now be determined by going through each of the five ways in which quantum mechanics is new in terms of *physics*. In each case we can ask: Does *this* new aspect of the way we understand *matter* imply anything about how we should understand *mind* and its relation to matter?

3. The Role of Mind in the New Physics

Our original and perhaps murky question has been reduced to five somewhat clearer questions. Each of these, I think, has a clear and relatively short answer.

A. Does the wave-particle duality of quantum mechanics have any implications for the mind-matter distinction of metaphysics? None whatsoever. We now know that certain classical views about *subcategories of matter* are wrong—or at best are mere approximations for certain circumstances. *Matter* does *not* consist of classical discrete particles and classical waves. It consists of entities which, though *still physical*, are more subtle than that— entities which cannot be understood in terms of these concepts of classical physics. Nothing is said or implied about the nature or existence of *mind* or the relation of mind to matter. This new feature of quantum mechanics is as mute on these issues as is the now falsified classical view.

B. Does the fact that we now believe the world to be statistical in nature— not fully determinate—have any implications for the mind-matter distinction? There were once some philosophers and physicists who took this new "openness" in nature as providing a foot in the door for mind to influence matter, but this view is no longer seriously entertained. And rightly so. Indeterminacy does *not* leave room for free will any more than Descartes' purposeful omission of direction of motion from his conservation laws gave mind a material foothold or something to do. Mind is not just a more ethereal kind of matter working its will when matter can't decide what to do or where to go. If it were, the statistics would disappear! But as far as I know, no one has offered a "more complete" theory of the universe in which *mental* laws help to fill in the statistical gaps left by *physical* laws.

C. Does the essential incompleteness of quantum mechanics have any mind-matter implications? Here again the answer is "no", for reasons similar to those just given. No one has seriously entertained the idea, for example, that quantum mechanical descriptions would one day be "completed" by inserting *mental* events into the lacunae left by the *physical* descriptions. The hidden variables, if such there be, are not mental in nature. De Broglie's hidden "guiding waves were not mind waves," *whatever* those might be.

D. Does non-separability imply anything about the relation of mind to matter? No again. Non-separability certainly forces upon us a radically new view of the nature of matter—as do all of the new characteristics I have been discussing—and indeed perhaps a radically new view of space and time. The notion of space and time, or even "space-time" within Special Relativity Theory, as a kind of stage upon which independent material entities generally go their own way, influencing one another only if "close enough," is no longer acceptable. General Relativity Theory already undermined aspects of this view, but non-separability adds a new wrinkle to the demise of this aspect of the classical world view. Once again the new physics—the new science of matter—presents us with a radically new view of *matter* (and indeed of space-time) but not of *mind* or of mind's relation to matter.

On this point, however—finally—there is some controversy. Consider, for example, Mermin's "We now know that the moon is demonstrably not there when nobody looks." This is best discussed, however, in the context of the final new feature of quantum mechanics, where non-separability is connected with the new view of measurement, observation, and detection within quantum mechanics. After all, Mermin's "looking" is certainly an *observation* and a kind of "measurement." We now get to the nub.

E. Our original question boils down to asking whether the mind-matter distinction is affected by the fact that the way we choose to observe, measure, or detect subatomic entities affects the nature of those entities. It is primarily *this* new feature of quantum mechanics that generates the startling quotations with which I began this discussion and which seems to have far-reaching consequences for the nature of the mind-matter relation. But here I will again argue that this feature of quantum mechanics, when properly understood, implies nothing differentially about the mind-matter relation—nothing that classical mechanics doesn't imply.

The plausibility—indeed attractiveness—of the contrary view rests upon two main errors, in my view, plus a third subsidiary error. The first is a serious misunderstanding of *observation, measurement*, and *detection* in quantum mechanics—indeed, in classical physics too—as *mental phenomena*. The second error involves the misdescription referred to earlier of the *role* of observations, measurements, or detections in quantum mechanics: misdescription which illegitimately separates the *theory* on the one hand from the human *observer* (with his measuring or detecting equipment) on the other. Finally, there is the third error, centered on the notion of what we, as purposeful agents, "*choose*" to observe, measure, or detect. I shall go through these in order.

Let me first emphasize that "*observations,*" "*measurements,*" and "*detections*" *in quantum mechanics are not lookings, listenings, and the like understood as events in the minds of human observers*—or at least no more so than they were in classical physics. They are *physical* or *material* things— dark spots on photographic plates, tracks in cloud chambers or photographs thereof, positions of needles on meters, arrays of numbers on computer printouts, spikes on oscilloscope screens, and the like. It is sets of *objective material entities* such as these which we look at, consider, and then *interpret* when we apply the laws of quantum mechanics to the systems being "observed" *and* to the array of "measuring apparatus" on the laboratory bench. This interpretation leads us to draw a conclusion about some entity, property, or event at the subatomic level. It is this *use* of these objective material entities which *renders* them "data"—which renders them *observations*! Objective material events *become* data, "observations," or "measurements" when they are *taken* by us as indicative of *other* such events, including subatomic events, based on the connections which quantum mechanics draws—or which *we* draw *via* the theory—between the former and the latter.

Notice that all of this can be said about classical mechanics as well, however. Thus *mind* is *not* brought within the ambit of the new theory any more than it was in classical physics. Any conclusion to the contrary bespeaks a general predilection for subjective idealism, not differentially supported by quantum mechanics. Here one might fruitfully consider again the passage from Heisenberg quoted by Capra in my introduction: "Natural science does not simply describe and explain nature; it is part of the interplay between nature and ourselves."

Secondly, let me emphasize that it is a mistake to describe the new physical situation as a new relation between two conceptually distinct domains: quantum mechanics and quantum phenomena, on the one hand, and the measuring apparatus or instruments of "observation" on the other (usually described as always being classical). For then we get the second half of the paradox, in which the latter is thought of as *causing* or *bringing into being* features of the former. What we should rather say, as suggested earlier, is that *within* the new physics of quantum mechanics the *modes of observation*—not mental events, but the particular arrangement of the *measuring apparatus* as understood above—is brought within the theory's descriptive and explanatory net in a radically new *integrated* way, precluding the disjoint description of two independent parts of what is taking place in the complex subatomic-system-plus-measuring-apparatus. Thus *of course* the exact nature of the latter—the apparatus—mathematically "determines" the properties of the former in accordance with the theory. One can make *calculations* going both ways in this single complex system. Change the apparatus, thereby creating a new, complex, non-decomposable system, and the calculations run differently—you get different subatomic entities. *Particle plus spin-A detector* is one complex system. *Particle plus spin-B detector* is another. But it doesn't follow from this that measurements or observations of positive spin *create* it, or—bringing in the first error as well—that my mental looking for it creates it. The truth is the less astounding claim that a certain complex system is not the simple sum of two subsystems.

Finally, and briefly, let me mention the third error. Having committed the first two errors, in particular the second, one might go on to notice that we can *choose* which spin detector to use. By itself this is not a very deep observation, of course, but if we already mistakenly believe that instruments *create* what they detect, measure, or observe, then since *we* choose what instruments to use the conclusion follows that *we* are *creating* the subatomic entities. We are therefore "participators" rather than merely observers, as announced in the Wheeler quote above. But of course in classical physics we "participate" in this sense as well. We create new arrangements of matter every time we move a chair, after all. No differential implications here.

Putting all three errors together in one small package, we get the following *mistaken description*:

We create *spin*-A in the human mental act of looking for it. Therefore, according to the new physics, mind actively creates features of matter by its choices.

What a startling discovery! The new physics is astounding! "Oh Brave New World, that has such features in it." Without the three errors, however, we get the following *correct description*:

We construct a certain laboratory arrangement—just the sort of thing physicists have done for many generations. Using the current theory in accordance with which we did the construction, and in accordance with which we interpret what happens to it, we interpret the objective physical data collected—the "observations"—as indicating (among other things) 60% *spin*-A. A somewhat *different* overall arrangement, similarly analyzed, yields a calculation of 80% *spin*-B. This was to be expected, however, since the *overall arrangement* was indeed *different* in ways fully intelligible according to quantum mechanics. As good quantum physicists, after all, we have learned that a measurement is not an interpreted simple sum of two distinct systems, each of which has the same characteristics after their integration into one complex system as they did before.

The matter is worth further elaboration, for paradoxes are hard to shake even when unmasked.

Suppose that at a given time neither a *spin*-A detector nor a *spin*-B detector is "attached to" a given subatomic system. If we think of the atomic system and the spin detector as *distinct* systems, *each of which retains its full integrity and characteristics when hooked together*—what I have called the second main error—we'll be tempted to think about what the spin *really is, even when not being measured or observed.* If we also bring in the first error, we'll wonder what the spin really is even when "no one is *looking.*" And since the results of the measurements when we *do* look yield the strange

80 + 60 > 100% result when we think of these systems in the classical way, we find ourselves forced to say that quantum mechanics shows that spin does not exist when no one is looking—or when no one *chooses* to look (to bring in the third error as well).

But if indeed no spin detection apparatus *at all*, of *any type whatsoever*, is "attached" to the system, then of course it makes no sense to attribute "60% spin-A" to the system since this specific property is only defined for the more complex system of subatomic-system-plus-detector. That complex system, by hypothesis, *does not now exist*.

This now brings us to Mermin. Does the *moon* exist at a time at which no one is looking at it—indeed when no one is *choosing* to look at it?

Let's restate this without two of the errors I have been discussing. Does the moon exist when not *observed*—when no *moon detectors* (moon-measuring instruments) are "attached" to it? Here choice and "looking" are removed; I can, for example, use photographic evidence that the moon did exist at the time a photograph was made even if no person was actually looking in its direction at that time. But of course a *moon detector, observer, or measurer* is any apparatus (indeed any arrangement of matter, whether humanly constructed or not) which when properly interpreted in light of the relevant theories allows us to infer the moon's presence and characteristics. So, for example, tidal behavior can be interpreted as constituting a "moon detector," "observer," or "measurer"—one manner of "observing" the moon and allowing calculation of its characteristics. As long as "looking" isn't taken literally, it's tempting to say that tides provide a way of "looking" at the moon. So are we to imagine no moon detectors *at all*? No tides, for example?

But then our initial question boils down to this: Does the moon exist at a time at which *no objective physical events whatsoever*—photographic images, tides, or anything of the kind—can be interpreted as allowing us to conclude the presence of this allegedly massive, reflective object relatively near the Earth? No, under such circumstances it does not! And what a shock this would indeed be! But note that this is rather a far cry from the *apparent* claim that if we *humans* all *chose* to avert our *eyes* the moon would go out of existence, and is recreated by us (*each* of us) at the *instant* we choose to look up again. Mind has not created matter.

If my observations and arguments are correct then there is nothing about the new *physical* aspects of quantum mechanics—astounding as they may be *in physics*—that has any implications for the mind-matter distinction of the Western metaphysical tradition since Descartes. If we accept a mind-matter distinction, then quantum mechanics—the new science of matter—indeed has some revolutionary things to say about the nature of *matter*. But it says nothing about the relation between matter and mind, and certainly nothing about the nature of mind. *A fortiori*, it does not imply that matter does not exist, or that it exists only at our sufferance; it does not imply subjective idealism or anything like it. Physics has not become "a branch of psychology" (Zukav); it is not "the study of the structure of consciousness" (Zukav). There is nothing about quantum mechanics that differentially implies any of these claims; that is, nothing that implies or supports these claims in a manner in which classical physics does not.

4. Concluding Observations

It is tempting to try to understand why it has appeared to be otherwise to many physicists, many popularizers of physics, and some philosophers. I believe that, aside from the errors of interpretation of the theory that I have laid out, there are other kinds of reasons which more deeply motivate these idealist understandings of quantum mechanics.

In particular, there is what I take to be an idealist metaphysical predilection among the thinkers and writers in this area. Given such a predilection, it is no surprise to find bad arguments in favor of these views. For if you want to find something badly enough, you probably will. The arguments can be glued on later, as it were; one can always find some kind of argument, good or bad, for just about any strongly held antecedent view. That may not always be such a bad thing—but that's another story.

This is not the place to go into the history of quantum mechanics—which would include the sociology, psychology, and education of both its founders and its popularizers. Here I will simply state my belief—my hypothesis, if you will—that many of these thinkers and writers had a pre-existing tendency to accept a Cartesian mind-body distinction. They then offered *metaphysical* arguments in favor of matter being (at best) a manifestation of mind—arguments which were often independent of the advent of the new physics and which would go through just as well (or just as poorly) on the basis of a classical view of the "physical world." Berkeley's arguments, after all, did not depend *at all* on any particulars of the "theory of matter" held at the time he wrote.

The arguments are often somewhat incoherent and at best, let us say, idealist *in tone*. Sometimes one finds mere idealist manifestos, as if the view seems too obvious to *require* argument. One often finds a confusion between the claim that matter doesn't exist *at all* and the claim that mind brings matter and its properties, or some features of these, into existence. The former view is Berkeley's subjective idealism, or something like it—"Physics is the study of the structure of consciousness" (Zukav). This view does not follow differentially from the nature of quantum mechanics, although one can make an interesting case for it based on the nature of physical theories or theory construction *in general*. It was in fact a philosophical idea in the air when many of the founders of quantum mechanics did their early work.

The latter view, that quantum mechanics has shown within a Cartesian dualism (where matter *does* exist) that mind and matter interact in a manner not implied by classical physics—that features of matter are created by mind—I have shown to be false.

But it might be worth asking what such a view would have to provide if it were to be essentially correct. Put aside for a moment my arguments against it. How would one flesh out, in detail, the new metaphysics allegedly based on the new physics—allegedly based on the theory of measurement, in particular? Here we would need in essence an updated Cartesian interacting dualism involving two basic "substances": *matter*, now understood on the basis of quantum mechanics, and *mind*.

The first point to notice is that the new physics says nothing about the nature of mind by virtue of which one can understand *how* it interacts with the newly conceived realm of matter. But perhaps that's not a totally fair criticism, for neither did classical physics. This much *can* be said, however: the conceptual, metaphysical, and causal gulf which existed for a classical Cartesian dualism, remarked on in part 1, will also exist for any quantum mechanical mind-matter dualism. It doesn't help to be told, for example, that " . . . measurements are interactions which create 'sensations' in our consciousness—for example, the visual sensation of a flash of light, or a dark spot on a photographic plate . . ." (Capra).[7] For this is true within a classical world view as well. Even Descartes spoke of physical events giving rise to sensations in our consciousness. But he was at a loss to say *how* such a thing could happen—he could only remark that it somehow *did* happen. He had no detailed theoretical account of the intricacies of each of his two realms, and of the intricacies of their detailed interconnections, *by virtue of which one could understand the details of the causal activity alleged to be taking place in both directions.* Not only did he not have such an account, but his work posed conceptual problems for the very notion of the *possibility* of such a rich and detailed causal account, as I have already noted (and indeed as philosophers such as Leibniz and Spinoza noted long ago).

But all of this is equally true within a quantum mechanical world view! What we have again is the *bare assertion*—equally true (or false) in a classical world—that mind causes things to happen in matter, and matter causes things to happen in mind(s).

We have a new "theory of the physical"—quantum mechanics. But where is a similar "theory of the mental"? What would such a theory look like? What would be its fundamental concepts, and what basic laws would relate those concepts? What would be its "state functions" or "state equations"? How would it describe and explain the temporal unfolding of complex "mental systems" in accordance with applicable mental laws? What would its mathematical formalism look like? Finally, what will the nature of the *connections* be between the theory of the physical (quantum mechanics) and the theory of the mental—the "bridge laws" by virtue of which the two become integrated into one overarching theory which would yield an understanding of the complex causal interactions between mind and matter?

The kind of rich, coherent, testable (and yes, mathematical) theories of matter *and of mind* required to flesh out in an empirically rich manner the *bald claim* of causal interaction is as absent in the quantum mechanical view of the world as it was in the classical world of Descartes and those who came after him.

The heart of the difficulty, it seems to me, is what might be called the Humpty Dumpty problem. If one begins with a metaphysics which postulates a radical dualism of mind and matter—as do the thinkers we have been considering—then it will not be possible to coherently reintegrate these two realms into one universe.[8] The move from classical physics to quantum mechanics in the realm of matter won't do the trick, for the ways in which quantum mechanics differs from classical mechanics cannot bear the conceptual burden required of such a merging of these radically distinct realms.

Let me sum up, and tentatively offer a suggestion or two:

The physical world as newly depicted by quantum mechanics is indeed strange, given the expectations of the classical world view. These strange new features of the quantum mechanical world are indeed difficult to digest, if not also hard to swallow, and cannot be explained away—or so it now seems to the overwhelming majority of working physicists.

But nothing metaphysical follows from this about the age-old mind-matter problem, for a number of classical metaphysical views can equally well (or ill) embrace the new physics. As I have put it, quantum mechanics does not imply anything differentially about the nature or role of mind in the universe. Given the mind-matter distinction, all that follows is that matter is weirder than previously believed.

Quantum mechanics not only has not solved the mind-matter problem of traditional metaphysics. Given the problem as traditionally stated, quantum mechanics *cannot in principle* solve it—or even speak to it—any more than could classical physics.

Indeed, I suspect that this problem is insoluble as traditionally stated. There are, it seems to me, two viable possibilities. Either one must reject attempts to bridge these two realms in a *scientific-causal* manner, or—more radically—one must reject mind and matter as fundamental categories upon which to construct a metaphysics. Perhaps in a radically different metaphysics of this sort—whatever it might look like, and whatever categories it might take as fundamental—we would be able to reach an interpretation of what we today call matter and mind that allows for a coherent presentation of the interplay of what we today describe as *physical* and *mental* phenomena. Perhaps in such a world view a *re-interpreted* Western physics and a

re-interpreted Eastern world view will appear as mere subcases or as different aspects of deeper conceptual schemes. Perhaps. But before that happens, we'll need to see much more than the poor arguments, superficial
analogies, and misunderstandings of physics—both quantum and classical—
that today seem to generate so much of the discussion in this area.[9]

If this sounds overly harsh, let me say that I nevertheless hope these
speculations continue, though at a much more serious level. I want to
conclude with an anecdote from the history of science which may explain
that hope.

Toward the end of the 19th century, Lord Kelvin advanced an argument
based on undergraduate-level physics that the Earth could not possibly be
anywhere nearly as old as it would have to be if the forces of *evolution* had
in fact produced the current variety of life. The "residual heat" of the
planet was simply too high. Assuming some plausible initial temperature at
the time of its formation, the Earth would have cooled to absolute zero in
the time required to account for the variety of life in accordance with
Darwin's principles of chance variation and subsequent natural selection.
Evolutionists, including Darwin himself, were troubled by this very real
problem. Some of them, "Darwin's bulldog" T. H. Huxley, for example,
defensively resorted to sophistry in the face of Kelvin's irrefutable calculations.

As it turned out, of course, Kelvin was wrong. Darwin and Huxley were
right—or so it now appears. But who could have guessed at that time that
the heat of the Earth was *not* "residual" heat that remained after the gradual dissipation of the heat present at the Earth's formation? Who could
have known that this heat resulted instead from an ongoing process of
radioactive decay due to the uranium content of the Earth's crust?

This episode in the history of science (and others like it) raises some interesting methodological questions. Although Kelvin was mistaken in his *conclusion*, he was surely *rational* (or "scientific") in arriving at it. Huxley and
Darwin turned out to be right. But were their *reasons* for clinging to their
belief in the great antiquity of the Earth *rational* in view of the clarity and force
of Kelvin's calculations? What if they had given up in the face of these
calculations? Which (if any) of these thinkers proceeded in a *methodologically correct* manner in pursuing and developing their various hypotheses?

How, ultimately, are we to adjudicate between the following two methodological principles?:

(1) Reject a hypothesis that contradicts established theories, or has been
refuted by clear arguments.

(2) Pursue a [promising] hypothesis even if [for a while] one has only
analogies and weak arguments for maintaining it in the face of [apparently]
refuting evidence and arguments.

If we favor the second principle, moreover, how are we to understand
the bracketed adjectives?

I leave to the reader the pleasure of comparing the case of Kelvin to the case at hand. But let me emphasize that those whose views I have criticized will have to do more—if their views are to be taken seriously, much less accepted—than simply take comfort in this and similar anecdotes.

Notes

1. For a more detailed characterization of classical mechanics, see chapters 2 and 3 of my book, *Methodological Foundations of Relativistic Mechanics*, Notre Dame, Indiana: Notre Dame Press, 1972.

2. Although, as I have argued elsewhere, electrodynamics is actually a rather elaborate *branch* of Newtonian mechanics—but that's not important for present purposes. See my *Methodological Foundations of Relativistic Mechanics, op. cit.*, chapter 4.

3. An aside: If the "big bang" theory of the origin of the universe is correct (as seems to be the case), then it appears that all parts of the present universe *did* indeed once interact and therefore do interact now.

4. Again, if the big bang theory is correct, the universe would appear to constitute one colossal experiment of such a sort.

5. For further details, see Mermin's paper in this volume. One could also look at the 1935 Einstein-Podolsky-Rosen thought experiment (A. Einstein, B. Podolsky, and N. Rosen, *Physical Review*, XLVII, 777 (1935)). But a better place for the reader to look might be the exciting exchange between Bohr and Einstein in the Schilpp volume devoted to Einstein from which the opening quotations in this section of the present volume are taken (A. Schilpp, ed., *Albert Einstein: Philosopher-Scientist*, La Salle, Illinois: Open Court, 1969). That exchange is a marvelous *tour de force* of physical argumentation—an exhilarating performance to observe.

6. The choice of words here—"measure," "detect," and "observe"—is important, as we shall see in part 3. I will therefore continue to use all three together as a way of preparing the reader for what I shall have to say in the next section.

7. I will deal only with "the visual sensation of a flash of light," not the second part of the 'or'—a dark spot on a photographic plate"—for the latter is just a part of the objective *material* realm, as I argued in part 3. This is part of what I called the *correct* description of the relation between theory and measurement, and has no implication for the *mind*-matter distinction. The vacillation in the quote indicates how difficult it sometimes is to get a handle on some of these claims.

8. This is very similar to what I earlier called the "second error" in understanding the nature of observation in quantum mechanics.

9. To put it somewhat paradoxically, we'll need to see something even more startling than the quotations with which I began this paper, and we'll need to see it fleshed out in some detail.

Suggested Readings

As noted in the introduction, this is not intended as anything like a complete bibliography. The works mentioned are generally standard sources, for the most part easily available, which may be of help at the next stage of the reader's research.

Quantum mysticism
Bentor, Itzhak. *Stalking the Wild Pendulum.* New York: Bantam Books, 1979.
Capra, Fritjof. *The Tao of Physics.* New York: Bantam Books, 1983.
Davies, Paul. *God and the New Physics.* New York: Simon and Schuster, 1983.
Talbot, Michael. *Beyond the Quantum.* New York: Macmillan, 1986.
Walker, Evan Harris, "Consciousness and Quantum Theory," in Edgar D. Mitchell, ed. *Psychic Exploration: A Challenge for Science.* New York: Paragon Books, 1974.
Zukav, Gary. *The Dancing Wu Li Masters.* New York: William Morrow and Co., Inc., 1979.

Critiques of quantum mysticism
Crease, Robert P., and Charles C. Mann. "Physics for Mystics." *The Sciences*, July/August 1987, pp. 50-57.
Gardner, Martin. "Quantum Theory and Quack Theory." *New York Review of Books*, May 17, 1979. Reprinted in Martin Gardner, *Science: Good, Bad, and Bogus.* New York: Avon Books, 1981. This piece includes an exchange between John Wheeler and a number of physicists sympathetic to quantum mechanical treatments of parapsychology.
Restivo, Sal. *The Social Relations of Physics, Mysticism, and Mathematics: Studies in Social Structure, Interests, and Ideas.* Dordrecht, Holland: D. Reidel, 1985.

Popular introductions to quantum mechanics
Davies, Paul. *Other Worlds.* New York: Simon and Schuster, 1981.
D'Espagnat, Bernard. "Quantum Theory and Reality," *Scientific American*, vol. 241, no. 5 (November 1979), pp. 128-140.
Gibbin, John. *In Search of Shrödinger's Cat: Quantum Physics and Reality.* New York: Bantam Books, 1984.

Pagels, Heinz R. *The Cosmic Code: Quantum Physics as the Language of Nature.* New York: Bantam Books, 1982.

Wolf, Fred Alan. *Taking the Quantum Leap: The New Physics for Nonscientists.* San Francisco: Harper and Row, 1981.

Philosophical works on interpretations of quantum mechanics

Colodny, Robert G., ed. *Paradigms and Paradoxes: The Philosophical Challenge of the Quantum Domain.* Pittsburgh: University of Pittsburgh Press, 1972.

Fine, Arthur. *The Shaky Game: Einstein, Realism, and the Quantum Theory.* Chicago: Univ. of Chicago Press, 1986.

Margenau, H. "Advantages and Disadvantages of Various Interpretations of Quantum Mechanics." *Physics Today* 7 (1954), pp. 6-13.

Davies, P. C. W. and J. Brown, eds. *The Ghost in the Atom: A Discussion in the Mysteries of Quantum Physics.* New York: Cambridge Univ. Press, 1986. Based on a BBC radio documentary, this includes interviews with Alain Aspect, John Bell, John Wheeler, and David Bohm.

On Bohr, Einstein, and Others

Bohr, Neils. "On the Notions of Causality and Complexity," and Albert Einstein, "Quanten-Mechanik und Wirklichkeit" ("Quantum Mechanics and Reality"), *Dialectica*, 2 (1948), pp. 312-324. A translation of a form of Einstein's article here appears in Born, ed. *The Born-Einstein Letters, op. cit.*

Born, Max, ed. *The Born-Einstein Letters.* New York: Walker and Co., 1971.

Hooker, Clifford A. "The Nature of Quantum Mechanical Reality: Einstein Versus Bohr," in Colodny, ed. *Paradigms and Paradoxes*, op. cit., 67-302.

Schilpp, Paul Arthur, ed. *Albert Einstein: Philosopher-Scientist.* La Salle, Illinois: Open Court, 1969.

Crease, Robert P., and Charles C. Mann. *The Second Creation: Makers of the Revolution in 20th-Century Physics.* New York: Macmillan, 1986.

OTHER APPROACHES TO THE OCCULT

Introduction

The problem of demarcation—how science is to be distinguished from pseudoscience—has been a central issue in our examination of various topics that might loosely be included within "the occult." But it might be that in dealing with the occult in this way we have done at least some aspects of the occult a disservice, or have been unfair to some pursuits or traditions that it includes. This final section is intended to broach some more general questions of this kind regarding science and other endeavors.

A relatively weak claim that might be made in this regard is that to decide that something is not science is not necessarily to condemn it. Art, music, and aesthetic and ethical values clearly do not qualify as science, but are in no way inferior for that. There are things of value other than science, and there are perhaps pursuits of greater value. In dealing with the occult with scientific status so much in mind, we might be led to ignore other traditions valuable in other regards. Perhaps ethical and aesthetic values are embodied in some aspects of the occult to which scientific status or the lack of it is simply irrelevant.

Often, however, a stronger claim is made; that science has equivalent or superior rivals within the occult. In the first piece of this section, WILLIAM JAMES comes close to making such a claim in discussing mysticism and the effects of nitrous oxide. "... our normal waking consciousness," James maintains, "... is but one special type of consciousness, whilst all about it, parted from it by the filmiest of screens, there are potential forms of consciousness entirely different." James attempts to be scrupulously fair to the claims of mysticism, and concludes that although they can wield no special authority, they may nevertheless reveal "the truest of insights into the meaning of this life."

In the second piece, EDWARD CONZE makes a much more explicit attack on science in behalf of the occult. Conze condemns "science-bound philosophers" for their "ruthless will for boundless power," and argues that the methods of science are "useless for the exploration of two-thirds of the universe." Here the reader should attempt to separate carefully different arguments and appeals that Conze makes. It is also important to distinguish points at which Conze is arguing for values other than those exhibited in science from points at which he is arguing for the occult as in some other way a rival to science.

In the third piece of the section, I attempt to raise some questions concerning science and values and to reply in part to some common attacks on science made not only by Conze but by Carlos Castaneda, Alan Watts, Philip Slater, Colin Wilson, and others. Here I attempt to distinguish essential from non-essential values in science, and to argue that although science is "value-laden" it does not follow that science is on all fours with conflicting value systems or with rival "ways of knowledge" within the occult. Although this is the final piece of the collection, it is of course not intended as anything like the last word.

The discussion in this final section raises philosophical issues broader than those addressed in the volume as a whole. For the most part we have concentrated on issues within the philosophy of science. But James's and Conze's pieces raise the quite general issue of epistemological relativism, and my own piece touches on important questions concerning science and values. This is, I think, entirely appropriate as a brief indication of the broader philosophical settings of the particular issues with which we have dealt.

Notes on Mysticism
and Nitrous Oxide

WILLIAM JAMES

Nitrous oxide and ether, especially nitrous oxide, when sufficiently diluted with air, stimulate the mystical consciousness in an extraordinary degree. Depth beyond depth of truth seems revealed to the inhaler. This truth fades out, however, or escapes, at the moment of coming to; and if any words remain over in which it seemed to clothe itself, they prove to be the veriest nonsense. Nevertheless, the sense of a profound meaning having been there persists; and I know more than one person who is persuaded that in the nitrous oxide trance we have a genuine metaphysical revelation.

Some years ago I myself made some observations on this aspect of nitrous oxide intoxication, and reported them in print. One conclusion was forced upon my mind at the time, and my impression of its truth has ever since remained unshaken. It is that our normal waking consciousness, rational consciousness as we call it, is but one special type of consciousness, whilst all about it, parted from it by the filmiest of screens, there lie potential forms of consciousness entirely different. We may go through life without suspecting their existence; but apply the requisite stimulus, and at a touch they are there in all their completeness, definite types of mentality which probably somewhere have their field of application and adaptation. No account of the universe in its totality can be final which leaves these other forms of consciousness quite disregarded. How to regard them is the question—for they are so discontinuous with ordinary consciousness. Yet they may determine attitudes though they cannot furnish formulas, and open a region though they fail to give a map. At any rate, they forbid a premature closing of our accounts with reality. Looking back on my own experiences, they all converge towards a kind of insight to which I cannot help ascribing some metaphysical significance. The keynote of it is invariably a reconciliation. It is

Excerpted from "Mysticism," lectures XVI and XVII of William James, *The Varieties of Religious Experience* (New York: New American Library, 1958).

as if the opposites of the world, whose contradictoriness and conflict make all our difficulties and troubles, were melted into unity. Not only do they, as contrasted species, belong to one and the same genus, but *one of the species,* the nobler and better one, *is itself the genus, and so soaks up and absorbs its opposite into itself.* This is a dark saying, I know, when thus expressed in terms of common logic, but I cannot wholly escape from its authority. I feel as if it must mean something, something like what the Hegelian philosophy means, if one could only lay hold of it more clearly. Those who have ears to hear, let them hear; to me the living sense of its reality only comes in the artificial mystic state of mind. . . .

I have now sketched with extreme brevity and insufficiency, but as fairly as I am able in the time allowed, the general traits of the mystic range of consciousness. *It is on the whole pantheistic and optimistic, or at least the opposite of pessimistic. It is anti-naturalistic, and harmonizes best with twice-bornness and so-called other-wordly states of mind.*

My next task is to inquire whether we can invoke it as authoritative. Does it furnish any *warrant for the truth* of the twice-bornness and supernaturality and pantheism which it favors? I must give my answer to this question as concisely as I can.

In brief my answer is this—and I will divide it into three parts:—

(1) Mystical states, when well developed, usually are, and have the right to be, absolutely authoritative over the individuals to whom they come.

(2) No authority emanates from them which should make it a duty for those who stand outside of them to accept their revelations uncritically.

(3) They break down the authority of the non-mystical or rationalistic consciousness, based upon the understanding and the senses alone. They show it to be only one kind of consciousness. They open out the possibility of other orders of truth, in which, so far as anything in us vitally responds to them, we may freely continue to have faith.

I will take up these points one by one.

1

As a matter of psychological fact, mystical states of a well-pronounced and emphatic sort *are* usually authoritative over those who have them.[1] They have been "there," and know. It is vain for rationalism to grumble about this. If the mystical truth that comes to a man proves to be a force that he can live by, what mandate have we of the majority to order him to live in another way? We can throw him into a prison or a madhouse, but we cannot change his mind—we commonly attach it only the more stubbornly to its beliefs.[2] It mocks our utmost efforts, as a matter of fact, and in point of logic it absolutely escapes our jurisdiction. Our own more "rational" beliefs are based on evidence exactly similar in nature to that which mystics quote for theirs. Our senses, namely, have assured us of certain states of fact; but mystical experiences are as direct perceptions of fact for those who have them as any sensations ever were for us. The records show that even though the five senses be in abeyance in them, they are absolutely sensational in their epistemological quality, if I may be pardoned the barbarous expression—that is, they are face to face presentations of what seems immediately to exist.

The mystic is, in short, *invulnerable*, and must be left, whether we relish it or not, in undisturbed enjoyment of his creed. Faith, says Tolstoy, is that by which men live. And faith-state and mystic state are practically convertible terms.

2

But I now proceed to add that mystics have no right to claim that we ought to accept the deliverance of their peculiar experiences, if we are ourselves outsiders and feel no private call thereto. The utmost they can ever ask of us in this life is to admit that they establish a presumption. They form a consensus and have an unequivocal outcome; and it would be odd, mystics might say, if such a unanimous type of experience should prove to be altogether wrong. At bottom, however, this would only be an appeal to numbers, like the appeal of rationalism the other way; and the appeal to numbers has no logical force. If we acknowledge it, it is for "suggestive," not for logical reasons: we follow the majority because to do so suits our life.

But even this presumption from the unanimity of mystics is far from being strong. In characterizing mystic states as pantheistic, optimistic, etc., I am afraid I over-simplified the truth. I did so for expository reasons, and to keep the closer to the classic mystical tradition. The classic religious mysticism, it now must be confessed, is only a "privileged case." It is an *extract*, kept true to type by the selection of the fittest specimens and their preservation in "schools." It is carved out from a much larger mass; and if we take the larger mass as seriously as religious mysticism has historically taken itself, we find that the supposed unanimity largely disappears. To begin with, even religious mysticism itself, the kind that accumulates traditions and makes schools, is much less unanimous than I have allowed. It has been both ascetic and antinomianly self-indulgent within the Christian church.[3] It is dualistic in Sankhya, and monistic in Vedanta philosophy. I called it pantheistic; but the great Spanish mystics are anything but pantheists. They are with few exceptions non-metaphysical minds, for whom "the category of personality" is absolute. The "union" of man with God is for them much more like an occasional miracle than like an original identity.[4] How different again, apart from the happiness common to all, is the mysticism of Walt Whitman, Edward Carpenter, Richard Jefferies, and other naturalistic pantheists, from the more distinctively Christian sort.[5]

The fact is that the mystical feeling of enlargement, union, and emancipation has no specific intellectual content whatever of its own. It is capable of forming matrimonial alliances with material furnished by the most diverse philosophies and theologies, provided only they can find a place in their framework for its peculiar emotional mood. We have no right, therefore, to invoke its prestige as distinctively in favor of any special belief, such as that in absolute idealism, or in the absolute monistic identity, or in the absolute goodness, of the world. It is only relatively in favor of all these things—it passes out of common human consciousness in the direction in which they lie.

So much for religious mysticism proper. But more remains to be told, for religious mysticism is only one half of mysticism. The other half has no accumulated traditions except those which the text-books on insanity supply. Open any one of these, and you will find abundant cases in which "mystical ideas" are cited as characteristic symptoms of enfeebled or deluded states of mind. In delusional insanity, paranoia, as they sometimes call it, we may have a *diabolical* mysticism, a sort of religious mysticism turned upside down. The same sense of ineffable importance in the smallest events, the same texts and words coming with new meanings, the same voices and visions and leadings and missions, the same controlling by extraneous powers; only this time the emotion is pessimistic: instead of consolations we have desolations; the meanings are dreadful; and the powers are enemies to life. It is evident that from the point of view of their psychological mechanism, the classic mysticism and these lower mysticisms spring from the same mental level, from that great subliminal or transmarginal region of which science is beginning to admit the existence, but of which so little is really known. That region contains every kind of matter: "seraph and snake" abide there side by side. To come from thence is no infallible credential. What comes must be sifted and tested, and run the gauntlet of confrontation with the total context of experience, just like what comes from the outer world of sense. Its value must be ascertained by empirical methods, so long as we are not mystics ourselves.

Once more, then, I repeat that non-mystics are under no obligation to acknowledge in mystical states a superior authority conferred on them by their intrinsic nature.[6]

3

Yet, I repeat once more, the existence of mystical states absolutely overthrows the pretension of non-mystical states to be the sole and ultimate dictators of what we may believe. As a rule, mystical states merely add a supersensuous meaning to the ordinary outward data of consciousness. They are excitements like the emotions of love or ambition, gifts to our spirit by means of which facts already objectively before us fall into a new expressiveness and make a new connection with our active life. They do not contradict these facts as such, or deny anything that our senses have immediately seized.[7] It is the rationalistic critic rather who plays the part of denier in the controversy, and his denials have no strength, for there never can be a state of facts to which new meaning may not truthfully be added, provided the mind ascend to a more enveloping point of view. It must always remain an open question whether mystical states may not possibly be such superior points of view, windows through which the mind looks out upon a more extensive and inclusive world. The difference of the views seen from the different mystical windows need not prevent us from entertaining this supposition. The wider world would in that case prove to have a mixed constitution like that of this world, that is all. It would have its celestial and its infernal regions, its tempting and its saving moments, its valid experiences and its counterfeit ones, just as our world has them; but it would be a wider world all the same. We should have to use its experiences by selecting and subordinating and substituting just as is our custom in this ordinary naturalistic world; we should be liable to error just as we are now; yet the counting in of that wider world of meanings, and the serious dealing with it, might, in spite of all the perplexity, be indispensable stages in our approach to the final fullness of the truth.

In this shape, I think, we have to leave the subject. Mystical states indeed wield no authority due simply to their being mystical states. But the higher ones among them point in directions to which the religious sentiments even of non-mystical men incline. They tell of the supremacy of the ideal, of vastness, of union, of safety, and of rest. They offer us *hypotheses*, hypotheses which we may voluntarily ignore, but which as thinkers we cannot possibly upset. The supernaturalism and optimism to which they would persuade us may, interpreted in one way or another, be after all the truest of insights into the meaning of this life.

Notes

1. I abstract from weaker states, and from those cases of which the books are full, where the director (but usually not the subject) remains in doubt whether the experience may not have proceeded from the demon.

2. Example: Mr. John Nelson writes of his imprisonment for preaching Methodism: "My soul was as a watered garden, and I could sing praises to God all day long; for he turned my captivity into joy, and gave me to rest as well on the boards, as if I had been on a bed of down. Now could I say, 'God's service is perfect freedom,' and I was carried out much in prayer that my enemies might drink of the same river of peace which my God gave so largely to me." Journal, London, no date, p. 172.

3. Ruysbroeck, in the work which Maeterlinck has translated, has a chapter against the antinomianism of disciples. H. Delacroix's book (Essai sur le mysticisme spéculatif en Allemagne au XIVme Siècle, Paris, 1900) is full of antinomian material. Compare also A. Jundt: Les Amis de Dieu au XIV Siècle, Thèse de Strasbourg, 1879.

4. Compare Paul Rousselot: Les Mystiques Espagnols, Paris, 1869, ch. xii.

5. See Carpenter's Towards Democracy, especially the latter parts, and Jefferies's wonderful and splendid mystic rhapsody, The Story of My Heart.

6. In chapter i. of book ii. of his work Degeneration, "Max Nordau" seeks to undermine all mysticism by exposing the weakness of the lower kinds. Mysticism for him means any sudden perception of hidden significance in things. He explains such perception by the abundant uncompleted associations which experiences may arouse in a degenerate brain. These give to him who has the experience a vague and vast sense of its leading further, yet they awaken no definite or useful consequent in his thought. The explanation is a plausible one for certain sorts of feeling of significance; and other alienists (Wernicke, for example, in his Grundriss der Psychiatrie, Theil ii., Leipzig, 1896) have explained "paranoiac" conditions by a laming of the association-organ. But the higher mystical flights, with their positiveness and abruptness, are surely products of no such merely negative condition. It seems far more reasonable to ascribe them to inroads from the subconscious life, of the cerebral activity correlative to which we as yet know nothing.

7. They sometimes add subjective *audita et visa* to the facts, but as these are usually interpreted as transmundane, they oblige no alteration in the facts of sense.

Tacit Assumptions

Edward Conze

The mutual incomprehension of Eastern and Western philosophy has often been deplored. If there is even nó contact between "empiricist" European philosophy on the one side, and that of the Vedānta and Mahāyāna on the other, it may be because they presuppose two different systems of practice as their unquestioned foundations—science the one, and yogic meditation the other. From the outset all philosophers must take for granted some set of practices, with specific rules and aims of their own, which they regard both as efficacious and as avenues to worthwhile reality.

It is, of course, essential to grasp clearly the difference between sets of practices, or "bags of tricks" which regularly produce certain results, and the theoretical superstructures which try to justify, explain and systematize them. The techniques concern what happens when this or that is done. The theories deal with the reasons why that should be so, and the meaning of what happens. However gullible and credulous human beings may be about speculative tenets, about practical issues they are fairly hard-headed, and unlikely to persuade themselves over any length of time that some technique "works" when it does not.

Yogic meditation, to begin with, demands that certain things should be *done*. There are the well-known breathing exercises, which must be performed in certain definite bodily postures. Certain foods and drugs must be avoided. One must renounce nearly all private possessions, and shun the company of others. After a prolonged period of physical drill has made the body ready for the tasks ahead, and after some degree of contentment with the conditions of a solitary, beggarly and homeless life has been achieved, the mind is at last capable of doing its proper yogic work. This consists in systematically withdrawing attention from the objects of the senses.[1] And what could be the aim and outcome of this act of sustained introversion—so strikingly dramatized by Bodhidharma sitting for nine years cross-legged and immobile in front of a grey wall? All the adepts of Yoga, whatever their theological or philosophical differences, agree that these practices result in a state of inward tranquility (*samatha*).

Many of our contemporaries, imprisoned in what they describe as "common sense," quite gratuitously assume, as "self-evident," that all the contents of mental life are derived from contact with external sense-data. They are therefore convinced that the radical withdrawal from those sense-data can but lead to some kind of vague vacuity almost indistinguishable from sleep or coma. More than common sense is needed to discover that it leads to a state which the Indian yogins, who under the influence of Sanskrit grammar were almost obsessed with a desire for terminological precision, called one of "tranquillity," full of ease, bliss and happiness. Likewise a Bornean Dayak must find it difficult to believe that hard, black coal can be changed into bright light within an electric bulb. There is ultimately only one way open to those who do not believe the accounts of the yogins. They will have to repeat the experiment—in the forest, not the laboratory—they will have to do what the yogins say should be done, and see what happens. Until this is done, disbelief is quite idle, and on a level with a pygmy's disbelief in Battersea power station, maintained by a stubborn refusal to leave the Congo basin, and to see for himself whether it exists and what it does. In other words, it seems to me quite unworthy of educated people to deny that there exists a series of technical practices, known as Yoga, which, if applied intelligently according to the rules, produces a state of tranquillity.[2]

So much about the technical substructure. The ideological superstructure, in its turn, consists of a number of theoretical systems, by no means always consonant with each other. Theologically they are Hindu, Buddhist or Jain. Some are atheistic, some polytheistic, others again henotheistic. Philosophically some, like Vaiśeṣika and Abhidharma, are pluralistic, others, like Vedānta and Mādhyamikas, monistic. These two monistic systems, again, seem to be diametrically opposed in their most fundamental tenets—the one claiming that the Self (*ātman*) is the only reality, the other that it is just the absence of a self (*nairātmya*) which distinguishes true reality from false appearance.

On closer study these disagreements do, however, turn out to be fairly superficial. All these "yogic" philosophies differ less among themselves than they differ from the non-yogic ones. They not only agree that yogic practices are valid, but in addition postulate that these practices are the avenues to the most worthwhile knowledge of true reality, as well as a basis for the most praiseworthy conduct, and that, as the source of ultimate certainty, the yogic vision itself requires no justification. Only in a state of yogic receptivity are we fit and able to become the recipients of ultimate truth. Observations made in any other condition concern an illusory world, largely false and fabricated, which cannot provide a standard for judging the deliverances of the yogic consciousness.

A closely analogous situation prevails in Western Europe with regard to science. In this field also we can distinguish between the technology itself and its theoretical developments. The prestige of the scientific approach among our modern philosophers seems to me entirely due to its applications. If a philosopher assures us that all the "real" knowledge we possess is due to science, that science alone gives us "news about the universe"—what can have led him to such a belief? He must surely have been dazzled by the practical results, by the enormous increase in power which has sprung from the particular kind of knowledge scientists have evolved. Without these practical consequences, what would all these scientific theories be? An airy bubble, a diversion of otherwise unoccupied mathematicians, a fanciful mirage on a level with *Alice in Wonderland*. As a result of science, considerable changes have recently occurred in the material universe. Although by no means "more enduring than brass," the monuments to science are nevertheless rather imposing—acres of masonry, countless machines of startling efficiency, travel speeded up, masses of animals wiped out, illnesses shortened, deaths postponed or accelerated, and so on. This scientific method demonstrably "works," though not in the sense that it increases our "tranquillity"—far from it. All that it does is to increase "man's" power to control his "material environment," and that is something which the yogic method never even attempted. Scientific technology indeed promises limitless power, unlimited in the sense that by itself it places no limitations, moral or otherwise, on the range of its conquests. Very little notice would presumably be taken of the thought-constructions of our scientists if it were not for their impressive practical results. Dean Swift's *Voyage to Laputa* would then voice the general attitude, including that of the majority of philosophers.

As with Yoga, the bare technology is also here clothed in numerous theories, hypotheses, concepts and philosophical systems, capable of considerable disagreement among themselves. But all scientific philosophies agree that scientific research, based on the experimental observation of external objects,[3] is the key to all worthwhile knowledge and to a rational mode of life.

But though I were to speak with the tongues of angels, my "empiricist" friends will continue to shrug their shoulders at the suggestion that Yoga and other non-scientific techniques should be taken seriously. As professed "humanists" they might be expected to have a greater faith in the depth and breadth of the human spirit and its modalities. As "empiricists" they might have a more catholic notion of "experience," and as "positivists" a clearer conception of what is, and what is not, a "verifiable" fact. And even as

363

"scientists" they ought to have some doubts as to whether the world of sense-bound consciousness is really the whole of reality. But alas, a staggering hypertrophy of the critical faculties has choked all the other virtues. Contemporary empiricist and positivist philosophers, in their exclusive reliance on scientific knowledge, are guilty of what Whitehead has charitably called a "narrow provincialism." Usually unfamiliar with the traditional non-scientific techniques of mankind, they are also, what is worse, quite incurious about them. At best these techniques, if noticed at all, are hastily interpreted as approximations to scientific ones, worked out by ignorant and bungling natives groping in the dark. On the wilder shores of rationalism it is even rumoured that "the poet was the primitive physicist."[4] With a shudder we pass on.

To judge all human techniques by the amount of bare "control" or "power" they produce is patently unfair. Other goals may be equally worth striving for, and men wiser than we may deliberately have turned away from the pursuit of measureless power, not as unattainable, but as inherently undesirable. A graceful submission to the inevitable is not without its attractions, either. A great deal might be said, perhaps, for not wanting more power than can be used wisely, and it is much to be feared that the "captors of an unwilling universe"[5] may end as many lion tamers have ended before them.

Of all the infinite facets of the universe, science-bound philosophers will come to know only those which are disclosed to scientific methods, with their ruthless will for boundless power and their disregard for everything except the presumed convenience of the human race, and they cannot prove, or even plausibly suggest, that this small fraction of the truth about reality is the one most worth knowing about. As for the vast potentialities of the human mind, they will bring out only those which have a survival value in modern technical civilization. Not only is it a mere fraction of the human mind that is being used, but we may well wonder whether it is the most valuable section—once we consider the ugliness, noisiness and restlessness of our cities, or the effects which the handling of machines has on workers, that of scientific tools on scientists. At present it looks as if this mode of life were sweeping everything before it. It also demonstrably sweeps away much that is valuable.

2a. Turning now to the "triple world," we find that the unanimous tradition of the Perennial Philosophy distinguishes three layers of qualitatively different facts—natural, magical, and spiritual. The constitution of man is accordingly composed of three parts, reality presents itself on three levels, and threefold is the attitude we can adopt towards events.

In man we have body-mind as the first constituent, the "soul" as the second and the "spirit" as the third. In the objective world, the first level is the body of facts which are disclosed by the senses and scientific observation, and arranged by common sense and scientific theory. The second comprises a great variety of facts which with some justice are called "occult," because they tend to hide from our gaze. They weighed heavily with our forefathers, but are now widely derided. An example is astrology, or the study of the correspondences which may exist between the position of the celestial bodies on the one hand and the character, destiny, affinities and potentialities of people on the other. In addition this second level includes the activities of the psychic senses, such as clairvoyance, clairaudience, pre-cognition, thought transference, etc., the huge field of myths and mythical figures, the lore about ghosts and the spirits of the departed, and the working of "magic," which is said to cause effects in the physical world by means of spells and the evocation of "spirits." Thirdly, the spiritual world is an intangible, non-sensuous and disembodied reality, both one and multiple, both transcending the natural universe and immanent in it, at the same time nothing and everything, quite non-sensory as a datum and rather nonsensical as a concept. Indescribable by any of the attributes taken from sensory experience, and gained only by the extinguishing of separate individuality, it is known as "Spirit" to Christians, as "emptiness" to Buddhists, as the "Absolute" to philosophers. Here our senses are blinded, our reason baffled, and our self-interest defeated.

The three worlds can be discerned easily in our attitudes, say, to cold weather. The common-sense reaction is to light a fire, to wear warm clothing, or to take a walk. The magician relies on methods like the *gtum-mo* of the Tibetans, which are claimed to generate internal heat by means of occult procedures. They are based on a physiology which differs totally from that taught in scientific textbooks, and depend on the manipulation of three mystic "arteries" (*nadis*), which are described as channels of psychic energy, but which ordinary observation fails to detect, since they are "devoid of any physical reality."[6] Finally, the spiritual man either ignores the cold, as an unimportant, transitory and illusory phenomenon, or welcomes it, as a means of penance or of training in self-control.

Technical progress and scientific habits of thought increasingly restrict us to the natural level. Magical events and spiritual experiences have ceased to be familiar, and many people do not admit them as facts in their own right. By their own inner constitution the three realms differ in their accessibility to experience, the rules of evidence are by no means the same in all three, and each has a logic of its own. In the infinitude of the spiritual realm no particular fact can be seized upon by natural means, and everything in the magical world is marked by a certain indefiniteness, a nebulousness which springs partly from the way in which the intermediary world presents itself and partly from the uncertainties of its relation to the familiar data of the bright daylight world of natural fact. Every student of the occult knows that in this field the facts are inherently and irremediably obscure. It is impossible to come across even one magical fact which could be established in the way in which natural facts can be verified. There is a twilight about the magical world. It is neither quite light nor quite dark, it cannot be seen distinctly, and, like a shy beast when you point a torch at it, the phenomenon vanishes when the full light is turned on.[7]

The situation becomes more desperate still when we consider the spiritual. Here it is quite impossible to ever establish any fact beyond the possibility of doubt. The Buddhists express this by saying that Nirvana is "sign-less," i.e. it is of such a nature that it cannot be recognized as such. This is really a most disconcerting thought. Spirit is non-sensuous and we have no sense-data to work on. In addition, spiritual actions are disintegrated when reflected upon. If they are not to lose their bloom, they must be performed unconsciously and automatically. Further, to be spiritual, an action must be "unselfish." It is in the nature of things quite impossible ever to prove with mathematical certainty that an action has been unselfish, because selfishness is so skillful in hiding itself, because insight into human motives is marred by self-deception, and, in any case, at any given time the motives are so numerous that no one can be sure of having got hold of all of them. I. Kant has spoken the last word on this subject when he points out that "in fact it is absolutely impossible to make out by experience with complete certainty a single case in which the maxim of an action, however right in itself, rested simply on moral grounds and on the conception of duty. Sometimes it happens that with the sharpest self-examination we find nothing beside the moral principle of duty which could have been powerful enough to move us to this or that action and to a great sacrifice; yet we cannot infer from that with certainty that it was not really some secret impulse of self-love, under the false appearance of that idea, that was the actual determining cause of the will. We like then to flatter ourselves by falsely taking credit for a noble motive, whereas in fact we can never, even by the strictest examination, get completely behind the secret springs of action; since, when the question is of moral worth, it is not with the actions which we see that we are concerned, but with those inward principles of them which we do not see." [8]

Here is one of the inescapable difficulties of the human situation. All the meaning that life may have derives from contact with the magical and spiritual world, and without such contact it ceases to be worth while, fruitful and invested with beauty. It seems rather stupid to discard the life-giving qualities of these realms simply because they do not conform to a standard of truth suited only to the natural world,[9] where to the scientist phenomena appear worthy of notice only if they are capable of repetition, public observation, and measurement. They are naturally more inaccessible to natural experience than natural things are. The methods of science, mighty and effective though they be, are useless for the exploration of two-thirds of the universe, and the psychic and spiritual worlds are quite beyond them. Other faculties within us may well reveal that which the senses fail to see. In

Buddhism faith, mystical intuition, trance and the power of transcendental wisdom are held to disclose the structure of the spiritual and intermediary worlds. No one can be said to give Buddhist thinking a fair chance if he persists in condemning these sources of knowledge out of hand as completely futile and nugatory.

2b. Next, the perennial philosophy assumes that there are definite "degrees of reality." In this book we will be told that "dharmas" are "more real" than things, the images seen in trance "more real" than the objects of sense-perception, and the Unconditioned "more real" than the conditioned. People at present can understand the difference between facts which exist and "non-facts" which do not exist. But they believe that facts, if real, are all equally real, and that qualitative distinctions between them give no sense. This is the "democratic" viewpoint in vogue at the present time, which treats all facts as equal, just as all men are said to be equal.[10] In science nothing has any "meaning," and "facts" are all you ever have.

At the time when Buddhism flourished, this would have seemed the height of absurdity. Also the leading European systems of that time, like those of Aristotle and Plotinus, took the hierarchy of levels of reality quite for granted, and were indeed entirely based upon it. The lowest degree of reality is "pure matter," the highest "pure form," and everything else lies somewhere in between. The higher degrees of reality are more solid and reliable, more intellectually satisfying, and, chief of all, they are objectively "better" than the lower, and much more worth while. *Ens et bonum convertuntur.* In consequence contact with the higher degrees of reality entails a life which is qualitatively superior to one based on contact with the lower degrees. This is what sticks in the throat of the present generation. For here we affirm that "judgments of value" are not just subjective opinions, which vary with the moods of people, or their tastes or social conditions, but that they are rooted in the structure and order of objective reality itself.[11]

If the value of life depends on contact with a high level of reality, it becomes, of course, important to ascertain what reality is in its own-being (*svabhāva*), and to be able to distinguish that from the lesser realities of comparative fiction which constitute our normal world of half-socialized experience which we have made ourselves so as to suit our own ends. To establish contact with worthwhile reality has always been the concern of the exponents of the "perennial" philosophy, i.e. of most reputable philosophers of both Europe and Asia up to about AD 1450.

About this time there began in Europe that estrangement from reality which is the starting-point of most modern European philosophy. Epistemology took the place of ontology. Where ontology was concerned with the difference between reality and appearance, epistemology concentrated on that between valid and invalid knowledge. The Occamists who set the tone for all later phases of modern philosophy asserted that things by themselves have no relations to one another, and that a mind external and unrelated to them establishes all relations between them. Ontology as a rational discipline then lost its object and all questions concerning being *qua* being seemed to be merely verbal. Science should not concern itself with the things themselves, but with their signs and symbols, and its task is to give an account of appearances (*salvare apparentias*), without bothering about the existence *in esse et secundum rem* of its hypothetical constructions.[12] In consequence, thinkers seek for "successful fictions" and "reality" has become a mere word.

It is remarkable that 1,400 years before the Mahāyāna Buddhists had taught almost exactly the same. When they realized their estrangement from reality, they looked for a reality more real than they found around them, i.e. for the "Dharma-element" itself. Modern philosophy concludes that it is better for us to turn our backs on nebulous ideas about reality as such, and to concentrate on gaining power over the environment as it appears. Power by whom, and for whom? Here a philosophy which teaches that the particular alone exists and that universals are mere words, finds refuge in an abstraction called "man," who is somehow regarded as the highest form of rational being, and for whose benefit all these developments are said to take place. To Nāgārjuna and his followers this by itself would seem to indicate a serious logical flaw at the very basis of such doctrines.

2c. Finally, and that is much easier to understand, the hierarchical structure of reality is duplicated by and reflected in a hierarchy among the persons who seek contact with it. Like is known by like, and only the spirit can know spiritual things. In an effort to commend Buddhism to the present age, some propagandists have overstressed its rationality and its kinship with modern science. They often quote a saying of the Buddha who told the Kalamas that they should not accept anything on his authority alone, but examine and test it for themselves, and accept it only when they had themselves cognized, seen and felt it.[13] In this way the Lord Buddha finds himself conscripted as a supporter of the British philosopical tradition of "empiricism." But who can do the testing? Some aspects of the doctrine are obviously verifiable only by people who have certain rather rare qualifications. To actually verify the teaching on rebirth by direct observation, one would have to actually remember one's own previous births, an ability which presupposes the achievement of the fourth *dhyāna*, a state of trance extremely scarce and rarefied. And what width and maturity of insight would be needed to actually "know" that the decisive factor in every event is a "moral" one, or that Nirvana means the end of *all* ill! The qualifications are moreover existential, and not merely intellectual. Buddhism has much to say about the spiritual hierarchy of persons, for what someone can know and see depends on what he is. So the saint knows more than the ordinary person, and among the saints each higher grade more than the lower. In consequence, the opinions and experiences of ordinary worldlings are of little account, on a level with the mutterings of housepainters laying down the law about Leonardo da Vinci's "Virgin of the Rocks."

Notes

1. This saying from Bhagavadgīta II 58 may well be regarded as the clue to all Yoga:

yadā smharate cāyam kūrmo'ṅgānīva sarvaśah
indriyānīndriyārthebhyas tasya prajñā pratiṣṭihitā.

"He who draws away the senses from the objects of sense on every side, as a tortoise draws in his limbs (into the shell), his intelligence is firmly set (in wisdom)" (Radhakrishnan).

2. The most comprehensive and authoritative textbook is M. Eliade, *Yoga. Immortality and Freedom*, 1958.

3. The data of introspection have given rise to much uneasiness in this scheme of things. The most logical solution seems to be that of Behaviourism, which transforms psychic events into externally observable objects.

4. K. Nott, *The Emperor's Clothes*, 1953, p. 248.

5. Quis neget esse nefas invitum prendere mundum
Et velut in semet captum deducere in orbem? (Manilius II 127–8).

6. So A. David-Neel, *With Mystics and Magicians in Tibet*, p. 203. For further information about the nadīs see G. Tucci, *Tibetan Painted Scrolls*, 1949, and S.B. Dasgupta, *An Introduction to Tāntric Buddhism*, 1950.

7. For a further discussion of the difficulties of getting hold of the facts about the "intermediary world" see my article on "The Triple World" in *The Aryan Path*, xxv 5, 1954, pp. 201–2. Very instructive also is *Times Literary Supplement* 28.1.1956 on "New Concepts of Healing."

8. I. Kant, *Metaphysics of Morals*, trsl. T.H. Abbott, 1879, p. 33.

9. There is also something mean and timid about the caution of someone who wishes everything to be established beyond any reasonable doubt, and to have it inspected again and again with myopic and distrustful eyes.

10. The structure of the universe always reflects the structure of society. Likewise it is interesting to note that those who replace ontology by epistemology are Protestants who repudiate collective or corporate authority, whereas Roman Catholics, Marxists and Buddhists believe that meaningful statements can be made about the "real being" of things.

11. In addition, of course, the very assumption of qualitative differences in the worthwhileness of life has no scientific foundation, because "science," as we know it, has no eye for quality, but only for quantity. Likewise no moral qualifications are required of scientists, and the quality of their lives is unimportant when their findings are judged.

12. For further particulars see E. Conze, *Der Satz vom Widerspruch*, 1932, pp. 205–66, and the summary in English in the *Marxist Quarterly*, 1937, pp. 115–24.

13. Majjhima Nikāya I, p. 265; Taishō Issaikyō 26, k. 54, p. 769 b. Ch. A. Moore, Buddhism and Science: Both Sides, in "Buddhism and Culture," Kyoto, 1960, p. 94.

Scientific and Other Values

Patrick Grim

In that vague realm known as the popular imagination, science is quite frequently conceived of as something that stands apart from values. Science is thought of as value-free, and this supposed freedom from values is thought to be one of its prime virtues. On one side of a fact–value gap lies the quarrelsome morass of bickering values; arbitrary, indefinite, and perhaps merely subjective. On the other side of the gap, pristine and pure, untainted by mere values, stands the imposing edifice of science.

In what follows I hope to challenge this popular image of science and values. The claim that science is or even could be value-free, I shall argue, is simply false.

But at this point it is tempting to replace the first image of science and values with a second image, an image that itself boasts no small measure of current popularity. If science does not stand apart from the morass of values, it seems, it must be but one among many rival "value systems" bogged down in that morass. There is then nothing special to be said for science that could not be said with equal justice for any of its rivals. Any favoring of science over rival "systems" becomes a matter of personal and idiosyncratic taste, ultimately as arbitrary and subjective as any other choice of values.

This second image of science and values, I shall argue, is as wrong as the first. Science is shaped by and is essentially committed to certain values. But I hope to show that because of the *ways* in which science is shaped by values and because of the *particular* values to which it is committed, it does not follow that science is on all fours with any other "system" of values. Nor does it follow that the choice becomes arbitrary between science and Edward Conze's Buddhism, or William James's mysticism, or Carlos Castaneda's Yaqui sorcery.[1]

Is Science Value-Free?

In order to answer this question we need some sketch or outline or indication, however rough, of what we are to take as "science." But any attempt to define "science" with accuracy, or to offer necessary and sufficient conditions for properly scientific procedure, is bound to embroil us immediately in the devastating difficulties of demarcation. So let us settle, at least initially, for something less: a rough and ready characterization of science in terms of lists of claims.

We are all familiar with a large number of scientific claims; claims such as the following:

(1) Water freezes at 32° F.

(2) Most calico cats are female.

(3) Hydrogen is composed of one proton and one electron.

(4) Syphilis is caused by a spirochete.

(1) through (4) are claims currently accepted, on grounds of scientific study and research, within the scientific community. And at least one way to characterize the science of our times would be in terms of a list, indefinitely large or infinite, of all such scientific claims.

Let us thus envisage such a list of scientific claims. We need not hold that any science at any time will share these claims; it is enough that our current science can be so represented, whatever inevitable discoveries or refutations the future will bring.

With this rough and ready characterization of current science we have at least one way of trying to determine whether science is value-free. Are there any value judgments in the list we have envisaged? If so, it would appear, we have good grounds for concluding that at least *our* science is not value-free. If there are no value judgments in the list, we have at least some basis for claiming that science *is* value-free.

Consider in this regard (1) through (4) above. None of these, I think, is a very promising candidate for a "value judgment." And if our complete list of current scientific claims were composed entirely of claims like (1) through (4), we would have at least some grounds for maintaining that science *is* value-free.

But (1) through (4) are only some of the claims on our list. (5) and (6), for example, would appear to have an equal right to be included among current scientific claims:

(5) Smoking is hazardous to your health.

(6) Contemporary handling of nuclear wastes is unsafe.[2]

We can also expect to find on our list claims concerning other claims, such as (7) and (8):

(7) We have compelling evidence that smoking causes cancer.

(8) It has not yet been satisfactorily established that the use of high-sulfur coal poses a significant environmental risk.

Any list of currently accepted scientific claims that did not include claims such as (5) through (8), I think, would clearly be an inadequate and incomplete list. But it can at least be argued that claims such as (5) through (8) embody value judgments, and thus that at least our current science is not value-free.

Consider claims (5), (6), and (8). (5) invokes a notion of health and of hazards to health. But health, some have argued, is itself a term to which values are crucial. A healthy organism is one that is as it *should* be, and a hazard to health is a *threat*—something that at least prima facie ought to be avoided for the sake of *proper* functioning.[3] Consider also the notions of safety and of risk that appear in (6) and (8). To claim that a situation is unsafe is surely to claim that it jeopardizes something of *value*, and to speak of risk is to speak of probabilities of *loss*. The notions of safety and risk, then, like that of health, appear to assume a particular background of values. Were human lives held to be of no more value than lives of wood lice, we would not speak scientifically of risk and safety in the ways we do. Many of our scientific claims *do* concern risk and safety, and thus assume a background of values.

Consider also a slightly different argument. Claims (7) and (8) invoke notions of compelling evidence and of the satisfactory establishment of claims. And surely any science worth its salt will include claims of this sort concerning the establishment of other claims. But "compelling evidence" and "satisfactory establishment," it can be argued, assume the same sort of background of values as indicated above for "safety" and "risk." Richard Rudner has put this point quite elegantly:

> since no scientific hypothesis is ever completely verified, in accepting the hypothesis the scientist must make the decision that the evidence is *sufficiently* strong or that the probability is *sufficiently* high to warrant the acceptance of the hypothesis. Obviously our decision regarding the evidence and respecting how strong is "strong enough," is going to be a function of the *importance*, in the typically ethical sense, of making a mistake in accepting or rejecting the hypothesis. Thus, to take a crude but easily manageable example, if the hypothesis under consideration were to the effect that a toxic ingredient of a drug was not present in lethal quantity, we would require a relatively high degree of confirmation or confidence before accepting the hypothesis—for the consequences of making a mistake here are exceedingly grave by our moral standards. On the other hand, if, say, our hypothesis stated that, on the basis of a sample, a certain lot of machine stamped belt buckles was not defective, the degree of confidence we should require would be relatively not so high. *How sure we need to be before we accept a hypothesis will depend on how serious a mistake would be.*[4]

It thus appears, on several grounds, that science is not value-free.[5] Let us state our conclusions here in as strong a form as possible. Some scientific claims concern health, safety, hazard, and risk, and each of these seems to make sense only against a general background of values. Some scientific claims concern compelling evidence, adequate establishment, and sufficiently high probability for the acceptance of other claims. These claims, on Rudner's argument, rely on a similar background of values. But here we can also say more. The envisaged list of claims we have appealed to throughout is a list of claims currently accepted within the scientific community. If so, each claim is presumably on the list because it has been accepted on grounds of "compelling evidence," "adequate establishment," or "sufficiently high probability." Thus the inclusion of *each* item on the list, even of items as apparently dispassionate as (1) through (4), will reflect the general background of values that Rudner indicates. *Some* claims on the list will invoke values in virtue of their content; for example, those that address issues of health, risk, or sufficient establishment. But regardless of its content, the presence of *each* claim on the list will reflect its acceptability in terms of background values.

In Rudner's sense, then, *all* science is shaped by values. We might also note that in principle almost any background value might do the shaping. The scientific acceptability of a claim will depend on the prospects of accepting it or not, and on the relative values we place on those prospects. Almost any value concerns the relative merit of alternative prospects, and almost any value will rank alternatives in some way. Thus in principle almost any value might be of importance for the scientific acceptability of some claim, and in that sense almost any value could be of significance in shaping science.

We have presented in some detail an argument to the effect that science is not value-free. If this argument is adequate, it appears that the first popular image of science and values presented in our introduction is drastically misconceived.

What replies to the argument above might be made in behalf of a notion of "value-free" science? Let us briefly consider two possible replies.

First, it must be noted that the list of claims with which we have been working is merely a list of *current* scientific claims. But no argument on this basis alone, it might be argued, will suffice to show that science *could* not be value-free.

This reply does seem telling, as far as it goes, against the first part of the argument presented above. A science that did not address questions of health, hazard, safety, or risk would to that extent avoid certain questions of value. Thus science *could* escape values to some extent by confining its claims to those such as (1) through (4) and avoiding claims such as (5) through (8).[6]

But this reply does not seem adequate against the full force of Rudner's argument. Surely *any* science must accept and reject claims on the basis of adequate warrant, compelling evidence, sufficiently high probability, and the like. If any such acceptance and rejection of claims demands a background of values, then any science must assume some such background of values. In that repsect, although we have been working with a rough characterization of *current* science, our conclusions would appear to apply as well to any science at any time.

Consider also a second possible reply. What we have taken as "science," it might be argued, is not science *proper* or science *per se* but instead the everyday management, administration, and application of science. The management of science, it might be conceded, is shaped by background values. But science per se is nevertheless value-free. The scientist qua scientist neither accepts nor rejects hypotheses; he merely assigns probabilities to various hypotheses at issue.[7]

By carefully gerrymandering a distinction between "science per se" and the mere "management" of science one could, I think, carve out as "science per se" something that was value-free. But I seriously question the point of any stipulative linguistic exercise of this type. That which one would end up idolizing as "pure science" would be a barely recognizable amputation from current practice, consisting of little more than the mindless manipulation of test tubes and the mechanical marking of charts. It would not be what we think of as science and would not be science as we know it. It would fit best the work of those technicians who know least what they are doing, if it would fit the practice of any working scientist at all, and would relegate to a category of "mere management" the most important work of every Nobel Prize Winner we have. If that is the cost of maintaining that science is value-free, we had better simply concede that it is not.

Two Types of Scientific Values

Rudner's argument shows that standard scientific notions of "adequate establishment," "compelling evidence," and "sufficiently high probability" assume a background of values. In that sense we might conclude that science is "value-laden." But here it is important to consider precisely *what* values are at issue and precisely *how* science is "laden."

Rudner's argument shows that science is shaped by a background of values. But the argument does not show that any particular set of values must do the shaping. Our current science, perhaps, relies on a particular background of values. And all science, perhaps, must rely on *some* such background. But science *could* nevertheless assume a quite different background of values.

Because we value human life to the extent that we do, and because we do not value chimpanzee life more than we do, we are more wary of accepting claims regarding lethal quantities of food additives as "adequately established" where human lives, and not merely chimpanzees, are at stake. If we valued human life less, or chimpanzee life more, our list of "adequately established" claims could be expected to shift. Similarly, if we ranked Rudner's machine-stamped belt buckles as on a par with original Rembrandts, we could expect our list of acceptable claims to change. Different values would then bear on what establishment we took to be adequate establishment, and as a result different claims would be accepted as adequately established.

But would a science shaped by different values in this way be any less a science? I think not. Although *some* set of values must in this way serve as background for notions of sufficient evidence and adequate establishment, it need not be any particular set of values.[8] Were our values regarding human lives and chimpanzee lives to change tomorrow, which claims we found scientifically acceptable or sufficiently warranted might change as well. But that would not render our current work with either humans or chimpanzees unscientific.

One moral to draw from this is that changes in the ethical values of a society can change its science in subtle but important ways. We might also draw another moral: the fact that a particular set of values shapes scientific practice at any particular time is not necessarily any vindication of that set of values. Nazi science, for example, might well rely on Nazi values. Claims regarding lethal quantities of food additives might be differently treated where Aryans and non-Aryans are at stake much as we treat such claims differently where humans and chimpanzees are at stake. If some sets of values can be twisted or perverse, some forms of science—those forms of science that rely on such values—can be equally perverse.

The background of values that Rudner points out in science is a background that may shift. In that sense the background values that have so far been indicated for our science are what I will term *nonessential* values; science would still be science were these values to be replaced by others.

Are there also *essential* values within science; values such that any "science" that abandoned or replaced those values would no longer be science?[9] Here I want to consider first some values that might be proposed as essential to science, in order to reject them. I then hope to propose some central values that I think *are* genuinely essential to science.

In the preceding piece, Edward Conze portrays science as a naked manifestation of a mere will to power. He blasts "science-bound philosophers" for their "ruthless will for boundless power," and says of science that "all that it does is to increase 'man's' power to control his 'material environment,' and that is something the yogic method never even attempted." Conze concedes to science its success in this regard, but "to judge all human techniques by the amount of bare 'control' or 'power' they produce is patently unfair."

But must science be committed to a lust for this type of power, or is it always so committed? Does Conze's "ruthless will for boundless power" indicate a value *essential* to science? Certainly not. Science can be pursued, and often has been pursued, for its own sake and for the sake of mere knowledge. Paleontology and archaeology are dubious pursuits for any who lust for power, and not every form of scientific endeavor promises some high-powered technological application. It must be conceded that science *has* often been pursued with technological goals, and it is quite likely that our science would not have the shape it does today were that not the case. But it is also the case that religion has often been used for petty political ends. It no more follows that power is essential to science than that petty politics is essential to religion.

Consider also a second feature that is occasionally suggested as essential to science. It is sometimes proposed that science is in some way committed to, or is an elaborate apologetic for, an "everyday common-sense world of sense experience."[10] In this regard Conze speaks scathingly of the "narrow provincialism" of "sense-bound consciousness."

Is science essentially committed to "the everyday common-sense world of sense experience"? I am not sure what power this would have as a condemnation of science even if true. But in fact nothing could be much farther from the truth. Anyone who delves into contemporary physics expecting to find a familiar universe neatly expansive in absolute space and punctually sequential in time, filled with comfortably solid middle-size objects, will be very much surprised. From the perspective of common sense and everyday experience the claims of contemporary physics will seem among the most bizarre imaginable. The contemporary physicist seems as willing as any yogi to reject "the everyday common-sense world of sense experience" as illusion, and seems as insistent as any yogi that the more fundamental reality lies beyond it.

Neither a lust for power nor a commitment to an "everyday common-sense world of sense experience," then, appears to be essential to science. But there are, I think, values that *are* essential to science; values without which no pursuit could properly be considered science.

The first of these essential values involves an almost obsessive emphasis on truth. The aim of science is not to ennoble or to edify, to satisfy or sanctify or save. The aim of science is simply to sort claims on the basis of their truth or falsity. Within science no other criterion for sorting claims—on the basis, say, of accordance with religious tradition or political appeal—is even a close second. This is not, of course, to say that no claim has ever been accepted within a scientific community for baser motives than a sincere search for truth. But it is to say that any adoption of claims for other motives is to that extent *un*scientific, and that a "science" that abandoned this central emphasis on truth would no longer be science.[11]

Tied to this first essential characteristic is a second: science *demands* demonstration.[12] If a claim is to be adopted on truly scientific grounds, it must be openly *demonstrable*, and no other appeal—to venerated authority or to the dicta of the properly initiated—will do.

This emphasis on demonstration has sometimes been ridiculed as a myopic devotion to phenomena that fit neatly within the confines of a laboratory. That, I think, is a mistake; science, though essentially committed to demonstration, is committed to no particular type of demonstration. Field studies, expeditions, and the appearances of comets have played a major role in the history of science. Contemporary reliance on mathematics reflects a willingness to accept a priori deductive as well as inductive demonstration. And there are times when the course of science quite properly shifts on the basis of what appear to be almost purely philosophical arguments.[13]

It does not appear, then, that science is confined to any particular type of demonstration. Nevertheless commitment to demonstration, of one type or another, seems essential; a "science" that abandoned or even significantly weakened the demand for demonstration would no longer be recognizable as science.

Here, as with regard to the emphasis on truth, we must concede that the demand for demonstration is an ideal of scientific practice, not always fulfilled. Science has at times been swayed by misplaced authority and has settled too quickly for claims without demanding sufficient demonstration. But that an obsession with truth and a demand for demonstration are *ideals* of scientific practice does not mean that they can be waved cynically aside. That these *are* the ideals of scientific practice, even if "merely" ideals, is crucial; anything that did not value these as ideals would no longer be science.

It appears, then, that science is "value-laden" in two importantly different ways. As noted in the previous section, standard scientific judgments of "compelling evidence," "sufficiently high probability," and "adequate establishment" demand a background of values. These values, I have argued, can shift and hence are not *essential* values; though *some* set of background values is required by science in this regard, no *particular* set of values is demanded.

But there are also particular values within science that are essential. A purported lust for technological power and a supposed slavery to an "everyday common-sense world of sense experience" are not among these. But an obsession with truth and a commitment to demonstration are; each of these is a value essential to science in the sense that a "science" without them would not be science at all.

Science and Its Rivals

It appears that we must reject, then, the first image of science and values presented above. Science is far from value-free; it is both shaped by a general background of values and essentially committed to some particular values.

But at this point it is tempting to replace the first image of science and values with a second. It is quite often claimed, at least within some circles, that there are equally worthy alternatives to science. The general claim is that there are bodies of knowledge or types of pursuit that in some way conflict with science but that have an equal or greater claim to our serious attention. Conze appears to hold that forms of Buddhism are the equal or the superior of science, Alan Watts makes similar claims for similar religious traditions, Colin Wilson and Philip Slater hold that the occult somehow embodies a respectable rival to science, and Carlos Castaneda portrays the "world" of a fictitious Yaqui sorcerer as an equally acceptable alternative to the "world" of science.[14]

In the general spirit of this general claim, however, quite different particular claims have been made by different authors, or even by the same author at different times.[15] A complete treatment of all such claims would involve carefully disentangling and clarifying each claim, seeing what support can be offered in its behalf and what can be said against it. That complete task is well beyond us here. Instead I shall concentrate on only a sampling of such claims; those that rely on the "value-ladenness" of science.

If science is "value-laden," as we have argued that it is, is it not then on all fours with any other value system? Does not devotion to science then amount to dogmatic adherence to but one value system among many? Is not then the choice of those values allied with science as ultimately arbitrary as the choice of any alternative set of values?

These are, of course, rhetorical questions; those who pose them expect us to answer "yes" and to concede that science can claim no special status over rival values; that the adoption of scientific values is on a par with the adoption of those values evident in Christianity or Buddhism or Hinduism, or for that matter in Nazism or the fantasies of Carlos Castaneda.

The force of such an attack should not be underestimated. In this regard it may help to use a distinction hinted at above: a distinction between the current practice of what we think of as our scientific community on the one hand, and what we might call *truly* scientific practice on the other. The actual practice of those who staff our laboratories and conduct our various research programs may not always be fully or genuinely scientific. As you read this someone somewhere may be fudging a result, or misrecording an observation so as to support a pet theory, or doctoring the data. There also be claims applauded by our current scientific community that are not genuinely scientific claims, backed by truly scientific evidence and subjected to genuinely scientific scrutiny. Current practice has its share (although I doubt that it has more than its share) of frauds and foibles and human frailties. And current practice may be *un*scientific to a greater or lesser degree because of such things.

But those who press the type of attack at issue are not merely criticizing "science," in the sense of current practice, for the extent to which it is *un*scientific. We would not need them for that, and they are out for bigger game. To the extent that current practice is *un*scientific, we hardly need an appeal to "rivals" or even a concentrated examination of values to condemn it. And to the extent that claims applauded by the scientific community are not genuinely scientific claims, we hardly need an appeal to "rivals" or a concentrated examination of values to condemn them. *Un*scientific aspects of current practice and claims currently accepted without genuinely scientific warrant are clearly to be condemned on purely scientific grounds, and would be whether or not science had any "rivals" in the sense at issue.

Those who press such an attack are instead questioning current practice precisely to the extent that it *is* scientific. Were it *perfectly* scientific they would still urge Buddhism and the occult and Yaqui sorcery as equivalent rivals or alternatives. Thus the full force of the attack is against science itself, in no matter how pure a form, rather than merely against the variously corrupted forms of current practice.

How good is the attack? *Does* it follow from the fact that science is "value-laden" that it is on all fours with any rival set of values?

As noted, science is "value-laden" in two importantly different ways: it is shaped by a background of inessential values, and it is also committed to some particular values. Consider first, then, those values that function as background to our science, and that shape our notions of "compelling evidence," "sufficiently high probability," and the like. To these values—the relative values we place on human and animal lives, for instance—there are clearly alternatives. Instead of *these* values we might adopt others, and if some religious tradition emphasizes importantly different values, we might come to adopt those values instead. We might decide that all life is as precious as human life, or that Brahmin life is of greater value than the life of lower castes.

But are these rivals to the background values of our science really rivals to *science*? I think not. As we have seen, science demands *some* background of values of the sort Rudner indicates. But it does not demand any *particular* set of values. A form of science shaped in this way by other values than those that shape our science would nevertheless be a form of *science*. So the choice here is not between science and those other values; it is merely a choice between a science shaped by the values by which our science is shaped and a science shaped by other values. If we are to set up a genuine conflict between science and its supposed rivals, then, it must be a conflict regarding something other than merely the background of non-essential scientific values.

Consider then the essential values we have claimed for science, which would seem much more likely candidates for a genuine clash between scientific and other values. If science is *essentially* committed to certain values, is it not *then* on all fours with its alternatives? Have we not *then* shown that the choice between science and its rivals is simply a choice of values?

In order to answer this question carefully we must pay closer attention to the different *types* of things that may be at issue as "rivals" or "alternatives" to science. The term "science" is itself used ambiguously enough that what is meant is sometimes the *activity* of science and sometimes a set of *claims* accepted on scientific grounds. Hence science is sometimes contrasted with other activities or pursuits, and in considering rivals or alternatives we are considering rival or alternative activities or pursuits. But science is also sometimes contrasted with rival sets of claims—the claims of science versus the claims of Genesis, for example.

Let us take these one at a time. Does science, with its essential values, clash with other activities? Not, I think, in the sense that objectors to science commonly have in mind. Essential to the activity of science is an obsession with truth and a commitment to demonstration. And clearly this is not the case with all activities; art, music, and poetry, as well as certain forms of asceticism and meditation, are quite different. Does that mean that one must pursue science, devote oneself to truth and demonstration, and abandon these other activities? Not at all. There are other ways to live one's life, and in some respects—or for some people—these other ways may be infinitely preferable. But that there are other activities that are rivals to science in *this* sense is surely no condemnation of science, any more than one's choice to pursue a career in business is any condemnation of poetry (or vice versa). If this is what objectors to science have meant by portraying science as on all fours with other endeavors, they are certainly right; there is nothing in science or its essential values that absolutely compels us to do science or to emphasize those values. But it would seem odd to characterize this as in any significant respect a *challenge* to science.

So far, then, we have not found a genuine clash between science and its supposed rivals, and thus have not found the clash between science and "equally worthy rivals" suggested by its critics. Alternatives to the inessential background values of science are not alternatives to *science*, because science could assume any of various sets of background values. And alternatives to the activity of science, despite its essential values, are merely other worthy pursuits rather than "rivals" in any more threatening sense.

Finally, however, science is often set in contrast to other sets of claims. And here, unlike in the previous two cases, we do seem to have a genuine conflict. It must be remembered that we are dealing with *genuinely* scientific claims, and not all and only claims accepted by those who staff our laboratories may qualify as genuinely scientific. But even so, there are undoubtedly some claims current within some religious traditions, and some claims accepted within some cultural groups, that will simply contradict genuinely scientific claims. On pain of contradiction, one cannot accept both.

Does it follow from this, together with the fact that science is essentially "value-laden" in the ways specified, that the claims of science are on no firmer ground than any of a number of rival sets of claims, or that each is equally worthy or unworthy of our attention? Certainly not. The mere fact that two sets of claims conflict in no way forces us to throw up our hands, or to suspend judgment, or to pick a set at random. There may be much that can be said for one set rather than another. And the particular values essential to science, far from relegating it to the level of any of various alternatives, may offer a significant argument in its behalf.

At least the following can be said for the claims of science. Science is essentially committed to truth and demonstration, and genuinely scientific claims will reflect those values. They will have been selected with an obsessive ideal of truth and through the patient toil of dogged demonstration. This does not, of course, guarantee their truth; genuinely scientific claims may nevertheless be false and claims adopted without scientific warrant may nevertheless be true. But it does mean that scientific claims will be backed by a wealth of demonstration, evidence, and argument. Only further demonstration, better evidence, and stronger argument can rationally force us to abandon them; and then, of course, we will be replacing some scientific claims with other scientific claims. None of this can be said for most sets of claims that appear as religious or cultural tradition.

The essential values of science, then, are scientific *virtues*, and as genuine virtues are not to be flippantly waved aside. Science *is* essentially committed to particular values, but the values at issue are ideals of truth and demonstration. Those are ideals that are not to be taken lightly and are not to be lightly put aside.

On what basis, then, might one choose to abandon scientific claims and to adopt some other set of claims instead? One might do so by abandoning the essential values of science, or by relegating them to a secondary position. One might take something else as more important than truth, and choose claims contradictory to those of science on the basis of some other value; aesthetic or (broadly) political or personal appeal, perhaps. Or one might take something to have precedence over demonstration; the words of sacred authority or of sacred texts, perhaps, or the counsels of some form of mystical experience. But here other claims will have been accepted on the basis of *other* values, not on the basis merely of truth and demonstration. One will be able to appeal for one's chosen claims on the basis of whatever values they embody. But if they are to be chosen on grounds other than merely ideals of truth and demonstration, one cannot expect to be able to defend them solely with an eye to truth and solely on the basis of demonstration.[16]

One might also, of course, choose to adopt some set of claims—however unpopular—and to demonstrate their truth. This is a very different case. If science can adopt any form of demonstration, as argued above, one will not then have abandoned science. One will be defending one's claims as rival scientific claims, rather than as rivals to scientific claims, and will be contributing to the scientific project rather than renouncing it. One may, indeed, be contesting the accepted claims of current practice. But that is how science proceeds.

None of the arguments considered above, then, shows that science has equal rivals in the sense sometimes proposed by its opponents. Science *does* rely on a background of values, and there are rival values to those on which our current science relies. But this does not mean that those rival values are rivals to science itself. Science *is* one activity among many, and others may for various reasons be rightly pursued instead of science. But these other activities are not *rivals* in the threatening sense that critics of science generally have in mind. Finally, scientific claims will be contradicted by other sets of claims. But it does not follow that these need be *equal* rivals. Genuinely scientific claims, *because* of the essential values of science, have ideals of truth and demonstration on their side.

We have, then, found reason to question each of the images of science and values with which we began. Science is not value-free, and is moreover essentially committed to some particular values. At least with an eye to the arguments considered, however, it need not follow that science, because "value-laden," is on all fours with any of various "rivals" or "alternatives." The essential values of science, far from indicating a fatal weakness, may be its greatest strength.

Notes

1. It is perhaps appropriate to mention also two general topics that I will not attempt to deal with in their entirety. The first is value-relativism. Implicit in both images of science and values above is an assumption that all values corrupt, or that any introduction of values must represent capricious dogmatism, or that all "systems" of value are equally worthy—and hence equally unworthy—of our allegiance. I do not think any of these relativistic assumptions is true, but I will not here be able to argue fully against any of them. Second, it must be said that arguments based on the premise that science is value-laden are only some among the many arguments, not usually distinguished, that attempt to show science to be no better than various rivals. I do not think any of the other arguments are any better than those explicitly addressed here, but in what follows I will concentrate only on arguments that in one way or another turn on the issue of values.

Some of the claims of this essay—particularly those in the final section—are posed in a slightly stronger form than I might choose in another context. For present purposes, and because my intent is to offer a corrective to exaggerated views on the other side, I think this is relatively harmless.

Conze and James are represented by the two preceding papers. Carlos Castaneda's views are scattered throughout *The Teachings of Don Juan: A Yaqui Way of Knowledge* (Berkeley: University of California Press, 1968), *A Separate Reality: Further Conversations with Don Juan* (New York: Simon & Schuster, 1971), *Journey to Ixtlan: The Lessons of Don Juan* (New York: Simon & Schuster, 1972), *Tales of*

Power (New York: Simon & Schuster, 1974), and *The Second Ring of Power* (New York: Simon & Schuster, 1977). See also Richard De Mille, *Castaneda's Journey: The Power and the Allegory* (Santa Barbara, Calif.: Capra Press, 1976), and Richard De Mille, ed., *The Don Juan Papers: Further Castaneda Controversies* (Santa Barbara, Calif.: Ross-Erikson, 1980).

2. That (6) is perhaps controversial is of no importance to the argument here; "Contemporary handling of nuclear wastes is perfectly safe" or "It has not yet been clearly determined whether or not contemporary handling of nuclear wastes is unsafe" would serve our purposes equally well. The same applies to (8).

3. It has become quite widely accepted that the notion of health is value-laden. See Joel Feinberg, *Doing and Deserving* (Princeton: Princeton University Press, 1970), pp. 253–255; and H. Tristram Engelhardt, Jr., "Human Well-Being and Medicine: Some Basic Value-Judgments in the Biomedical Sciences," in *Science, Ethics, and Medicine*, ed. H. Tristram Engelhardt and Daniel Callahan (Hastings, N.Y.: Institute of Society, Ethics, and the Life Sciences, 1976). Both of these are reprinted in Thomas A. Mappes and Jane S. Zembaty, *Biomedical Ethics* (New York: McGraw-Hill, 1981).

4. Richard Rudner, "The Scientist *qua* Scientist Makes Value Judgments," *Philosophy of Science* 20 (1953): 1–6; reprinted in Baruch Brody, ed., *Readings in the Philosophy of Science* (Englewood Cliffs, N.J.: Prentice-Hall, 1970).

5. For the sake of simplicity I have not dealt here with a complex argument to the effect that "the 'softness' of social facts may affect the 'hard' notions of truth and reference" (p. 46), which appears in Hilary Putnam's *Meaning and the Moral Sciences* (Boston: Routledge & Kegan Paul, 1978).

6. A science that scrupulously avoided values in this way, of course, would be a science that was itself of little value.

7. This view appears in Richard C. Jeffrey, "Valuation and Acceptance of Scientific Hypotheses," *Philosophy of Science* 23 (1956): 237–246; reprinted in Brody. But Jeffrey concedes, in the spirit of the following paragraph, that "this account bears no resemblance to our ordinary conception of science."

8. To choose to value human and chimpanzee lives equally, or Rembrandts and belt buckles equally, would of course not be to abandon a background of values but to choose a particular background. As Rudner notes, science could attempt to escape *any* background of values only by abandoning notions of "adequate establishment," "sufficient evidence," and the like.

9. This is all too brief as an introduction of essential properties, let alone of essential values. I hope that it will nevertheless prove adequate for the task at hand.

10. This phrase appears in scare quotes because I am not sure precisely what a "world" is here supposed to include, nor am I sure—even when most sympathetic—that there is only one such "world."

11. This claim does not, I think, conflict with Rudner's conclusion, appealed to above. The fact that various values may be crucial to what we accept as "adequate evidence" for the truth of a claim does not mean that we are ultimately concerned with anything less than the *truth* of the claim.

It might also be noted that even claims that have slipped into currency by illegitimate means may later be offered proper scientific support. I am indebted to David Pomerantz for discussion on this point.

12. The tie between an emphasis on truth and a demand for demonstration is as subtle as it is intimate. The more sincerely one is concerned with truth, the more emphatically one will demand demonstration. And the more one demands demonstration, the better our reason to believe that one is committed to truth. Both of these

claims hold only ceteris paribus. But the ceteris-paribus status of each need not indicate that the tie between truth and demonstration is therefore a weak one.

13. Many of the arguments for behaviorism of some forms, against evolutionary "emergence" in biology, for important claims regarding quantum mechanics, and regarding tachyons traveling backward in time seem to be fully "philosophical" arguments.

14. Conze is represented by the preceding piece, and Castaneda's work is mentioned in footnote 1. For other authors mentioned, see Colin Wilson, *The Occult* (New York: Random House, 1971), quoted in the general introduction; Philip Slater, *The Wayward Gate* (Boston: Beacon Press, 1977); and Alan Watts, especially *Psychotherapy East and West* (New York: Vintage Books, 1975).

15. Here Conze might serve as a convenient example. At one point Conze criticizes science for ignoring the accounts of yogis: "There is ultimately only one way open to those who do not believe the accounts of the yogins. They will have to repeat the experiment . . . they will have to do what the yogins say should be done and see what happens" (p. 300). If this is an attack on science at all, it is a pretty weak attack. What Conze calls for here is *more* science—a scientific investigation of the claims of yogis—rather than for some alternative to science. Yet a few pages later Conze claims that the world of the yogi is one incapable of scientific exploration; within the magical world "the phenomenon vanishes when the full light is turned on" (p. 304), and within the spiritual world "it is quite impossible to ever establish any fact beyond the possibility of doubt" (p. 304). Here Conze changes his tune and presses a quite different attack: "The methods of science, mighty and effective though they may be, are useless for the exploration of two-thirds of the universe, and the psychic and spiritual worlds are quite beyond them" (p. 304). This is a quite different claim from that proposed earlier, and calls for quite different argumentative support. Can it be *demonstrated* that "two-thirds of the universe" is beyond the reach of science, and that claims made by Conze or anyone else about this major portion of the universe are true? If claims about this portion *can* be demonstrated, and if—as argued in the previous section—science can adopt any form of demonstration, then these claims are within the potential grasp of science after all. If these claims *cannot* be demonstrated, of course, we lose at least one good ground for believing that they are true at all, or for believing Conze's claim that we have overlooked a major portion of the universe.

16. Here it might be claimed that yogic meditation, for example, involves its own unique form of demonstration. I do not think that is true; what is at issue seems to me entirely unlike anything we normally consider "demonstration." But if there *is* a legitimate form of demonstration buried here, then, as argued above, it is grist for the mills of science. What remains to be demonstrated (and this is by no means a unique matter in the history of science) is that there *is* a form of demonstration at issue.

Suggested Readings

As noted in the general introduction, this is not intended as anything like a complete bibliography. The works mentioned are generally standard sources, for the most part easily available, which may be of help at the next stage of the reader's research.

The attack on science evident in Edward Conze's "Tacit Assumptions" appears in various forms in the following:
Castaneda, Carlos. *The Teachings of Don Juan: A Yaqui Way of Knowledge* (Berkeley: University of California Press, 1968).
Pearce, Joseph C. *The Crack in the Cosmic Egg.* New York: Pocket Books, 1973.
Pirsig, Robert M. *Zen and the Art of Motorcycle Maintenance.* New York: Bantam Books, 1976.
Slater, Philip. *The Wayward Gate.* Boston: Beacon Press, 1977.
Watts, Alan. *Psychotherapy East and West.* New York: Vintage Books, 1975.
Wilson, Colin. *The Occult: A History.* New York: Random House, 1971.

Critical works:
Evans, Christopher. *Cults of Unreason.* New York: Dell, 1973.
De Mille, Richard, ed. *The Don Juan Papers: Further Castaneda Controversies.* Santa Barbara, Calif.: Ross-Erikson, 1980.
Gardner, Martin. *Science: Good, Bad, and Bogus.* Buffalo, N.Y.: Prometheus, 1981.
Sagan, Carl. "Night Walkers and Mystery Mongers: Sense and Nonsense at the Edge of Science." In *Broca's Brain.* New York: Random House, 1979.
Sladek, John. *The New Apocrypha: A Guide to Strange Science and Occult Beliefs.* New York: Stein & Day, 1973.
General journal sources include *The Humanist*, which is overwhelmingly critical in tone, and *Zetetic Scholar*, published by Marcello Truzzi of the Department of Sociology, Eastern Michigan University, Ypsilanti, MI 48197, which attempts to be a less biased forum.

On science and values, see:
The section on "The Acceptance of Scientific Theories" in Baruch Brody, ed. *Readings in the Philosophy of Science*. Englewood Cliffs, N.J.: Prentice-Hall, 1970, pp. 539-570.
Grim, Patrick. "Meaning, Morality, and the Moral Sciences." *Philosophical Studies* 43 (1983), 397-408.
Hempel, Carl G. "Science and Human Values." In R. E. Spiller, ed. *Social Control in a Free Society*. Philadelphia: Univ. of Pennsylvania Press, 1960, pp. 39-64. Reprinted in E. D. Klemke, Robert Hollinger, and A. David Kline, eds. *Introductory Readings in the Philosophy of Science*. Buffalo, N.Y.: Prometheus Books, 1980, pp. 254-268.
Martin, Michael. "Value Judgments and the Acceptance of Hypotheses in Science and Science Education." *Philosophical Exchange* 1 (1973): 83-100.
Putnam, Hilary. *Meaning and the Moral Sciences*. Boston: Routledge & Kegan Paul, 1978.
Rescher, Nicholas, "The Ethical Dimension of Scientific Research," in Robert G. Colodny, ed. *Beyond the Edge of Certainty*. Englewood Cliffs, N.J.: Prentice-Hall, 1964, pp. 261-267. Reprinted in Klemke, Hollinger, and Kline, eds. *Introductory Readings in the Philosophy of Science, op. cit.*, pp. 238-253.
Scriven, Michael. "The Exact Role of Value Judgments in Science." In R. S. Cohen and K. Schaffner, *Proceedings of the 1972 Biennial Meeting of the Philosophy of Science Association*. Dordrecht, Holland: D. Reidel, 1974, pp. 219-247. Reprinted in Klemke, Hollinger, and Kline, eds. *Introductory Readings in the Philosophy of Science, op. cit.*, pp. 269-291.

On ethical and epistemological relativism:
Brandt, Richard. "Ethical Relativism." In *The Encyclopedia of Philosophy*, ed. Paul Edwards. New York: Macmillan and the Free Press, 1967.
Gardiner, Patrick. "Irrationalism." In Edwards, *Encyclopedia of Philosophy*.
Ladd, John, ed. *Ethical Relativism*. Belmont, Calif.: Wadsworth, 1973.
Meiland, Jack, and Krausz, Michael, eds. *Relativism: Cognitive and Moral*. Notre Dame: Univ. of Notre Dame Press, 1982.
Montague, Phillip. "Are There Objective and Absolute Moral Standards?" in *Reason and Responsibility*, ed. Joel Feinberg. Fifth edition. Belmont, California: Wadsworth, 1981, pp. 480-491.
Stace, Walter T. "Ethical Relativism." Reprinted from *The Concept of Morals*. New York: Macmillan, 1937. In Feinberg, pp. 492-497.
Wilson, Bryan, ed. *Rationality*. Evanston, Ill.: Harper & Row, 1970.

Name Index

Abel, Reuben, 51, 71
Abell, George O., 19, 83
Addey, John, 66ff, 75ff
Adler, Alfred, 88, 105ff
Agassi, Joseph, 121n
Alexander The Great, 2
Allen, George, 11
Aller, Lawrence H., 19
Althusser, Louis, 165n
Amaldi, Edorado, 19
Anderson, Margaret, 192
Anglade, H., 44
Aquinas, St. Thomas, 35
Aristotle, 35, 118, 146, 329, 367
Arons, M. E., 250ff
Asimov, Isaac, 36n
Ayer, A. J., 121n, 230n

Bacon, Francis, 161, 165n
Barnes, Barry, 157
Bastien, J., 44
Batteux, A., 44
Baum, Robert F., 7, 11, 90
Bell, J. S., 310, 322ff
Bell, Charles, 160
Ben-Shakhar, G., 61
Berendzen, Richard, 19
Berg, Larry, 17
Berkeley, George, 270, 328, 345
Bethe, Hans A., 19
Bidelman, William P., 19

Bigeleisen, Jacob, 19
Bigot, J., 44
Birge, William R., 202
Birney, D. Scott, 19
Blackmore, S. J., 58
Blackstone, W. T., 70ff
Blum, Jeffrey, 157ff, 165n
Bodmer, W., 157
Bohm, David, 280, 324, 329
Bohm, Karl-Heinz, 19
Bohr, Niels, 1, 10, 268, 272ff, 279, 289, 296ff, 301n, 304ff, 322ff, 329
Bok, Bart J., 19, 24-25
Born, Max, 121n, 272, 303ff, 323, 335
Borst, Lyle B., 19
Bostock, John, 160
Bouet, Michel, 42, 45
Bouma, 27
Boyce, Peter B., 19
Boyle, Robert, 294
Brahe, Tyco, 125, 127
Braud, L. W., 196
Braud, W. G., 201
Brier, Bob, 5, 9, 185
Broad, C. D., 184ff, 187, 215, 218, 222ff, 227, 228n, 231n, 232ff
Brooks, Harvey, 19
Brown, Thomas, 160
Brown, F. R., 25
Brush, Steven, 312
Buddha, 368

Name Index

Burckhart, Jacob, 176
Burr, H. S., 27
Burtt, 137n
Buscombe, William, 19
Busnel, R., 44

Cannon, Susan, 165n
Capra, Fritjof, 1, 5, 10, 267ff, 306, 308ff, 326, 346
Capriotti, Eugene R., 19
Cardan, Jerome, 53
Carington, Whately, 226
Carlson, S., 74
Carnap, Rudolf, 135ff
Carnegie, Andrew, 60
Carpenter, Edward, 357, 360n
Carter, H. E., 19
Castaneda, Carlos, 10, 354, 370, 378, 382n
Cavalli-Sforza, L., 157
Chamberlain, Von Del, 20
Chamberlain, Wilt, 95ff
Chamberlain, J. W., 19
Chambers, Robert, 159
Chandrasekhar, S., 20
Chaplin, J. P., 43
Chartrand, Mark R. III, 20
Chiu, Hong-Yee, 20
Chuang Tzu, 283
Cloud, Preston, 20
Codos, P., 44
Columbus, Christopher, 2
Comfort, Alex, 313n
Comte, August, 159
Conti, Peter S., 20
Conze, Edward, 10, 353ff, 370, 375ff, 384n
Conze, Muriel, 11
Cook, Alan F. II, 20
Cooper, Joe, 53
Cooter, Roger, 7, 90
Copernicus, 25, 125, 128, 170, 172ff, 175, 311
Copland, James, 160
Cottrell, Alan, 20
Cratty, B., 45
Crawford, Bryce, Jr., 20
Crease, Robert P., 1, 5, 10, 17n, 269ff
Crick, Sir Francis, 19
Crombie, A. C., 119
Csonka, Paul L., 251
Cudaback, David, D., 20
Culver, Roger G., 54, 72, 78
Curry et al, 66
Curry, Patrick, 66, 78, 83

da Vinci, Leonardo, 369
Dalgarno, A., 20
Dallenbach, Karl, 159

Darnton, Robert, 159
Darwin, Charles, 77, 115, 348
David, James W., 253
Davis, Morris S., 20
Davis, Hallowell, 20
Davis, T. Patrick, 17
de Grazia, 27
Dean, Geoffrey, 54, 59, 61
Dean, G. A., 1, 8, 16, 54, 57ff, 62ff, 72ff
DeBroglie, Louis, 340
Deglane, H., 44
Delsemme, A. H., 20
Democritus, 279
Descartes, 329ff, 344
Dewaele, M., 44
Dicke, Robert H., 20
Dickson, David, 165n
Dobyns, Zipporah, 71
Donn, Bertrum, 20
Doty, Paul, 20
Douglas, Mary, 166n
Drake, Frank D., 20
Drigny, G., 44
DuBridge, Lee A., 20
Ducasse, C. J., 187, 234, 236, 239
Duhem, Pierre, 115, 121n, 151, 171
Duran, Jane, 9, 184ff
Dutton, 58

Eastman, George, 166n
Eccles, Sir John, 19
Eddington, Sir Arthur, 105ff, 228
Edison, Thomas Alva, 60
Eichhorn von-Wormb, H. K., 20
Einstein, Albert, 1, 10, 35, 105ff, 115, 124ff, 131, 138n, 170ff, 177n, 268, 271ff, 285ff, 310, 322ff, 329, 297ff, 301n, 304ff, 329
Elliot, 75
Elwell, Dennis, 62ff, 65, 70, 73, 75ff
Emberson, R. M., 20
Emerson, Ralph Waldo, 94
Emmons, Howard W., 20
Empedocles, 109
Epstein, Eugene E., 20, 195
Erhard, Werner, 92
Ertel, Suitbert, 68ff
Estabrooks, G. H., 259ff
Estebany, 198
Eyring, Henry, 20
Eysenck, Hans, 44, 54, 55, 65, 68, 72, 159

Famechon, E., 44
Faraday, Michael, 115
Faucheux, L., 44
Federer, Charles A., 20
Feinberg, Gerald, 250, 309

388

Feleppa, Robert, 89ff
Feuerbach, Ludwig, 167n
Feyerabend, Paul K., 5, 8, 15, 27n, 116, 122n, 137n, 138n, 150, 174, 176
Feynman, Richard, 309
Fleischer, Robert, 20
Flew, Anthony, 5, 9, 184ff, 234
Fliegel, Henry F., 20
Flourens, J. P. M., 160
Flugel, J. C., 61
Fournier, A., 44
Fowler, William A., 20
Franklin, Fred A., 20
Fraser, Sir James, 166n
Fredrick, Lawrence W., 20
Freud, Sigmund, 53, 72, 88, 105ff, 299ff
Fulton, Robert, 2

Gale, Richard, 245
Galileo, 35, 94, 110, 115, 118, 175, 205, 291, 329ff
Gall, Franz Joseph, 159ff
Galton, 72
Gardner, Martin, 112, 118, 210
Gauquelin, F., 40, 63ff
Gauquelin, Michel, 5, 8, 16ff, 33, 36n, 40, 54, 63ff
Gauthier, B., 44
Geertz, Clifford, 165n
Gell-Mann, Murray, 309
Gerardin, L., 44
Gerbault, A., 44
Gettier, Edmund, 239ff
Giacconi, Riccardo, 20
Gingerich, Owen, 20
Gittelson, Bernard, 92
Gladstone, William E., 60
Glymour, Clark, 5, 7, 88
Godbey, John W., Jr., 235ff
Gold, Thomas, 20
Goldberg, Leo, 20
Goldhaber, Maurice, 20
Goodman, Nelson, 90, 141, 145ff, 153n
Gordon, Mark A., 20
Grad, Bernard, 198
Graule, V., 44
Greenstein, Jesse L., 20
Greisen, Kenneth, 20
Greyber, Howard D., 20
Grim, Elgas, 11
Guest, J. Z., 307
Gursky, Herbert, 20

Hafner, 129n
Hagan, John P., 20
Hamaker, 75

Hamilton, Sir William, 160
Hand, 75ff
Handler, Philip, 20
Hansel, C. E. M., 201ff, 257
Hanson, N. R., 113
Hartmann, William K., 20
Harvey, 74ff
Hausman, Daniel M., 11n
Haworth, Leland J., 20
Heelan, Patrick A., 314n
Hegel, G. W. F., 356
Heiles, Carl, 20
Heisenberg, Werner, 115, 268ff, 272, 275, 281ff, 287, 289, 303ff, 326, 329, 342
Heiser, A., 20
Heitler, 272
Helfer, H. L., 20
Hempel, Carl G., 90, 144ff, 151, 153n
Herbig, George H., 20
Hergenhahn, B. R., 62
Herz, Norbert, 27n
Herzberg, Gerhard, 19
Hesse, Mary, 165n
Hilsz, M., 44
Hipparchus, 30
Hitler, Adolf, 98
Hoag, Arthur A., 20
Hobbes, Thomas, 217
Hodge, Paul W., 20
Hodgkin, Luke, 165n
Hoffleit, Dorrit, 20
Home, Sir Edward, 160
Homer, 109
Honorton, Charles, 196ff, 201
Houk, Nancy, 20
Hövelmann, G.H., 77
Howard, William E., III, 20
Hoyle, Fred, 20
Hubbard, L., Ron, 103
Huber, L., 75
Huber, B., 75
Hume, David, 9, 35, 115, 135, 171, 184, 221ff, 230n
Humphrey, Betty, 260ff
Huxley, T. H., 348

Ianna, Philip A., 54, 72, 78
Iben, Icko, Jr., 20

James, Edward W., 8, 15, 16
James, William, 10, 187, 301n, 353ff, 370
Jefferies, Richard, 357, 360n
Jefferies, John T., 20
Jeffrey, Francis, 160
Jeffrey, Richard C., 383n
Jensen, Arthur, 157
Jerome, Lawrence E., 19, 36n, 112

Jettner, Frank C., 20
Jokipii, J. R., 20
Jonas, 57
Joplin, Janis, 95ff
Jouvet, Louis, 46ff
Jung, C. G., 27n, 292ff, 327

Kant, Immanuel, 171, 366
Kellerman, Kenneth, 21
Kelly, I. W., 1, 8, 16, 54, 57, 62, 72
Kelvin, 348ff
Kennedy, Jim, 253, 255
Kepler, J., 26, 27n, 32, 36n, 37, 110, 118, 127, 174
Kiewiet de Jonge, Joost H., 20
King, Ivan R., 21
Kompfner, Rudolf, 21
Kovach, William S., 21
Koyré, Alexander, 137n
Kretschmer, Ernst, 48
Krippner, S., 77
Krutzen, R., 72
Kuhn, Thomas S., 1, 5, 7, 8, 16, 28-29, 32ff, 89ff, 131ff, 140ff, 170ff, 279, 286ff, 310, 329ff
Kundu, M. R., 21
Kurtz, Paul, 11, 19, 83, 228

Lakatos, Imre, 1, 7, 36n, 89ff, 115, 137n, 143ff, 147, 151, 170ff, 174ff
Lama Anagarika Govinda, 283
Larmore, Lewis, 21
Larrain, Jorge, 165n
Laudan, Larry, 115
Lavoisier, 124
Leboyer, Claude Levy, 42
Lee and Yang, 124
Leek, Sybil, 92
Lefebvre, Henri, 165n
Leibniz, 329f, 348
Leontief, Wassily, 19
Leung, Kam-Ching, 21
Levin, Jerry, 254ff
Levitt, I. M., 21
Levy, Walter J., 9, 194, 185, 253ff
Liggins, C. G., 49
Lin, C. C., 21
Linell, Albert P., 21
Livingston, M. Stanley, 21
Locke, John, 35
Loptson, P., 72, 80
Lorentz, 125
Lorenz, Konrad, 19
Low, Frank J., 21
Lucretius, 35
Lukács, Georg, 158
Luyten, Willem J., 21

Lwoff, Andre M., 19
Lynes, B., 58, 75

MacDougall, William, 189
Mach, E., 329
MacLaine, Shirley, 307, 311
Magendie, Francois, 160
Maharishi Mahesh Yogi, 306
Mann, Charles C., 1, 5, 10, 74ff, 269ff
Margenau, 272
Marshall, Ninian, 226
Martens, R., 57
Martin, Michael, 11n
Marvel, Carl S., 21
Marx, Karl, 66, 88, 105ff, 159, 167n, 215
Mather, Arthur, 54, 59, 61, 66
Mauriac, François, 43
Maxwell, J. Clerk, 121n
Maxwell, Sir Robert, 170, 332ff
Mayall, Margaret W., 21
Mayall, Nicholas U., 21
McCrosky, Richard E., 21
McDougall, 256
McElroy, W. D., 21
McMahan, Elizabeth A., 202
Medawar, P. D., 19, 159
Meehl, Paul, 223
Meinecke, Friedrich, 177n
Mendel, G., 115
Mendelsohn, Everett, 165n
Menzel, Donald H., 21
Mermin, N. David, 1, 10, 270, 327, 331, 341ff
Michard, L., 44
Michelson, 223
Mikesell, Alfred H., 21
Mill, John Stuart, 9, 185, 247ff
Miller, Freeman D., 21
Mitchell, John, 27n
Moffet, Alan T., 21
Mook, Delo E., 21
Morley, 223
Morse, Marston, 21
Mulders, G. F. W., 21
Müller, A., 66
Mulliken, Robert S., 19
Munch, Guido, 21
Mundle, C. W. K., 226, 234
Murphy, Gardner, 201, 211
Musgrave, Alan, 11, 36n, 113, 137n, 143ff
Myers, Frederic, 225

Nagarjuna, 268, 368
Nauman, Eileen, 17n
Nelson, J. H., 27n, 359
Newton, Sir Isaac, 30, 105, 110, 115, 138n, 157, 170, 173, 176,
Ney, Edward P., 21

Neyman, J., 21
Nias, D.K.B., 54, 55, 62, 65, 68, 77
Niehenke, P., 54, 71
Nielson, K., 70
Nixon, Richard M., 98
Noonan, George, 75
Nordau, Max, 360n

O'Dell, C. R., 21
O'Keefe, John A., 21
Oedipus, 109
Oken, 75
Omarr, Sydney, 92
Oort, J. H., 21
Oppenheimer, Robert, 306
Owen, Tobias C., 21
Ownbey, 190

Pais, A., 315
Paracelcus, 70
Parker, Eugene N., 21
Parmenides, 109
Pasteur, Louis, 115
Pauli, W., 30, 272, 315, 323, 327
Pauling, Linus C., 19
Pearce, 189ff, 201
Penzias, Arno A., 21
Piccardi, G., 24, 27n
Pierce, A. Keith, 21
Planck, Max, 35, 170ff, 329
Plato, 35, 125, 151
Pletcher, Galen K., 9, 184
Plotinus, 367
Podolsky, B., 322ff, 337
Poe, Edgar Allen, 61
Poincaré, Henri, 171
Polanyi, Michael, 136
Pomerantz, David, 383n
Pope Innocent VIII, 24, 178n
Popper, Daniel M., 21
Popper, Karl, 1, 5, 7, 8, 15, 28-29, 32f, 88ff,
  111ff, 123ff, 131ff, 135, 140ff, 170ff
Post, H. R., 116, 121n
Pratt, J. G., 189ff, 201, 207ff, 261
Press, Frank, 21
Presswood, 129n
Price, G. R., 221, 223, 256ff, 261
Price, H. H., 187
Price, Harry, 221
Price, R. M., 21
Pritchard, James Cowles, 160
Protheroe, William M., 21
Ptolemy, 25-26, 118, 127, 173, 175ff
Purcell, Edward M., 19
Putnam, Hillary, 115, 383n

Rachleff, Owen, 36n

Rael, 75
Ramtha, 307
Rather, John D. G., 21
Rawlins, Dennis, 83
Rawls, John, 90, 141, 146ff, 151, 154n,
  155n
Reinsel, Ruth, 9, 183ff
Restivo, Sal, 269, 308, 311ff
Rhine, L. E., 11, 210ff, 239
Rhine, J. B., 5, 9, 185ff, 189ff, 201, 221, 225,
  255ff
Richardson, Robert S., 21
Rimet, J., 44
Risley, A. Marguerite, 21
Roach, Franklin E., 21
Roberts, Walter Orr, 21
Roberts, William W., 21
Robertson, R. N., 21
Robinson, Richard, 229n
Rodman, James P., 21
Roe, Ann, 42
Roget, Peter Mark, 160, 168n
Roll, W. G., 201
Rose, Steve, 159
Rosen, N., 322ff, 337
Rossi, Bruno, 21
Rothbart, Daniel, 7, 88
Rotton, J., 57
Rudhyar, Dane, 16, 72ff
Rudner, Richard, 154n, 372ff, 383n
Rudolphi, Karl, 160
Russell, Bertrand, 35
Rutherford, Ernest, 115

Sagan, Carl, 2
Saklofske, D. H., 1, 8, 16, 57
Salpeter, E. E., 21
Samuelson, Paul A., 19
Santo, Gregory, 71
Schaffner, Kenneth, 114
Scharff-Goldhaber, Gertrude, 21
Schmeidler, Gertrude R., 199, 204, 229n
Schmidt, Helmut, 197ff
Schmidt-Raghavan, Maithili, 9, 185
Schneider and Gauquelin, 67
Schopp, John D., 21
Schreur, Julian J., 21
Schrödinger, Erwin, 302ff, 329, 335
Schwinger, Julian, 19
Scott, E. L., 21
Scriven, Michael, 223
Seaborg, Glenn T., 19
Seitz, Frederick, 21
Severn, J. Milton, 60
Sewall, Thomas, 160
Shackleton, Basil, 211
Shaffner, Kenneth, 122n

Shane, C. D., 21
Shapere, Dudley, 155, 178n
Shapin, Steven, 168n
Shapley, Alan H., 21
Sheldon, W.H., 48
Sherrington, Charles, 216
Shu, Frank H., 21
Sidgwick, Henry, 187, 225, 256
Sitterly, Bancroft W., 21
Sitterly, Charlotte M., 21
Skinner, B. F., 21, 210, 229n
Slater, Philip, 354, 378
Smart, Ninian, 11
Smiles, Samuel, 159
Smith, Harlan J., 21
Smith, M. J., 198
Smythies, J. R., 231
Soal, S. G., 194, 211, 229n
Socrates, 201
Sorm, Frantisek, 21
Soubirois, Bernadette, 214
Spector, Marshall, 1, 5, 10, 270ff
Spencer, Herbert, 159
Spinoza, 346
Stalker, Douglas, 7, 88, 90
Stanford, Rex G., 201
Stapp, Henry Pierce, 275, 278, 298ff, 301n
Startup, M. J., 54, 55, 65, 67
Stebbins, G. Ledyard, 22
Stephenson, C. Bruce, 22
Stewart, Dugald, 160
Stewart, Gloria, 215
Stockmayer, Walter H., 22
Stone, Marshall H., 22
Storer, N. Wyman, 22
Sudarshan, E. C. G., 250ff
Suess, Hans E., 22
Swift, Dean, 363
Swihart, T. L., 22
Swings, Pol, 22
Szanto, G., 81
Szentagotai, J., 22

Taylor, Joseph H., 22
Terman, Frederick E., 22
Terzian, Yervant, 22
Thaddeus, Patrick, 22
Thagard, Paul, 11n
Thales, 125
Thoret, J., 44
Thorne, Kip S., 22
Tinbergen, J., 19
Tinbergen, N., 19
Tolstoy, Leo, 357
Toomre, Alan, 22

Toulmin, Steven, 174ff
Trollope, Anthony, 104
Tromp, W. W., 27
Turner, 190
Tuve, Merle A., 22

Urey, Harold C., 19

van Hayek, F. A., 165n
van de Kamp, Peter, 20
Vanek, M., 45
Vasilevskis, S., 22
Velikovsky, Immanuel, 22, 27n, 92, 112, 162, 312
Verfaillie, G. R. M., 27n
Visscher, Maurice B., 22
von Daniken, Erich, 92, 99
Vorhpal, Joan, 22

Wade, Campbell M., 22
Wagman, N. E., 22
Wald, George, 19
Wallace, A. R. 159
Wallace, Alfred Russell, 60, 77
Wallerstein, George, 22
Watkins, Anita M., 199
Watkins, Graham K., 199
Watkins, John, 135ff, 137n, 174ff
Watson, A., 57
Watson, H. C., 60
Watson, Lyall, 27n
Watts, Alan, 10, 354, 378
Weingarten, Henry, 36n
Wheeler, John, 282, 290ff, 327
Whewell, William, 121n
Whipple, Fred L., 22
White, Morton, 154n
Whitehead, A. N., 329, 364
Whitman, Walt, 357
Whitney, Hassler, 22
Wilson, Colin, 2, 354, 378
Witt, Adolf, 22
Wittgenstein, Ludwig, 35, 219
Wood, Frank Bradshaw, 22
Worley, Charles E., 22
Wright Brothers, 2

Young, Bob, 165n
Yuan, Chi, 22

Zelen, M., 83
Zirkle, George, 190
Zukav, Gary, 1, 5, 9, 10, 267, 269, 307, 309ff, 313, 327ff, 345

Subject Index

ad hoc assumptions, 108
Adler, 105
Alchemy, 157
alphabetology, 97 ff.
American Association for the Advancement
    of Sciences, 224
ancient astronauts, 1
anomalies, 134
astral projection, 92
astrology, 7 ff., 15 ff., 51, 54, 71, 92, 105, 108,
    110, 111, 112, 118, 120, 126, 141, 142, 187,
    214, 267, 365
astronomy, 24, 111, 127, 128, 142
Atlantis, 92, 183

background knowledge, role of in theory
    selection 117
basic limiting principles, 185, 222, 232
Bell's theorem, 270, 324
Bermuda Triangle, 183, 214
big bang, 349
Bigfoot, 92, 183
biorhythms, 92
bootstrap model, 308, 310
Boyle's Law, 294, 295
Buddhism, 267, 274, 281, 306, 310, 311, 367

Cartesian dualism, and QM, 329 ff., 346
Cartesian dualism, and parapsychology, 226
    ff.
causality, 185, 228, 232; paradoxes of 251

clairvoyance, 183 ff., 188, 191, 201, 205, 206,
    209, 232 ff., 235, 365
coincidence, 93, 184, 188
Comite Para, 39, 83
confirmation, 106, 107
consciousness, 355 ff.; and QM 326 ff., 341
    ff., 345
conservation of momentum, 330
conventionalism, 171, 172
Copenhagen Interpretation, 268, 275, 279,
    297, 298, 300, 312, 305, 306, 335
cosmology, 176; Aristotelian and Ptolemaic,
    118
crucial experiments, 115, 124, 134
CSICOP, 57, 58, 83, 214

Darwinian evolution, 111, 153, 348
decline effect, in parapsychological data,
    261
demarcation, 87 ff., 91, 109, 142, 186, 267, 353
demonstration, role of in science, 376 ff.,
    381
double-blind, 254; and parapsychology 194
dreams, 205, 206, 214
Duke University parapsychology laboratory
    189 ff., 192, 225, 243, 253, 261

Eastern mysticism, 268, 271, 303, 306, 308
eclipse observations, and relativity theory,
    105, 107
effect sizes, and astrology, 54 ff.

393

ego, super-ego, and id, 109
Einstein-Podolsky-Rosen experiment, 270, 322, 323
empiricism, 368; and mysticism, 361 ff.
endomorphism, mesomorphism and ectomorphism 48
engineering, 127
epicycles, 127
ESP, 92, 189, 190 ff., 193, 201, 217, 243
essential values in science, 375 ff., 381
essentialism, 149, 379
experimental replication, in parapsychology, 184, 195, 215, 221 ff., 224
experimental methodology, in parapsychology, 193 ff., 200
explanation, 106; and causality, 249; and coincidence, 205 ff.; and theory choice, 116, 133
Eysenck personality inventory, 47

faith healing, 73
falsifiability, 31, 28, 88, 89, 108, 112, 132, 134, 135, 138, 140, 141, 148, 149, 154
fascinating rhythms, 95 ff.
fraud, 184, 185, 194, 223, 253 ff., 378
fraud-proof evidence, need for parapsychology, 258 ff.
Freud, 105, 109, 131

ganzfeld technique, and parapsychology, 196, 229

half-life, 296
health, and values, 372, 383
Heisenberg uncertainty principle, 282, 289
heredity, and astrology, 48, 65
hidden variables in QM, 304
Hinduism, 267, 274, 311
hypothesis, 123 ff., 308; hypothesis selection for testing, 113 ff.

ideology, and science, 157 ff., 166, 177
incommensurability, 174, 175
incompleteness of quantum mechanics, 268, 270, 272, 304, 334, 335, 336, 340
induction, 104, 170, 172
Institute for Parapsychology, 259
instrumentalism, 172
interference pattern, 334
interpretations, 110
introversion/extroversion, and astrology, 47
IQ 56, 157, 159, 164
irrationalism, 28, 29, 34, 89, 132, 136

Jupiter, and soldiers, 39; and actors, 41

kinetic theory of gases, 295

knowledge, and parapsychology, 238 ff.

logic of discovery, 89, 132
logical positivism, 172
lunar effects, and astrology, 57

magic, 2, 18, 19, 25, 114, 119, 187, 365, 366
Mahayana Buddhism, 281, 368
Malleus Maleficarum, 23
Mars effect, 82
Mars, and sports champions, 39, 42
Marxism, 90, 105, 108, 131, 135, 311
materialism, 226, 331
meaningfulness, 109
medicine, 127
metaphysics, 328
meteorology, 127
method of difference (J. S. Mill), 247 ff.
methodology, 120
mind-brain identify theory, 235 ff.
miracles, 184, 221 ff., 224
mystical experience, 275
mysticism, 9, 10, 280, 283, 355 ff.

negative definitions, in parapsychology, 215, 218 ff.
Newtonian mechanics, 131, 267, 286, 290, 300, 338, 349
nitrous oxide, 355 ff.
non-natural births, and astrology, 49
non-separability, and QM 270, 274 ff., 324, 336 ff., 340
normal science, 89, 124, 125, 128, 131, 142, 143, 144, 153, 173
normative vs. descriptive, 90, 140, 144, 145
numerology, 92, 187

objectivity, and QM, 292 ff., 315, 327, 344
observation, 112, 339, 341 ff.
observer, role of, in QM, 270, 276 ff., 291, 304, 306, 326, 337, 344
ovulation, and astrology, 57

paradigm, 77, 111, 175, 132
parallelism, 269, 270, 308 ff.
paranormal, 187, 188, 205, 236
parapsychology, 8 ff., 183 ff.
particle accelerator, 278
Peruvian pick-up sticks, 100
phrenology, 60 ff., 77, 156, 158 ff.
physics, 24
PK, 192, 198, 199, 220
positivism, 363
pragmatism, 298
pre-testing standards, in science, 113
precession of the equinoxes, 30
precognition, 183 ff., 191, 197, 212, 227 ff., 232, 235, 244, 365

prediction, 107, 288
preparation and measurement, in QM, 277 ff.
Presocratics, 125
principle of sufficient reason, 250
probabilities, 335, 374; and QM 287, 289
probability theory, and astrology, 38 ff.
probability-density function, 335
progressive and degenerating problemshifts, 133, 137
pseudo-science, 6, 12, 87, 90, 104, 157, 312
psi, 186, 189, 195, 200, 201, 208, 216, 262
psi-gamma, 184, 216
psi-kappa, 216
psychic surgeons, 92
psychoanalysis, 26, 105, 118, 127
psychokinesis (PK), 183 ff., 191, 201, 212
psychological factors, and belief in astrology, 58
psychological measures, 57
psychology, 26
purpose, 93
puzzle-solving, 29, 31, 33, 126, 128, 173, 177
puzzles, as normal science, 124
pyramid power, 92
quanta, 170
quantum mechanics, 9, 267 ff., 286
quarks, 308, 310
Quine-Duhem thesis, 115, 154

radio propagation, and astrology, 57
radioactive decay, 268, 296, 302, 305; and parapsychology, 197
random event generators, 197
random phenomena, and parapsychology, 193 ff.
rationalism, and astrology, 28 ff.
rationality, 153; and mystical experience, 356 ff.
reality, degrees of, 367
reflective equilibrium, 90, 140, 141, 146, 147
relativism, 118, 151, 158 ff., 167, 174, 175, 176, 354, 370 ff., 382, 386
relativity, 118
religious mysticism, 357, 358
repeatability, 195
retrocognition, 216
reverse causation, 9, 135, 228, 230, 234, 243 ff., 250, 265
revolutionary science, 89, 124 ff., 131, 173, 223

S-matrix theory, 309

Saturn; and Jupiter, 46; and scientists, 39
Schrodinger wave equation, 303, 304, 305
Schrodinger's cat, 302 ff., 305, 307, 310, 311
science and values, 386
scientific methodology, 113
scientific progress, 144
scientism, 158, 165
scientology, 92
self-fulfilling prophecy, and astrology, 65
sheep-goat effect, 195, 229
simplicity, 114
social sciences, 33
Society for Psychical Research, 187, 188
solar activity, 24
spectral lines, 297
spiritual astrology, 70, 82
split-brain analysis, 299
statistical laws, 268, 270, 271, 272, 273, 287, 296, 304, 305, 336, 340
statistical character of parapsychological data, 219
statistics, 295, 320, 334; and astrology, 37 ff; and parasychology, 194
subjective idealism, 270, 328, 331, 342, 345

tachyons, 250
tacit knowledge, 152
Taoism, 267, 274, 311
telepathy, 183 ff., 188, 191, 201, 232 ff., 235, 365
the occult, 4, 6, 12, 119, 187, 267, 353, 365
theory, 104
theory, lack of in parapsychology, 216 ff.
translation, and parallelism, 309
truth, 171, 298, 310, 356, 376, 381, 383; and scientific status of theories, 104, 105

UFOlogy, 92
UFOs, 1, 87, 214
unconscious, and astrology, 71

vagueness, and astrology, 62
value judgments, 152
values, and science, 225, 354, 370 ff.
verification, 106, 107, 115
vitamins, 216

wave-particle duality, 270, 332 ff., 334, 336
witchcraft, 23, 114

yoga, 361 ff., 362

Zen, 306, 310, 311
Zener cards, 189, 219